The Eugenics Movement

The Eugenics Movement

An Encyclopedia

Ruth Clifford Engs

GREENWOOD PRESS

Westport, Connecticut • London

Library of Congress Cataloging-in-Publication Data

Engs, Ruth C.
    The eugenics movement : an encyclopedia / Ruth Clifford Engs.
       p. cm.
    Includes bibliographical references and index.
    ISBN 0–313–32791–2 (alk. paper)
      1. Eugenics—Encyclopedias. 2. Eugenics—United States. 3. Eugenics—Great Britain.
    4. Eugenics—Germany. I. Title.
    HQ751.E64 2005
    363.9'2'03—dc22   2005003391

British Library Cataloguing in Publication Data is available.

Library of Congress Catalog Card Number: 2005003391
ISBN: 0–313–32791–2

First published in 2005

Greenwood Press, 88 Post Road West, Westport, CT 06881
An imprint of Greenwood Publishing Group, Inc.
www.greenwood.com

Printed in the United States of America

The paper used in this book complies with the
Permanent Paper Standard issued by the National
Information Standards Organization (Z39.48–1984).

10 9 8 7 6 5 4 3 2 1

Contents

Preface

This reference work provides information concerning prominent individuals, organizations, publications, conferences, and concepts involved in the eugenics movements from the mid-nineteenth century to 2005. Although the focus is on the early-twentieth-century movement in the United States, links between the past and present and with Great Britain and Germany are also covered. These three nations were among the countries most involved with eugenic measures and debate. In the 1990s eugenics became an increasingly popular and scholarly topic among researchers, students, and the educated public. This volume has been produced to supply a ready source of information regarding the hereditarian and eugenics movements of the past two centuries and with some discussion of the current new movement. Although many authors imply that we are amid the surge of another eugenics movement, the term *eugenics* is generally not used. Today's terms include *cloning, family balancing, genetic engineering, in vitro fertilization, the new genetics,* and *genetic screening and testing.* On the other hand, others argue that some of these techniques do not even qualify as eugenics. In the past few years a multitude of articles, books, and Web sites have been produced with *eugenics* in the title. They present many interpretations of the early twentieth-century eugenics movement and its impact upon current trends and technological advancements. References for these divergent opinions on the meaning and concept of eugenics and of the hereditarian, eugenics, and "new eugenics" movements are provided in the selected bibliography. They range from academic to "yellow journalism."

Organization and Information Concerning the Structure of Entries

Entries contain numerous cross-references, in **bold**, providing links to related content and general references for further reading. The entries themselves range in length from a short paragraph to several pages. Due to size limitations of this work some eugenic leaders and topics were unable to be included. Likewise, it has not been practical to address the eugenics movements in other countries, including Argentina, Austria, Brazil, Canada, China, Finland, France, India, Italy, Japan, Mexico, Norway, and Sweden. There is not necessarily a correlation between the length and the importance of an item. For well-known personalities or subjects for which there are numerous other sources, such as inventor Alexander Graham Bell, short entries have generally been compiled, with emphasis on their connection to eugenics. Subjects with secondary sources that are limited or written in languages other than English have sometimes been described in more detail. Biographical entries present academic and overall career in the first paragraph, followed by contributions to eugenics or leadership in the movement. Since many eugenic supporters listed their family pedigree in their writings, this information is briefly noted. Book or periodical entries include sample chapter or article titles. Entries are in alphabetical order.

In the early twentieth century practitioners or supporters of eugenics were called "eugenicists" or "eugenists" in the United States, "eugenists" in Britain, and "eugenicists" or "race hygienists" in Germany. For the sake of clarity, the term *eugenicist* is used throughout this work. Entries are listed by the most common term or phrase in use during the period of time addressed. Some outdated terms used in entries, or some entries themselves, may be considered offensive by some, but it is not the intention of the author to offend anyone. The terms are included to promote understanding of the thought processes of an era. In addition, when searching for further reading from the past, the reader will need to look for information under the old terms. For example, the labels *feebleminded* and *moron* were considered acceptable and scientific terms through the mid-twentieth century to describe people with mental disabilities or learning disorders. The "menace of the feebleminded" was an important concept integral to the eugenics movement, particularly in the United States, as eugenicists and social reformers thought these individuals should be sterilized or segregated in custodial care to prevent "more of their kind." In several cases, a modern term, such as *sexually transmitted diseases,* may refer to the older term, for example *venereal disease.* The index points to other eugenics leaders and subjects discussed in the context of an entry.

Acknowledgments

This encyclopedia could not have been written without a host of individuals and institutions who provided material, time, and effort. Most of the resources for this work were found within the immense Indiana University Library system. These included the Special Collections Section of the Medical School Library in Indianapolis, and the Lilly Library, the Kinsey Institute Library, and the vast holdings of the Main Library, located in Bloomington. Librarians who have been most helpful include Frank Quinn and his staff from the Reference Department, among them Mark Day, David Frasier, Anne Graham, Luis Gonzalez, Anne Haynes, Jian Liu, Gabriel Swift, and Celestina Wroth. Patricia Steele and Randy Lent and their staff in the Circulation Department, including John Pate, Keith Welch, and Jane Tribe Al-Hashemi, were also very helpful. I am indebted to Rhonda Long, Rita Rogers, and Ron Luedemann of the Inter-Library Loan Department for finding obscure material. Other helpful library staff include Diana Hanson, Anne Haines, Charla Lancaster, Rita Barsun, Misty Walter, and Ron Hafft, for numerous items, and Roger Beckman in the Life Sciences Library. Special thanks go to Nancy S. Boerner for translating German material. For photographs, I am appreciative of the assistance of Bradley Cook, at the Indiana University Archives, Breon Mitchell of the Lilly Library, and Kristine R. Brancolini and her assistant Shana Berger of the IU Digital Library Program, who scanned most of the images in this book. I also appreciate the assistance of Clare Bunce and Sue Lauter from the Cold Spring Harbor Laboratory, Rob Cox of the American Philosophical Society, Nancy Dosch of the National Library of Medicine, and the Archive of the Max Planck Society (*Archiv zur Geschichte der Max-Planck-Gesellschaft*, Berlin-Dahlem), the Oneida Community Mansion House, and the Pickler Memorial Library, Truman State University.

A major task in the production of a resource of this nature is peer review by other scholars. I am most grateful for the review of the entire manuscript by Jessica Baldanzi, Nancy L. Gallagher, Diane Paul, Paul Weindling and Eugene Weinberg. Special thanks goes to Lindsay Franz and Jennifer Steinbachs for their review of certain subject areas; any errors are mine. The help of Robert W. Baird, who critiqued details of the manuscript, was most beneficial. The main "footwork" for this dictionary has come from my student assistants who have spent countless hours on this project and include Doris Burton Ronald C. A. Glass, Carolina Maria Guzmán C., Jenica Weiss Schultz, and Tony Sams. I would also like to thank my academic unit, the Department of Applied Health Science, and the School of Health Physical Education and Recreation for continued support of my research and writing. I am also appreciative of the many suggestions from my editor at Greenwood Press, Kevin Downing, and of Elizabeth Kincaid for her enhancement of the images and Carol Frenier for compiling the index. Most importantly, I am indebted to "my two Jeffs." Jeffrey Graf, a member of the Indiana University Main Library Reference Department, not only critiqued the entire manuscript but also compiled the chronology. I am especially indebted to my husband, Jeffrey Franz, for his help and constant support, without which this work would not have been completed.

Introduction

Eugenics is the science which deals with all influences that improve the inborn qualities of a race; also with those that develop them to the utmost advantage. . . . The aim of eugenics is to bring as many influences as can be reasonably employed, to cause the useful classes in the community to contribute more than their proportion to the next generation.

Francis Galton, "Eugenics: Its Definition, Scope, and Aims,"
The American Journal of Sociology 10 (July 1904)

Background Information Concerning the Concept of Eugenics

Eugenics, as noted by Charles B. Davenport (1911, 1), a prime mover of the American eugenics movement in the first third of the twentieth century, is the "science of the improvement of the human race by better breeding." Other terms for eugenics since the mid-nineteenth century have included *inherited realities, race betterment, race improvement, race culture, race regeneration, racial hygiene, sanitary marriages*, and *stirpiculture*, and in the late twentieth century the *new genetics*, the *new eugenics*, and *genetic engineering.* The idea that physical, mental, and moral characteristics ran in families has been discussed since antiquity. In this light families encouraged, or arranged for, their children to make the "best marriages" with the fittest, healthiest, and wealthiest individuals for producing the "best" children.

The term for encouraging marriage between those from "good stock" as a method for improving the race was not coined, however, until 1883. Francis Galton, a British naturalist, in *Inquiries into Human Faculty and Its Development* (1883, 24) derived *eugenics* from the Greek *eugenes*, meaning "good in stock, hereditarily endowed with noble qualities," after noting that wealth, ability, and intelligence appeared to run in certain families. He argued that the theory of evolution implies that "it would be quite practical to produce a highly gifted race of men by judicious marriages during several consecutive generations" (1869, 1). This classic concept that encouraged the middle classes and those with "good heredity" to reproduce became known as *positive eugenics*. Over his lifetime Galton continued to refine the concept of eugenics. Near the end of his life, in *Essays on Eugenics*, he suggested that eugenics is the study of agencies under social control that may improve or impair the racial qualities of future generations, either physically or mentally (1909, 81). The opposite of eugenics was termed *dysgenics* or breeding among "degenerates," and the "unfit," leading to "racial degeneracy," the decay of the health and efficiency of the population, and the decline of civilization. *Negative eugenics*, coined in 1909, defined the means of discouraging the unfit—including criminals, the mentally ill and disabled, prostitutes, and the poor—from reproducing.

Modern eugenic concepts that developed in the mid- to late nineteenth century resulted from the merging of several theories. A historian of the British eugenics movement, G. R. Searle (1976, 4), notes that

> eugenics in its modern form originates with Charles Darwin. The notion that man is part of the world of organic life and subject to the same natural process of evolution and decay was essential to the development of the new subject. The hypothesis of natural selection . . . raises . . . the possibility that . . . men might learn to control their own evolution; in place of the blind processes of natural selection a deliberate effort might be made to improve the species by attending in a scientific way the production of offspring.

In the late-nineteenth century, Darwinism became entwined with French naturalist Jean-Baptist Lamarck's theory of the inheritance of acquired characteristics. Lamarck's theory by the mid-nineteenth century had become the accepted theory of inheritance and was a foundation for degeneracy theory, in which acquired negative characteristics, such as poverty and alcoholism, were thought to be passed to offspring. This idea was also discussed as "inherited realities" by the 1840s in the United States. Darwin's theory of natural selection, sometimes called "survival of the fittest" led to social Darwinism, or the survival of the best social, economic, and political systems. Arthur de Gobineau described a "hierarchy of races" in the mid-nineteenth century in terms of intelligence and the advancement of civilization. He classified people from northern European ancestry at the top of the ladder and those with African ancestry at the bottom. In light of de Gobineau's theory and others, the fear of higher birthrates among "inferior races" became a concern on both sides of the Atlantic. A historian of the German eugenics movement, Sheila Faith Weiss, for example, suggests that these thoughts led the great majority of educated whites in Europe and North America to accept "the racial and cultural superiority of the Caucasians as a matter of course" (1987, 194). In the United States this belief manifested itself in fears of immigrants from eastern and southern Europe, and in Germany, in concerns about Jews and eastern Europeans. The degeneracy, inheritance, eugenic, and racial theories spawned different factions of the overall worldwide eugenics movement that included a social welfare, public health, and racist/nativist emphasis. In the early twentieth century eugenics was also considered an emerging science, as were genetics, sociology, and psychology. Eugenics supporters Paul Popenoe and Roswell H. Johnson, in *Applied Eugenics*, for example, note that "the science of eugenics consists of a foundation of biology and a superstructure of sociology" (1918, v).

In the United States eugenics became an underlying theme in many health-reform crusades, including the prohibition, sexual purity, birth control, antiprostitution, pure food and drug, and anti–venereal disease campaigns of the Progressive Era. However, "unlike other health movements of this era, such as prohibition and tuberculosis, the eugenics movement never became a crusade of the masses. Eugenics largely remained a matter of concern with the upper middle class, supported by leaders in biology, psychology, criminology, social work, sociology, liberal religion, and medicine" (Engs 2003, 115). In Germany and Britain eugenics became entwined with the public health and social welfare movements aimed at improving national vitality and health.

Background Information Concerning the Eugenics Movement

The importance of the "rights of society" versus the "rights of the individual" goes in and out of fashion and has been reflected in the trends of hereditarian and eugenics movements. The American academic Louis Menard, in *The Metaphysical Club* (2001, 441), argues that the "good of society" was more important than the "rights of the individual" in the early twentieth century. This attitude changed as "the great movement to secure civil liberties in the United States during the Cold War . . . was founded on the belief that every individual has an inalienable right to those freedoms by virtue of being human." Eugenics supporters in the early decades of the century, including prominent educators, physicians, scientists, and social workers, were convinced that eugenic practices would benefit humanity by curtailing reproduction among the unfit, thereby reducing disease and the cost of charity and public welfare. At that time, legislation for mandatory sterilization of the mentally ill and disabled, criminals, prostitutes, and even the poor was considered a humanitarian effort for the common good of society. By the late twentieth century, the rights of the individual, even if harmful to society or that individual, were considered more important. The mentally ill, for example, had the "right to be homeless" and to not be forced into custodial care. Reflecting this attitude, the new eugenics at the turn of the twenty-first century was driven not by a mass movement or by social reformers but by consumer demands for the "best product." In other words, by their individual "right" to have a mentally and physically healthy baby.

In the midst of the worldwide depression of the 1930s, although the number of sterilizations peaked, the eugenics movement shifted emphasis and the influence of some of its organizations waned. However, there are contested interpretations regarding the movement's change and decline. Genetic research suggested that environment, not just heredity, was important in molding human characteristics such as social achievement and intelligence, and even health leading to increased interest in environmental solutions. In the United States, where the focus was on negative eugenics, many eugenic measures such as sterilization, immigration restriction, and marriage license requirements had already been legislated, lessening interest in the crusade. In Britain laws regarding negative eugenics procedures, such as sterilization, were never passed because most Britons considered them an abridgement of individual rights. Germany, however, passed sterilization laws and legislation forbidding Germans to marry "undesirable races," especially Jews. In the aftermath of World War II, public awareness of the extremist and convoluted form of negative eugenics—genocide—that had occurred under the Nazi regime caused all eugenic ideals, supporters, and organizations, particularly in the United States, to be discredited, rejected, and ultimately demonized. The term *eugenics* became associated with the Holocaust and racism. Some claimed it had been a pseudoscience and many considered it taboo. Even scholars considered the topic unworthy of study until the 1960s, when historian Mark H. Haller published *Eugenics: Hereditarian Attitudes in American Thought* (1963), followed by Donald K. Pickens's *Eugenics and the Progressives* (1968).

In the immediate post–World War II era and in response to social and political changes, components of the eugenics movement, like some other early-twentieth-century health and social reform movements, shifted focus. For example, in the

United States the National Tuberculosis Association and the tuberculosis movement by the 1950s had evolved into the American Lung Association and focused upon the elimination of smoking. Many eugenics organizations and the eugenics movement had became an aspect of population studies, sociology, anthropology, family planning, psychology, medical genetics, genetic counseling, and social biology. In the postwar era, environmental theories of human behaviors came to the forefront, and heredity was believed to have little effect on differences in intelligence or achievement. Research concerning inherited differences of personality traits and abilities among population groups all but disappeared in the United States.

By the turn of the twenty-first century, contested interpretations of eugenics and the early-twentieth-century movement had been offered. A few suggested that it was primarily a racist campaign based upon pseudoscience, while others argued that it was a humanitarian effort and part of the public health, hygiene, physical fitness, and social welfare reform crusades of the era. The worldwide movement varied and embraced all these points of views, depending upon the country. Haller (1984, 5) argues that "eugenics at first was closely related to the other reform movements of the progressive era and drew its early support from many of the same persons. It began as a scientific reform in an age of reform." Political scientist Diane B. Paul (1995, 17) suggests it was a diverse movement reflecting scientific and social beliefs of the era, but notes that "many eugenicists were racist and reactionary, even by the standards of their own times, for some of them, the individual counted for nothing, the larger community all; only the 'fit' had the right to survive, or at least reproduce." Many current interpretations of the movement have focused only upon negative eugenics and have neglected other aspects of the movement that evolved into modern statistics, genetics, psychological testing, anthropology, medical genetics, and other applied sciences. Positive eugenics programs, such as encouraging women to have adequate diets and abstain from alcohol and tobacco during pregnancy, prenatal care to increase the probability of having a healthy child, and well-baby clinics to discover health problems for early correction, are universally part of health care and can be attributed to the eugenics movement, but they have rarely been discussed in detail.

Some works published over the past few years have compared the earlier eugenics movement with the current interest in genetics, genetic engineering, and assisted-reproduction methods. Some condemn the contemporary "new eugenics" movement, fearing the specter of a *Brave New World*, while others suggest that genetic manipulation is critical for the future of a healthy population. Still others argue that the techniques have little in common with eugenics. The current interest in eugenics is expressed in works that range from *The Perfect Baby* (1997), Glenn McGee's discussion of medical-technical choices for achieving a healthy infant, to the controversial *The Bell Curve* (1994) by American researchers Richard J. Herrnstein and Charles Murray, and *Eugenics* (2001) by British psychologist Richard Lynn, in which the authors examine intelligence-test score differences among socioeconomic classes and ethnic groups. The many interpretations of the eugenics movements and their leaders, results from psychological and social sciences research studies, and the advancement in genetic and bioengineering technology will likely lead to continued discussion and more controversy concerning the role of eugenics in the past, present, and future. As suggested by science historian Daniel J. Kevles (1985, 300), "How the public, or politically powerful public coalitions, will respond to the steady pressure of problems raised by the advance of genetics depends upon what reconciliation society chooses to make between . . . social obligations as against individual rights, and reproductive freedom and privacy as against the requirements of public health and welfare."

List of Entries

Whose heart does not beat high at the bare possibility of becoming the progenitor of a world, as it were, of pure holy, healthy, and greatly elevated beings—a race worthy of emerging from the fall—and enstamping [sic] on it a species of immortality.

William Alcott, *The Physiology of Marriage*, 1866

ABORTION

The termination of a pregnancy, generally during the first trimester, is termed *abortion*. Elective abortion has been an extremely contentious subject, particularly in the United States, since the latter half of the twentieth century. Before that time it was hardly discussed and carried out only in secret. The procedure is often accomplished when a fetus has been found to be afflicted with an incurable **genetic disease** or other condition. It is sometimes used in **assisted reproduction** techniques such as **in vitro fertilization** to reduce the number of implanted embryos. Abortion is rarely used in Western cultures for **family balancing** (selecting an embryo of the desired gender for non-medical reasons). In the 1960s the women's rights movement lobbied for safe and legal abortions. Legislative reform was first enacted in Colorado in 1967, and by 1970 eleven other states had passed abortion reform laws. These laws followed the guidelines of the America Law Institute and were endorsed in 1967 by the American Medical Association. They recommended that abortion should be available: (1) when continuation of pregnancy would threaten either the life or health of the mother; (2) when the child might be born with a grave physical or mental defect; and (3) when pregnancy resulted from rape, incest, or other felonious intercourse, and threatened the mental or physical health of the mother. In January 1973 the Supreme Court handed down a landmark decision in *Roe v. Wade*, affirming the legal right for women anywhere in the United States to legally receive an abortion in the first trimester of pregnancy.

The Supreme Court decision precipitated a strong backlash against abortion. Grassroots groups calling themselves "right-to-life" or "pro-life" organizations formed under the authority and support of the Roman **Catholic Church** and some **Protestant** denominations. The National Right to Life Committee (NRLC), one of the largest pro-life associations, was founded in 1973 as an umbrella organization. Its stated goal is the "protection of human life from abortion, euthanasia, and infanticide." This and other national organizations formed a pro-life coalition that gained power on the legislative front over the rest of the century in an effort to prevent all forms of pregnancy termination. In opposition to this campaign, several grassroots organizations coalesced to form the National Abortion Reproductive Action League (NARAL). The purpose of this "pro-choice" organization is to "maintain the right to legal abortion for all women." Abortion is an entity of the **new eugenics**.

FURTHER READING: Jacoby, Kerry N., *Souls, Bodies, Spirits* (1998); Segers, Mary C., and Timothy A. Byrnes, (eds.), *Abortion Politics in American States* (1995); Sloan, Don M., with Paula Hartz, *Abortion* (1992).

ACQUIRED CHARACTERISTICS

See Lamarckian Inheritance of Acquired Characteristics

AFRICANS AND AFRICAN AMERICANS

People with ancestors from sub-Saharan Africa are generally referred to as *African*. Although people from Egypt, Algeria, Morocco, Libya, and other countries around the southern Mediterranean are technically African, they tend to be known as *North Africans*. Native-born Americans with sub-Saharan ancestry were referred to as *African Americans*, *blacks*, and sometimes *Afro Americans* in the late twentieth century. *African American* could include peoples in all of the Americas, but the term usually refers to those living in the United States. From the mid-nineteenth until the mid-twentieth century, *Negro* or *colored* was commonly used; *Negroid* was the scientific classification. Nineteenth-century racial theorists, such as **Arthur de Gobineau**, classified Africans, or the "black **races**," as being on the bottom of a **"hierarchy of races"** in terms of ambition, **intelligence**, advancement in civilization, creativity, and "racial worth." In many Western cultures, this theory reinforced negative attitudes regarding the intermarriage of blacks with other races and fears of **racial degeneracy** and the ultimate decline of Western civilization from racial mixing.

The ancestors of most African Americans came from west Africa as slaves or indentured servants, beginning in the early seventeenth and ending in the mid-nineteenth century. At the turn of the twentieth century, more than 80 percent of African Americans lived in rural southern areas. Beginning around **World War I** and continuing at an increased pace during and after **World War II**, millions of blacks migrated from southern farms and rural areas to cities for better job opportunities. This migration led to African Americans becoming the majority population in many urban areas by the late twentieth century. However, by the 1990s a reverse migration to the South emerged among middle-class blacks. According to the United States census, blacks made up 11.6 percent and 12.9 percent, respectively, of the population in 1900 and 2000. During the early twentieth-century eugenics movement, African Americans were often socially invisible or ignored unless a health or social problem threatened to spill out into the dominant middle-class **Anglo-Saxon** culture.

FURTHER READING: Cavalli-Sforza, L. L., *Genes, Peoples, and Languages* (2000); Russell, Cheryl, *Racial and Ethnic Diversity* (2000); Salzman, Jack, David Lionel Smith, and Cornel West, (eds.), *Encyclopedia of African-American Culture and History* (1996).

ALCOHOL

Ethanol, commonly called *alcohol* or *beverage alcohol*, has been used by most cultures throughout history in one form or another. Types of beverages have included ales and beer, wine, mead, distilled spirits, and hard cider. Alcohol is produced from the fermentation of sugars in crushed fruits or grains by yeast. In different geographic regions of the world drinking preferences, attitudes, and behaviors were established in the distant past. In Europe two distinct drinking cultures developed, ranging from consumption of grain-based beverages, often to intoxication, in the north to moderate consumption of wine in the south. These drinking norms accompanied immigrant groups when they came to the New World. Overuse of alcohol—generally more than four or five drinks per day over time—can lead to physical and psychological damage and chronic **alcoholism**. In pregnant women it can cause physical and mental damage to the developing fetus. Some reformers in the nineteenth century saw alcohol as a major cause of **degeneracy**, leading to a decline in civilization. It was thought that damage to a parent could be passed to offspring through the mechanism of **Lamarckian inheritance** of acquired characteristics. In the first part of the twentieth century, alcohol was classified as a **racial poison** by some eugenicists. In the United States the Eighteenth Amendment to the Constitution went into effect in 1920 to eliminate all alcohol from society; it was repealed in 1932 as unworkable. Later, concerns about negative behaviors surrounding youthful drinking resulted in prohibition for people under age twenty-one in 1987.

FURTHER READING: Engs, Ruth Clifford, *Clean Living Movements* (2000); Engs, "Do Traditional Western European Drinking Patterns have Roots in Antiquity?" (1995); Wiseman, James, "In Vino Veritas" (1997).

ALCOHOLISM AND ALCOHOLICS

Heavy drinking of alcoholic beverages that causes severe physical, psychological, and/or social problems for an individual, one's family, and society has been termed *alcoholism, alcohol addiction, alcohol dependance*, or *problem drinking*. For centuries the problem has been noted to run in families. In the mid-nineteenth century the condition was termed *inebriation, drunkenness*, or *intemperance* and was regarded as being inherited through the mechanism of **Lamarckian inheritance** of acquired characteristics. Health and social reformers of the nineteenth and early twentieth centuries viewed intemperance as the root cause of **crime**, **insanity**, **pauperism**, **prostitution**, **racial degeneracy**, **venereal disease**, and most other problems of society. Some eugenicists in the early twentieth-century eugenics movement encouraged **eugenic sterilization** of alcoholics in order to prevent this trait from being passed to offspring. By the end of the century alcohol dependence was considered multifaceted and caused by a combination of hereditary and environmental factors. Around 10 to 15 percent of individuals who consume alcohol are likely to become addicted to it. From a worldwide perspective there is no consensus as to the cause, prevention, treatment, or even what constitutes alcoholism.

FURTHER READING: Begleiter, Henri, and Benjamin Kissin, *The Genetics of Alcoholism* (1995); Engs, Ruth C., (ed.), *Controversies in the Addiction Field* (1990); Levin, Jerome D., *Alcoholism* (1990).

ALCOTT, WILLIAM A. (August 6, 1798–March 29, 1859)

An American pioneer of health and physical education, Alcott promoted a healthy lifestyle and "hygienic marriages" to improve the human race. He was a leader of the **Jacksonian Era** hereditarian and **clean living movements** (1830s–1860). Alcott was born in Wolcott, Connecticut, of an old New England **Anglo-Saxon** family, was educated locally, and began to teach at age eighteen. Declining health due to **tuberculosis** inspired him to read medical books and enter Yale Medical School to gain a knowledge of physiology and the **laws of health**. He received a medical diploma to practice medicine and surgery (1826) and practiced briefly, but his health declined. He found that abstinence from **alcohol**, a near-vegetarian diet, and going outdoors helped restore his health. When further attempts at teaching again resulted in a decline of health, he left teaching and published educational tracts. In 1831 Alcott moved to Boston to edit an educational magazine and a periodical for children. In 1834 a series of his children's articles, called "The House I Live In," was published as a book designed to provide physiological and anatomical information for youngsters. He also wrote numerous health and educational works based upon his experiences and became nationally known as an education and health reformer.

Alcott promoted hygienic marriages to improve the human race in how-to manuals such as *The Young Man's Guide* (1835). Like other American **Protestant** reformers of the era, he advocated the position that the human race had degenerated from its "Biblical origins." In *Physiology of Marriage* (1866) Alcott suggested that once laws of health such as a semivegetarian diet and abstinence from alcohol were universally adopted, each generation would be sturdier and longer-lived than the previous one until the original long-life expectancy found in the Garden of Eden was regained. He was one of the first reformers to address **inherited realities**, which held that **alcoholism**, lack of exercise, and promiscuity could cause disease and debility to the "third and fourth generation," as hypothesized by **Lamarckian inheritance** of acquired characteristics. The use of Alcott's recommendations by phrenologists, including **Orson** and **Lydia Folger Fowler** in their **phrenology** publications, helped spread the protoeugenics hereditarian concept among the middle class. Alcott married Phebe Bronson (1836), with whom he had two children. His last years were spent in Newton, Massachusetts.

FURTHER READING: Walters, Ronald C., *American Reformers 1815–1860* (1978); Whorton, James C., *Crusaders for Fitness* (1982).

AMERICAN BIRTH CONTROL LEAGUE (1921–1939; Planned Parenthood Federation of America 1942–present)

The main educational and lobbying body of the birth-control movement, the American Birth Control League (ABCL) provided contraceptives primarily for poor and immigrant women not otherwise served by the medical community. The organization and its activities were considered controversial. In December 1916 **birth control** leader **Margaret Sanger** founded the Birth Control League of New York, which the following year launched the *Birth Control Review* as its official publication. On November 10, 1921, the ABCL was founded "to enlighten and educate all sections of the American public in the various aspects of the dangers of uncontrolled procreation and the imperative necessity of a world program of Birth Control." The typical member was a mainline **Protestant**, **Anglo American**, upper-middle-class, married, and educated woman with fewer than three children. The organization's board was composed of physicians, scientists, and prominent New York society women. Board members included eugenics supporters **Clarence C. Little**, **Lothrop Stoddard**, and **Lydia De Vilbiss**. In 1923 the first legal birth control clinic in the nation, the Clinical Research Bureau, was opened under the league's auspices.

Sanger led the ABCL until June 12, 1928, when clashes with the board forced her to resign. The league was largely funded by Sanger's husband, James Noah Henry

Slee (1860–1943), who in 1929 terminated his financial support of it. Without Sanger's notoriety and Slee's money, the league lost its stature in the 1930s. After her resignation Sanger assumed full control of another clinic she had founded and renamed it Birth Control Clinical Research Bureau. In 1937 the Birth Control Council of America was founded to coordinate the work of the old league and Sanger's bureau. In 1939 the two organizations merged to form the Birth Control Federation of America, which in 1942 changed its name to the Planned Parenthood Federation of America. The organization remains a major community force in reproductive health services for both women and men. However, like its early predecessors, the federation has been considered controversial by some **Protestant** groups and the Roman **Catholic Church** because some of its centers began to offer **abortion** services in the late twentieth century.

FURTHER READING: Chesler, Ellen, *Woman of Valor* (1992); Gordon, Linda, *Women's Body, Women's Right* (1990); Kennedy, David M., *Birth Control in America* (1970); Reed, W. James, *From Private Vice to Public Virtue* (1978).

AMERICAN BREEDERS ASSOCIATION (1903–1913; American Genetics Association, 1914–present)

Created during the first decade of the twentieth century, the American Breeders Association (ABA) was the first professional society to focus on the scientific study of heredity and eugenics in the United States. The ABA was founded December 29, 1903, as a spinoff of the Association of American Agriculture Colleges, meeting in St. Louis. The aim of the new organization was to further the practical side of heredity of all organisms, based upon the newly rediscovered rules of **Mendelian inheritance**. The Committee on Eugenics was formed in 1906 at the association's second meeting to "investigate and report on heredity in the human race." The committee was upgraded to the **American Breeders Association Eugenics Section** in 1910, with educator **David Starr Jordan** as chair and **Charles Davenport**, director of the **Eugenics Record Office**, as secretary. It became a powerful component of the association and greatly influenced the direction of the eugenics movement.

Reflecting the growth of genetics and eugenics study and the fact that many members of the organization were not livestock breeders, the association voted on November 15, 1913, to change its name to the American Genetic Association. It published the *Journal of Heredity*, and **Paul Popenoe**, later involved with the **Human Betterment Foundation**, served as editor. By the 1930s the association and its journal focused on **genetics**. In the late twentieth century it focused on various aspects of molecular biology, **genomics**, and **genetic engineering**.

FURTHER READING: Haller, Mark H., *Eugenics* (1984); Kevles, Daniel J., *In the Name of Eugenics* (1985); Paul, Diane B., *Controlling Human Heredity* (1995).

AMERICAN BREEDERS ASSOCIATION EUGENICS SECTION (1910–c. 1939)

A division of the **American Breeders Association** (ABA), the eugenics section greatly influenced the direction of the eugenics movement in the United States. Its membership overlapped with many eugenic groups, and it constituted the administrative board of the **Eugenics Record Office** (ERO). This section emerged out of the ABA's Committee on Eugenics, formed in 1906. The committee was upgraded to a section

PRICE, SINGLE COPY, 50 CENTS

WASHINGTON, D. C.
1910

AMERICAN BREEDERS MAGAZINE

Published by the American Breeders Association

The New Magazine has a Place
SEC'Y OF AGRICULTURE JAMES WILSON

Breeding Fat and Protein into Corn
DR. L. H. SMITH

Breeding the Army Horse
CARLOS GUERRERO

Poultry Breeding in Australia
D. F. LAURIE

Imperfection of Dominance
DR. C. B. DAVENPORT

The New Magazine—Editorials

First Quarter

Vol. I

Jan., Feb., Mar.

No. 1

Application for entry as second-class matter at the post-office at Washington, D. C., pending.

American Breeders Magazine, founded in 1910 and published by the American Breeders Association, was one of the first professional journals to publish eugenic research studies. (From: *American Breeders Magazine* [1910], Main Library, Indiana University, Bloomington, Indiana. Image courtesy of the Digital Library Program.)

in 1910 with the purpose of improving the **race** by "a better selection of marriage mates and the control of the reproduction of the defective classes." Many pioneer eugenicists were members of the section, including **David Starr Jordan**, **Charles Davenport**, **Alexander Graham Bell**, **Luther Burbank**, and **Roswell Johnson**.

The eugenics section, in conjunction with the ERO, conducted **family history studies** for characteristics believed to be inherited. It appointed subcommittees to study and make recommendations concerning **feeblemindedness**, **insanity**, **criminals**, deaf-mutism, inherited epilepsy, and immigration. Out of committee reports, the section advocated measures to eliminate **racial poisons** such as **alcohol**, admonished men to avoid **venereal diseases** that "decreased their value as parents," and supported "licensing only the fit to marry." It recommended **eugenic sterilization** and **immigration and marriage-restriction laws**. In the late 1920s, due to new findings in the field of **genetics**, environmental influences were also seen as a major determinant in human behavior. The American Genetics Association shifted to genetic research in the 1930s, and by 1939 the eugenics section had lost influence.

FURTHER READING: Haller, Mark H., *Eugenics* (1984); Kevles, Daniel J., *In the Name of Eugenics* (1985); Paul, Diane B., *Controlling Human Heredity* (1995).

AMERICAN BREEDERS MAGAZINE
See Journal of Heredity

AMERICAN EUGENICS SOCIETY (Eugenics Committee of the United States, 1922–1925; AES, 1925–1972; Society for the Study of Social Biology, 1973–present)

At the peak of the eugenics movement in the 1920s and early 1930s, the American Eugenics Society (AES) was one of the most important eugenic organizations in the **United States**. Membership included notable scientists, health reformers, and wealthy individuals. Its purpose was to promote eugenics on the popular level and to apply eugenic principles to the "improvement of the human race through education and legislation." It advocated **eugenic sterilization**, **segregation**, and **marriage-restriction laws** for the mentally ill and disabled. It proposed **immigration restriction laws**, a higher birth rate among the middle class, and the widespread use of **birth control** among the poor and **unfit**. Its primary activity was to sponsor **fitter family** contests at agricultural fairs in the Midwest and eugenics sermon contests. It set up exhibits at county fairs, schools, and libraries. The society participated in legislative initiatives and sent representatives to national and international conferences. It was the counterpart to the British **Eugenics Society** and the **German Society for Race Hygiene**. Its official journal was *Eugenics* (1928–1931) and it published a pamphlet, "A Eugenics Catechism" (1926) and it's replacement, *Tomorrows Children* (1935) in addition to other material.

The society emerged from the **Second International Congress of Eugenics**, held in 1921. At the conference an ad interim committee to guide and promote U.S. eugenics education was formed, with **Irving Fisher** as chairman. This committee evolved into the American Eugenics Society on October 31, 1925. It was officially incorporated January 30, 1926 by eugenics supporters Henry Crampton, Fisher, **Henry Fairfield Osborn**, **Madison Grant**, **Harry H. Laughlin**, **Clarence C. Little**, and **Harry Olson**. **Charles Davenport** was vice president and **Henry Fairchild**, secretary-treasurer. **Leon Whitney** was appointed full-time field secretary with an office staff of seven. There were 928 charter members. Presidents during its most active years were **Roswell Johnson** (1926–1927), Laughlin (1927–1929), Little (1929), Fairchild (1929–1931), **Henry F. Perkins** (1931–1934), **Ellsworth Huntington** (1934–1938), and **Samuel J. Holmes** (1938–1940).

By 1931 varied opinions within the AES leadership regarding sterilization, birth control, and **race**, due to increased information regarding **genetics** and environmental influences concerning human behaviors, spawned conflict within the society. The society was embarrassed in 1934 when Whitney, its executive secretary, publically supported **Nazi Germany**'s sterilization program. The organization began to disassociate itself from **nativistic** elements of the eugenics movement, replaced board members with scientists whom they felt had a more "balanced" view, and advocated environmental reform programs to raise the socioeconomic level of the poor. In the 1930s it forged alliances with other groups including the **American Birth Control League**, the American Statistical Association, the Population Council of America, and the **American Social Hygiene Association**. Maurice Bigelow, president from 1940 to 1946, carried the society through the **World War II** years, when it was largely inactive. **Frederick Osborn**, nephew of Henry Osborn, was secretary for much of the society's existence and the moving force for much of its later years. He was elected president in 1946 and served until 1952. The society increasingly turned its interest to the field of hereditary defects and diseases, population problems, and **genetic counseling**. *Eugenics Quarterly* was launched in 1954 to publish research in these areas. The society sponsored five conferences on population genetics and demography at Princeton University between 1964 and 1969. In response to the changing political climate, it changed its name to the Society for the Study of Social Biology in 1973. Its journal was renamed *Social Biology*. The society remains active.

FURTHER READING: Bigelow, Maurice A., "Brief History of the American Eugenics Society," (1946); Haller, Mark H., *Eugenics* (1984); Kevles, Daniel J., *In the Name of Eugenics* (1985); Mehler, Barry, "The history of the American Eugenics Society, 1921–1940," (1988).

AMERICAN GENETICS ASSOCIATION

See American Breeders Association

AMERICAN PHILOSOPHICAL SOCIETY (APS) (1743–present)

A repository of eugenics material, the American Philosophical Society, based in Philadelphia, Pennsylvania, was founded in 1743. Its aims were to pursue "all philosophical Experiments that let Light into the Nature of Things." It was the new nation's first scholarly organization, and over its more than 250 years of existence it has promoted information about the sciences and humanities through scholarly research, professional meetings, publications, library resources, archival material, grants, and scientific prizes. In the late twentieth century it became a repository for material from the eugenics movement in the **United States**. In 1944 documents from the **Eugenics Record Office** (ERO) were transferred to the Charles Fremont Dight Institute for the Promotion of Human Genetics at the University of Minnesota. They included its **family history and pedigree** records, trait cards, eugenic **field workers**' files, correspondence, photographs, and papers from the ERO's leaders **Charles Davenport** and **Harry H. Laughlin**. When the Dight Institute closed in 1991 many of the ERO records were given to the APS. From 1967 to 1995 former leaders of the **American Eugenics Society**, including **Frederick Osborn**, its moving force for most of its later history, and **Leon Whitney**, executive secretary during its peak years, donated their papers to the APS. These included committee records, educational programs from state fairs, and material from **fitter family** contests. The APS holds one of the largest consolidated eugenics collections anywhere, however, it is not a eugenics organization.

FURTHER READING: American Philosophical Society, *Aspects of American Liberty* (1977), http://www.amphilsoc.org/ (downloaded August 22, 2004).

AMERICAN SOCIAL HYGIENE ASSOCIATION (1913–1959; American Social Health Association, 1960–present)

The end product of a merger among several organizations that dealt with the legal, medical, religious, and social aspects of **prostitution**, **venereal disease**, and sexuality, the American Social Hygiene Association (ASHA) was concerned with the **dysgenic** effect of venereal disease and advocated sex education, rather than "prudery and secrecy," as the best method in preparing young people for life. In October 1913 leaders of two groups, the Federation for Sex Hygiene and the American Vigilance Association, met in Buffalo, New York, and formed the association. Its aims were to promote public health and morality and to act as "a clearinghouse for information." Educator **Charles Eliot** was elected its first president, and eugenics pioneer and educator **David Starr Jordan** was on its executive committee. The association published the *Journal of Social Hygiene*. As a result of ASHA agitation, red-light districts were closed adjoining armed-forces training camps in 1917 when the United States entered **World War I**. Working with eugenicists, the ASHA campaigned to prevent the spread of venereal disease to wives and the unborn. By 1920 thirteen states had **eugenic marriage-restriction laws** for untreated venereal diseases. The association changed its name to the American Social Health Association in 1960 and continues to educate concerning sexually transmitted diseases and sexuality.

FURTHER READING: Pivar, David J., *Purity Crusade* (1973); *Purity and Hygiene* (2001).

ANGLO OR ANGLO-SAXON AMERICANS

"Old-stock" Americans of British heritage who can trace their ancestry to the colonial period are termed *Anglo* or *Anglo–Saxon Americans*. This group includes people of English, Scotch-Irish, Scottish, and Welsh descent. Their ancestors emigrated to North America from the seventeenth through the mid-nineteenth century for better economic opportunities, religious freedom, or to escape legal prosecution. Some were indentured servants; others were ousted from small tenant farms by their landlords who wished to raise sheep in what is known as the Highland Clearances. Anglo-Saxon Americans formed the leadership cadre of the hereditarian and eugenics movements from the mid-nineteenth through the mid-twentieth centuries. These northern European **Protestants** took pride in their heritage and accomplishments and considered themselves the leaders of civilization, superior to other ethnic and **racial classifications**. They heralded their form of government, political freedoms, culture, economic system, religion, and good health, which had been achieved, in their view, through inborn superiority, strong moral values, hard work, and thrift. Nineteenth-century racial theorists, such as **Arthur de Gobineau**, classified the "white **races**" as being on top of a **hierarchy of races**. This theory reinforced middle-class bias against intermarriage with members of any group not from northern Europe.

American **nativist** eugenicists and health and social reformers of the **Jacksonian** and **Progressive Eras**' health reform **clean living movements** viewed newer immigrants who were not from northern European cultures as lacking high moral standards, being physically and mentally unhealthy, and genetically **unfit**, compared with themselves. A declining birth rate among the educated white middle class led many reformers to fear **race suicide** among Anglo Americans inasmuch as "low standard immigrants" from **eastern and southern Europe**, such as Polish and Italian **Catholics** and eastern European **Jews**, were reproducing much faster than "more valuable" old-stock Americans. They worried that the U.S. way of life and overall health were in decline and strove to institute programs and legislation to halt this trend. These included **eugenic sterilization**, **marriage-restriction**, and **segregation** laws and **immigration restrictions**. Almost all hereditarian, eugenic, and health reformers in the **United States** through the mid-twentieth century came from Anglo American Protestant backgrounds. The U.S. census of 2000 reports that around 13 percent of American population identified themselves as having British roots.

FURTHER READING: Bannister, Robert C., *Social Darwinism* (1988); Howe, Louise Kapp, (ed.), *The White Majority* (1971); Swain, Carol M., *The New White Nationalism in America* (2002).

ANNALS OF EUGENICS (UK, 1925–1954: *Annals of Human Genetics*, 1954–present)

The publication *Annals of Eugenics: A Journal for the Scientific Study of Racial Problems* was founded in 1925 by British statistician **Karl Pearson**, head of the **Galton Laboratory**. It was the only publication in Britain devoted to eugenics research. It was edited by Pearson, along with **Ethel M. Elderton**, until his retirement in 1933. In contrast, the British *Eugenics Review*, published by **Eugenics Society**, was aimed at eugenics education and promotion. On the front cover of the *Annals* was "I have no Faith in anything short of actual Measurement and the Rule of Three," quoted from **Charles Darwin**. The journal focused on research using the developing field of measurement and statistics. Papers were published by researchers in anthropology, sociology, genetics, population studies, and other areas. Titles ranged from "Studies of Paleolithic Man" to "On the Relative Value of the Factors which Influence Infant Welfare." From its beginnings the journal was a respected scientific publication.

After Pearson's retirement in 1933, new editor **R. A. Fisher** subtitled the publication (beginning with volume 6) *A Journal Devoted to the Genetic Study of Human Populations*. It was then published by both the Galton Laboratory and the Eugenics Society. In 1941 the Eugenics Society was omitted from its title page. S. S. Penrose (1898–1972) succeeded Fisher as editor for volume 13 in 1946, and the journal was then subtitled "A Journal of Human Genetics." Beginning in 1951 Cambridge University Press began to publish the *Annals* for the laboratory. In 1954 the journal's name was changed to *Annals of Human Genetics* with *Annals of Eugenics* printed below as a subtitle. Research reports included studies of inherited traits with titles such as "Sickle-cell Polymorphism" and "Muscular Dystrophy in Northern Ireland." In 1965, with volume 30, new editors were appointed and the eugenics subtitle was omitted. By the 1990s articles were written in the fields of human **genome** variation, statistical genetics, human population genetics, and **Mendelian inheritance** disorders, with titles such as "A Metric Linkage Disequilibrium Map of a Human Chromosome." In 1997 the bimonthly journal also began publishing in electronic format. It continues to be a rigorous journal for professionals in the field of genetics.

FURTHER READING: Keynes, Milo, (ed.), *Sir Francis Galton, FRS* (1993).

ARCHIVE FOR RACIAL AND SOCIAL BIOLOGY (*Archiv für Rassen- und Gesellschaftsbiologie*, 1904–1944)

In 1904 German physician and eugenics pioneer **Alfred Ploetz** founded *Archiv für Rassen- und Gesellschaftsbiologie*, which developed an international reputation as a respected scholarly journal. The *Archive*, a quarterly for most its lifetime, was the first journal in the world dedicated to eugenics. Its aims as stated on its title page were to investigate and publish scientific research on the interaction of race and society, conditions for the preservation and improving the health and efficiency of the human race, and the theory of evolution and how it related to race improvement. In its first few years the journal was funded by Ploetz and his associate founding editors, his brother-in-law Anastasius Nordenholz (b. 1862), ethnologist Richard Thurnwald (1869–1954), and zoology professor Ludwig Plate. Beginning in 1909 B. G. Teubner published the journal until 1921. With volume fourteen in 1922, J. F. Lehmann, publisher of medical and anthropology texts, took over publication of the Journal. Upon establishment of the **German Society for Race Hygiene** in 1905, the journal became its official voice. Notable German physicians, anthropologists, and eugenicists served on its editorial board. Among them were Max Von Gruber (1853–1927), an internationally known biologist and the first German professor of **public health**; psychiatrist Ernst Rüdin (1874–1952), who became editor-in-chief upon the death of Ploetz; anthropologist **Eugen Fischer**, who became director of the **Kaiser Wilhelm Institute**; and **Fritz Lenz**, the first **race hygiene** professor physician. German scholars contributed and so did leading researchers from both sides of the Atlantic, including British statistician **Karl Pearson** and American geneticist **Charles Davenport**. The journal also featured book reviews and abstracts of articles from the U.S. *Journal of Heredity* and *Eugenical News* and the British *Eugenics Review*. These served to inform German eugenicists about trends in other industrialized countries.

Articles published during the first two decades of the **German eugenics** movement were aimed at anthropologists, sociologists, and physicians. These technical reports included genetic research and **family history and pedigree studies**, which included genealogical reports of renowned families, such as the "The Fick Family" and "degenerate families" with several generations of **insanity, feeblemindedness**, and **alcoholism**. Articles appeared on the social and economic costs of protecting the **unfit**, the potentiality of **race suicide** of the Germanic people, along with anthropological and sociological investigations. Titles ranged from "Halting the Population Decline" to "Geological Forces and the Evolution of Life." Few contributions discussed differences between **races**. In the post–**World War I** Weimar Republic articles expressed concerns about the biological consequences of **war** due to the loss of fit young men, and the declining population. Technical studies examined the inheritance of mental illness and other genetic conditions. Titles included "The Consequences of Syphilis and Gonorrhea" and "Human Albinism." A few plant and animal genetic articles were also published. Under the control of **Nazi Germany**, the journal was mandated to limit its scope to human genetics in reports such as "The Reason for Twin Births." Biographies and information concerning organizations were allowed, such as "The Duties and Aims of the German Society for Race Hygiene." Near the end of **World War II**, articles focused on the genetics of mental illness, theoretical questions such as "The Evolution of Organisms," and birth-rate statistics. The final issue of the *Archiv* appeared in 1944.

The *Archive for Racial and Social Biology*, founded in 1904 by German physician Alfred Ploetz, was the first professional journal for eugenics. (From: *Archiv für Rassen- und Gesellschaftsbiologie* [1904], Main Library, Indiana University, Bloomington, Indiana. Image courtesy of the Digital Library Program.)

FURTHER READING: Weindling, Paul, *Health, Race, and German Politics between National Unification and Nazism, 1870–1945* (1989); Weingart, Peter, "German Eugenics between Science and Politics," (1989); Weiss, Sheila Faith, *Race Hygiene and National Efficiency* (1987); Weiss, "The Race Hygiene Movement in Germany," (1987).

ARTIFICIAL OR DONOR INSEMINATION (AI)

The technique by which sperm is clinically placed into the vagina, uterus, or fallopian tubes through a catheter or syringe to create a pregnancy is called *artificial* or *donor insemination*. This procedure is a facet of **assisted reproduction**. It was also known as "artificial impregnation" during the early to mid-twentieth century. American biologist **Hermann Muller** proposed AI as a **positive eugenics** method in 1935. The first documented success was in 1780 in Italy with a dog. A few sporadic successful human inseminations in Britain and France were reported in the late eighteenth century. In the United States the first human success was reported in 1866. However, for humans the procedure was not highly successful until the mid-1930s, when the female fertility cycle was better understood. From the beginning, legal and ethical issues surrounded AI. A report by British infertility researchers in 1945 discussing the procedure generated controversy. Some individuals and groups, including the Roman **Catholic Church**, considered it unnatural or immoral. The procedure continued to be used in the privacy of physicians' offices with both married couples and anonymous sperm donors. In 1953 successful human artificial insemination in the United States with frozen sperm was publicized. By the late 1950s it was estimated that several thousand children had been born as a result of the procedure, and the phrase **test tube baby** was popularized by the media. Livestock breeders also use AI. It did not become a widely accepted treatment for infertility until the 1970s. In the mid-1970s **sperm banks** were founded, where semen from healthy men was frozen in liquid nitrogen for storage for later use. The **Foundation for Germinal Choice** gathered semen of a Nobel laureate, athletes, and other talented men and made it available for married women desiring "superior sperm" as a positive eugenics technique. By the last decade of the twentieth century, single women and same sex couples who wanted children also used AI. They could select the sperm of attractive healthy anonymous donors known for intelligence, achievements, talents, and other characteristics, profiled in catalogs and made available through local or national sperm banks.

FURTHER READING: Foote, Robert H., *Artificial Insemination to Cloning* (1998); Vercollone, Carol, Heidi Moss, and Robert Moss, *Helping the Stork* (1997).

ARYANS OR ARYAN RACE

The term *aryan* originally denoted peoples speaking an ancient Indo-European language in what is now Iran and the Indus Valley. However, over the course of the twentieth century it became associated with northern European populations. The term originated from the ancient Sanskrit term *arya*, which means "noble" and of "good family lineage." It was first used in 1853 to denote the language spoken by *Caucasoids*, or whites, by **Arthur de Gobineau**, who constructed a **hierarchy of races**. Charles Morris (1833–1922) suggested in *The Aryan* (1888) that "the true Aryans" were blonds and that these "fair whites" originated the language. In *The Aryan and His Social Role* (1899), Georges Vacher de Lapouge (1854–1936), in classifying a "racial" hierarchy, designated Aryan as both a language and a racial group, at the top. The elite of the Aryans were tall, fair, Germanic "Teutons" or Scandinavians. By the end of the nineteenth century, ancient speakers of Indo-European languages were considered the ancestors of all European peoples. In the 1920s *Aryan* was used synonymously with *Nordic* for northwestern European populations. Some anthropologists and other scholars, however, in **Germany** and other countries, argued that *Aryan* designated a language, not a racial group, and that *Nordic* was the correct scientific term for the race. Under **Nazi Germany** in the 1930s, *Aryan* referred to non-**Jewish** Germans with northern European ancestry, and in particular, those with blond hair and light eyes. A search for the geographic origins of the "Aryan race" became part of Nazi ideology. After **World War II** the term fell out of general use. By the late twentieth century it was used primarily by white supremacist organizations to designate the "white race."

FURTHER READING: Cavalli-Sforza, L. L., *Genes, Peoples and Languages* (2000); Clay, Catrine, and Michael Leapman, *Master Race* (1995); Morris, Charles, *The Aryan Race* [1888] (1988).

ASIANS OR ASIAN AMERICANS

Populations from East Asia (China, Japan, and Korea), Southeast Asia (Vietnam, Laos, Cambodia, Thailand, and the Philippines), and South Asia (India, Bangladesh, and Pakistan), and certain other ethnic groups from the Asian continent are referred to as *Asians*. Until the late twentieth century this geographic region was referred to as *the Orient*. Asians in particular were referred to as *Oriental* or *Asiatic*, and *Mongoloid* was the scientific classification for this **race**. By the late twentieth century, the term *Asian American* was commonly used for individuals with Asian ancestors in the **United States**. Nineteenth-century racial theorists such as **Arthur de Gobineau** classified Asians, or the "yellow races," as being in the middle of a "**hierarchy of races**" in terms of ambition, **intelligence**, advancement in civilization, creativity, and "racial worth."

Immigration from East Asia to the west coast of North America began in the mid-nineteenth century. Most immigrants were single young men seeking economic opportunities. Because they planned to return home eventually, they did not readily assimilate. The majority were from China and Japan and lived on the West Coast. A few migrated to large eastern cities and worked as unskilled laborers; many built the railroad system across the nation. **Progressive Era** reformers at the turn of the twentieth century and **nativists**, alarmed by increased crime, opium smoking, importation of bubonic plague and cholera, and cheap "coolie" labor, felt threatened by these newcomers and launched virulent campaigns against the "yellow peril." **Immigration restriction laws** were passed to prevent Asians from entering the country, owning property, or gaining citizenship. Eugenic concerns played a part in these laws inasmuch as Asians were not considered as capable or intelligent as "old stock" northern European Americans. However, in spite of the Chinese Exclusion Act of 1882, the Immigration Act of 1917, and other laws, Asian immigration reached record levels between 1882 and 1920.

To enforce the immigration laws and to screen for disease and mental dysfunction, the Angel Island Inspection Station was opened in San Francisco Bay in May 1910. In 1940 a fire destroyed the island and all its files, dating back to 1850, resulting in the loss of most of the history of Asian immigration to the United States. Due to the laws few Asians immigrated from the 1920s until 1952, when the McCarran-Warren Act allowed immigration from South and East Asia and citizenship for immigrants. The Immigration and Nationality Act of 1965 repealed the older laws and facilitated more open immigration for all Asians, many of whom were technically trained or otherwise educated and came with their families. Asian Americans accounted for about 4 percent of the total U.S. population in the 2000 census. Of them, Chinese comprised 24 percent, Filipinos 20 percent, and Japanese 12 percent. More than 50 percent of Asians resided in three states: California, New York, and Hawaii. Most others lived near urban centers.

FURTHER READING: McClain, Charles J., *In Search of Equality* (1994); Ng, Franklin, *The Asian American Encyclopedia* (1995); Russell, Cheryl, *Racial and Ethnic Diversity* (2000).

ASSISTED REPRODUCTION

Technology to enhance the probability of a woman's becoming pregnant is termed *assisted reproduction*. Many forms of assisted reproduction have eugenic implications since eggs, sperm, and embryos can be selected for specific characteristics. Common forms of assisted reproduction include **artificial insemination** (AI), sometimes called *donor insemination*; **in vitro fertilization** (IVF); gamete in vitro fertilization (GIF); and **cloning**, which has yet to result in a live human birth and is fraught with ethical, moral, and religious questions. **Genetic engineering**, **genetic screening and testing**, and **gene therapy** also are important in the process of assisted reproduction. The ability to eliminate certain genetic traits and diseases or to select for gender and other characteristics through assisted reproduction are aspects of the **new eugenics**.

FURTHER READING: Duster, Troy, *Backdoor to Eugenics* (2003); Stacey, Meg, (ed.), *Changing Human Reproduction* (1992).

B

The rich get richer and the poor have children.

Anonymous folk saying

We hold that children should be: (1) Conceived in love; (2) Born of the mother's conscious desire; (3) And only begotten under conditions which render possible the heritage of health.

"The American Birth Control League, Margaret Sanger,
President," *Birth Control Review* 7 (June 1922)

BEING WELL-BORN: AN INTRODUCTION TO EUGENICS (1916)

In *Being Well-Born* geneticist **Michael F. Guyer** argues that both heredity and environment (**nature-nurture**) contributed to human traits and behaviors and that both **positive** and **negative eugenic** measures are necessary for "race improvement." Published in Indianapolis, Indiana, by the Bobbs-Merrill Company, the 374-page book contained many illustrations. For its second edition published in 1927, the subtitle was changed to *Introduction to Heredity and Eugenics* and 100 pages of new **genetics** information was added. The original publication contained ten chapters with titles including "The Bearers of the Heritage," "Mendelism," "Are Modifications Acquired Directly by the Body Inherited?" "Prenatal Influences," "Mental and Nervous Defects," "Crime and Delinquency," and "Race Betterment Through Heredity."

The book was intended to educate the middle class concerning the mechanism of heredity in order to encourage positive eugenics efforts such as wise selection of a marriage partner in order to improve the human race. The author argues that environmental influences such as **racial poisons** like **alcohol**, lead, and **venereal disease** could harm offspring. However, he suggests that "bad environment" alone cannot wholly explain "**degenerate** stocks" and that some individuals are born with "faulty **germ-plasm**," which can lead to **criminal** behavior, **insanity**, **feeblemindedness**, **pauperism**, and other health problems or negative social behaviors. Like other eugenicists, the author argues that since **natural selection** had been eliminated through **public health** and welfare programs, an increased number of "defectives" had entered society, reached maturity, and become capable of reproducing. Guyer also discusses the **dysgenic** effect of **war**, in which the "strong and brave" volunteered for military service during **World War I** while the "coward and weak" and those unfit for service remained at home to reproduce. To counter **racial degeneracy**, the author recommends **eugenic sterilization**, **segregation**, and **marriage-restriction laws** to prevent reproduction among the generically "**unfit**."

BELL, ALEXANDER GRAHAM

(March 3, 1847–August 2, 1922)

A Scottish inventor who moved to the United States, Bell was an early pioneer of the eugenics movement. He was the second of three sons born in Edinburgh, Scotland; both his father and grandfather taught and developed approaches to speech communication. Bell graduated at age thirteen from the Royal High School in Edinburgh, after which he was educated privately by his grandfather in London. At age sixteen he intermittently taught speech and music at preparatory schools and attended the University of London (1868–1870). Upon the death of his grandfather, in 1867 he became an assistant to his father, who was perfecting a method to help the deaf communicate. In 1870, with his parents, Bell emigrated to Canada. The following year he moved to Boston to teach his father's system to deaf-mutes. In order to train

Alexander Graham Bell was one of the first to use field workers for family pedigree studies in the early 1880s when he studied the genealogy of deaf-mute families in Martha's Vineyard. (From: *Journal of Heredity* [1922], Main Library, Indiana University, Bloomington, Indiana. Image courtesy of the Digital Library Program.)

teachers in his methods, he established a private school in Boston (1872) and served as professor of vocal physiology at the School of Oratory at Boston University (1873–1877). Simultaneously, he experimented with and invented a variety of instruments in an attempt to help deaf people hear or feel speech. In 1876 he developed the telephone.

Bell was among the first to use eugenic **field workers** to collect information for **family pedigrees**. He used this procedure in his study of the genealogy of deaf-mute families in Martha's Vineyard in the early 1880s. Influenced by British naturalist **Charles Darwin** and reflecting a growing interest in heredity, Bell began experiments with sheep-breeding at his summer home in Nova Scotia in 1889. These experiments led him into the **American Breeders Association** (ABA) and the foundation of the U.S. eugenics movement. In 1906 Bell became one of the original members of the ABA's Committee on Eugenics. He helped organize the **First International Eugenics Congress**, held in London in 1912. That year he also became chairman of the board of expert directors to guide the research of the **Eugenics Record Office** and chairman of the Committee on Eugenics for the ABA. He was made honorary president of the **Second International Congress of Eugenics** in 1921 and supported the foundation of a national eugenics society that culminated in the **American Eugenics Society**. Bell was in favor of **immigration restriction laws** to "eliminate undesirable ethnical elements" and was an early proponent of eugenics education. By the end of the twentieth century his contributions to the eugenics movement had been largely forgotten. During his lifetime Bell produced more than 500 publications on a variety of topics and was awarded many honors for his inventions. He married Mabel Gardiner, one of his deaf students, and fathered two daughters. He continued work on projects during his later years and died at his summer home in Baddeck, Nova Scotia.

FURTHER READING: Bruce, Robert V., *Bell* (1973); Kevles, Daniel J., *In the Name of Eugenics* (1985).

BELL CURVE, THE: INTELLIGENCE AND CLASS STRUCTURE IN AMERICAN LIFE (1994)

Written by Harvard psychologist Richard J. Herrnstein (1930–1994) and social scientist Charles Murray (b. 1943), *The Bell Curve* proposes that a strong correlation exists between **intelligence** and socioeconomic achievement, as well as between intelligence and "races." The authors project that an expanding "underclass" of less intelligent individuals with high birth rates will generate serious social and economic problems in growing technological societies. The title of this 845-page book refers to the normal distribution of characteristics such as height that are distributed evenly on either side of an average value to form a bell-like curve. Aimed at the educated middle class, the book is divided into four parts: "The Emergence of a Cognitive Elite," "Cognitive Classes and Social Behavior," "The National Context," and "Living Together." The work contains twenty-two chapters including "Cognitive Class and Education, 1900–1990," "Welfare Dependency," "Crime," "Ethnic Differences in Cognitive Ability," The Demography of Intelligence," "The Leveling of American Education," and "The Way We Are Headed."

Herrnstein and Murray present statistical analyses, graphs, and references to numerous studies both inside and outside the United States to support their premise that **intelligence** affects socioeconomic class structure. They argue that due to the increasingly technical nature of society, inherited intelligence is necessary for success and achievement. Because only the intelligent achieve social success, a "cognitive elite" is being produced at the top stratum and a poor underclass at the bottom. The authors contend that social problems in the United States including **crime**, poverty, public welfare, school failure, unemployment, and urban and rural violence are related to low intelligence. They advance the premise that innate intellectual differences exist between ethnic groups and races that cannot be accounted for by culturally or racially biased **intelligence or IQ tests** and that differences in mental ability account for inequalities in socioeconomic status among different groups. Because people with low intelligence have more children, and immigration policy allows immigrants with low intelligence to enter the United States, the authors claim, a growing underclass will lead to increased class conflicts and serious social and political problems. To remedy the situation and to increase the **IQ** of the nation, they recommend **positive eugenics** measures, such as encouraging "smarter women to have higher birth rates than duller women." They also suggest **negative eugenics** policies, such as ending welfare programs to discourage high birth rates among the less intelligent, and barring low-intelligent immigrants from entering the country. *The Bell Curve* reflects the **new eugenics** ideology, and it provoked considerable controversy, particularly in the **United States**, but it was highly popular. Numerous articles and books were published to refute its thesis. Herrnstein died a few weeks before its publication.

FURTHER READING: Herrnstein, Richard J., and Charles Murray, *The Bell Curve* (1994); Jacoby, Russell, and Naomi Glauberman, *The Bell Curve Debate* (1995).

"Better babies" contests, similar to livestock judging contests, were held in the pre–World War I era in the United States to encourage mothers to raise healthy infants as a positive eugenics measure. (From: *Physical Culture* [1916]. Courtesy of Lilly Library, Indiana University, Bloomington, Indiana.)

BETTER BABIES MOVEMENT

A **positive eugenics** and **public health** program, the *better babies movement* was established to educate parents in adequate child care, hygiene, and sanitation during the pre–**World War I** years. Its goal was to improve children's health and prevent **racial degeneracy**. Prevailing thought was that better care and feeding of infants would produce healthier babies, as had been found with domestic animals. The first Better Babies Contest was held in 1908 at the Louisiana State Fair and started by civic-minded leader Mrs. Frank (Mary) de Garmo (1865–1953). She introduced the idea to Mary T. Watts (d. 1926), who established a contest at the Iowa State Fair in 1911. Babies were judged not according to their beauty but, rather, according to their health and strength. In March 1913 the editors of *Women's Home Companion*, a popular women's magazine, developed a Better Babies Bureau after being influenced by de Garmo, headed by physician **Lydia De Vilbiss**. The bureau's goal was to educate mothers in "race betterment" methods through baby contests. It encouraged social workers and others to launch contests and educate parents about hygiene, diet, and positive child-rearing techniques for producing healthier children. Most competitions were at state and county fairs or urban settlement houses. These events encouraged mothers to obtain information on the care and feeding of their babies in order to bring them up to the standards of the prizewinners. When a physician examining a baby pointed out defects, parents were more receptive to health education messages. In 1915 the Children's Bureau a federal governmental agency founded in 1912 organized a "Better Babies Week," during which American mothers were encouraged to have their children weighed and measured. In local communities, club women, extension organizations, doctors, ministers, and others organized better baby clinics. Parents were urged to bring their children for free checkups. Better babies contests in the post–World War I years evolved into the **Fitter Family campaign**.

FURTHER READING: Dorey, Annette K. Vance, *Better Baby Contests* (1999); Holt, Marilyn Irvin, *Linoleum, Better Babies, and the Modern Farm Woman, 1890–1930* (1995).

BIOENGINEERING
See Genetic Engineering

BIRTH CONTROL
Birth control was considered a tool for both positive and negative eugenics. The term *birth control* for pregnancy prevention was popularized by **Margaret Sanger**, pivotal leader of the early twentieth-century birth control movement. Devices such as the diaphragm and the condom and information for preventing pregnancy were considered obscene under the Comstock laws of the **Progressive Era**; they could not be sent through the mails, given out by physicians, described in writings, or even publicly discussed. Contraception was not considered respectable. It was associated with libertine practices such as nonmarital sexuality. In reality, birth control was being used by the educated middle class. Information about methods was quietly passed among networks of women and between wives and husbands. Small families became increasingly fashionable among the middle class at the turn of the century, a trend that eugenics proponents considered a threat to the already declining birthrate among the "fit," leading to **race suicide**. However, by the mid-1920s some eugenicists saw birth control as a measure to reduce the high birthrate among **paupers**, immigrants, and the **unfit**. By the early 1960s new birth control methods such as the contraceptive pill and the intrauterine device (IUD) were developed, making it easier for all women to prevent pregnancy and determine the size of their families.

FURTHER READING: Chesler, Ellen, *Woman of Valor* (1992); Gordon, Linda, *Woman's Body, Woman's Right* (1990); Reed, W. James, *From Private Vice to Public Virtue* (1978).

BIRTH CONTROL REVIEW (1917–1940)

The monthly periodical *Birth Control Review* was the primary journal of the birth control movement and the official organ of the American Birth Control League, which also published material concerning eugenics. The journal was launched in New York City in February 1917 by **Margaret Sanger**, a pivotal leader of the birth control movement in the United States. The masthead read, "Dedicated to the principle of intelligent and voluntary motherhood." In the fall of 1921 Sanger founded the **American Birth Control League** to promote family planning. That November "Birth Control: To Create a Race of Thoroughbreds" was added to the masthead, reflecting increased focus on eugenics as an aspect of family planning. The *Review* published articles relating birth control to eugenics, immigration, temperance, **race suicide**, **social hygiene**, **tuberculosis**, and **venereal disease**. Both **positive eugenics**, encouraging the fit to reproduce, and **negative eugenics**, discouraging the unfit from reproducing, were discussed. Articles with eugenics themes ranged from "The Eugenic Conscience," "Birth Control or Racial Degeneration?" and "The Necessity of Population Control" to "Birth Control in its Relation to Tuberculosis." Over its lifetime the *Review* helped popularize the birth control movement among the middle class and also introduced the concept of eugenics and its relation to family planning. It chronicled activities and conferences and served as the chief source of information for supporters of the movement.

The format of the *Review* included debate of controversial issues, letters from women seeking birth control advice, book reviews, and feature articles. As editor, Sanger used the *Review* to publish many of her own speeches and articles on a wide variety of topics throughout the late 1910s and 1920s. Contributors included eugenics and birth control supporters **Robert Dickinson**, **Henry Fairchild**, **Paul Popenoe**, **Roswell H. Johnson**, **Clarence C. Little**, and **Lothrop Stoddard**. Vigorous street sales and circulation campaigns made the journal available in nearly every state and boosted circulation to about 10,000 copies per issue by 1922. However, due to internal conflicts, Sanger resigned as editor on January 31, 1929. The *Review* continued to appear until July 1933, after which it became a shorter monthly bulletin, lasting until January 1940.

FURTHER READING: Chesler, Ellen, *Woman of Valor* (1992); Reed, W. James, *From Private Vice to Public Virtue* (1978).

BLACK STORK, THE (1916; *Are You Fit to Marry?*, 1927)

A commercial film with a eugenics theme, *The Black Stork* was first released in 1916 and renamed *Are You Fit to Marry?* for the 1927 release. It was produced by newspaper magnate William Randolph Hearst. The silent black-and-white feature-length film aimed to educate the public about **social hygiene** (venereal disease prevention) and **negative eugenics**. It encouraged engaged couples to undergo premarital physical examinations to rule out the presence of **venereal diseases** or inherited conditions that could be passed to offspring. It also encouraged parents to consider allowing their mentally or physically disabled newborn to die rather than use heroic efforts to save the infants and sentence them to a potentially unhappy life. The movie shows individuals who should not marry, including **alcoholics**, the **feebleminded**, **paupers**, and "cripples." It also shows the "eugenically fit." The plot involves a physician who urges parents of a newborn with multiple disabilities, likely caused by congenital syphilis passed unknowingly by the father, to either let the child die or allow euthanasia. After a dream in which the mother sees the child living an unhappy life, passing the defect to many offspring, and murdering the doctor who allowed him to live, she decides to allow the infant to die rather than undergo surgery to save its life. The movie was based upon a real incident and the physician involved, Harry Haiselden, plays himself. The film was popular into the 1930s but was considered controversial because of the topic.

FURTHER READING: Pernick, Martin S., *The Black Stork* (1996).

BLACKWELL, ELIZABETH (February 3, 1821–May 31, 1910)

The first woman medical doctor in North America, Blackwell was a supporter of hereditarian and health reform practices. Born in Bristol, England, the daughter of a wealthy sugar refiner, she was one of nine children in a strongly religious and reform-minded Methodist family. As a child she emigrated with her family to New York, where she attended local schools. In 1838 the family moved to Cincinnati, Ohio, where her father soon died. She helped support her family by teaching school in Kentucky and then, for several years, in Asheville, North Carolina. Anxious to learn medicine, she saved her money, began to read medical texts, and studied with a professor at Charleston Medical College. After rejections from numerous medical schools, she was accepted in 1847 to the Geneva Medical School of Western New York, from which she received her M.D. degree in 1849. Fleeing hostility by the medical community because of her gender, she went to Paris to gain experience in obstetrics. She contracted gonorrhea, a **venereal disease**, in her eye from a patient, resulting in blindness in the eye. She then studied at St. Bartholomew's Hospital in London and returned to New York in 1850 to practice. Due to continuing hostility from the medical establishment and society in general, who mistrusted "lady doctors," with the help of her younger sister, also a physician, she opened the New York Dispensary for Poor Women and Children in 1853. In 1857 it was incorporated into the New York Infirmary and College for Women. The hospital was run entirely by women and supported by New York Quakers. In 1869 Blackwell moved back to England; established a practice; and lectured on hygiene, nutrition, family planning, and sex education. In 1875 she became professor of gynecology at the London School of Medicine for Women, where she remained until 1907.

Blackwell published numerous health education tracts including *Popular and Practical Science: The Laws of Life with Special Reference to the Physical Education of Girls, Vol. 2* (1852), which contained hereditarian—protoeugenics themes. She argued that the teaching of physiology, hygiene, and **physical culture** and exercise were needed to prevent "hereditary diseases," to halt **racial degeneracy**, and to promote steady progress in the improvement of the race. Based upon **Lamarckian inheritance** of acquired characteristics, Blackwell argued that since the time of Adam and Eve neglect of the **laws of health**, including intemperance and improper diet, had led to degeneracy, as evidenced by the disease and "ugliness" found among slum dwellers. Like others of the era she believed that perfect health was tied to physical beauty and considered it unthinkable to imagine "Adam suffering from dyspepsia, or Eve in a fit of hysterics. The thought shocks us—our Eden becomes a hospital." Although she never married, in 1854 she adopted a seven-year-old orphan named Kitty. Blackwell died of a stroke on May 31, 1910, in Hastings, England, after a long illness following a fall.

FURTHER READING: Blackwell, Elizabeth, *Pioneer Work in Opening the Medical Profession to Women* (1980).

BLOOD OF THE NATION, THE: A STUDY OF THE DECAY OF RACES THROUGH SURVIVAL OF THE UNFIT (1902)

An antiwar and pro-eugenics essay, *The Blood of the Nation* was written by educator and eugenics supporter **David Starr Jordan**. The 82-page book was published in Boston by the American Unitarian Association. The work is divided into two parts—"In Peace" and "In War." The author proposes that the "blood of a nation determines its history" and that the "history of a nation determines its blood." The term *blood* is defined as heredity. A major theme suggests that "a race of men or a herd of cattle are governed by the same laws of selection"and proposes that when the fit, brave, and strong are sent to battle to die, the weak and **unfit** remain home and reproduce. It is the descendants of these individuals who, in turn, make up the future character of the nation. Jordan argues that this factor was a cause of the decay of ancient Rome and Greece. He contends that the British **Anglo-Saxon** "race" developed a "superior" civilization due to its primogenitor laws (the oldest son was the sole heir), which constantly integrated younger sons and daughters back into the masses. These individuals with noble ancestry in turn colonized North America and brought democracy to the New World and Britain. The book, like other eugenics works of the day, decried the decrease in birthrate among the educated as leading to **race suicide**. It was favorably reviewed in popular periodicals.

BRITAIN AND THE BRITISH EUGENICS MOVEMENT

Although eugenics concepts were discussed, the British eugenics movement did not emerge until the turn of the twentieth century. The movement is divided into two phases, pre–**World War I** (1901–1914) and postwar (1920–1935). It focused on **positive eugenics** (encouraging the "fit" to reproduce). In Britain the terms *race culture, race improvement*, and **race regeneration** were also used for *eugenics*. In 1869 naturalist **Francis Galton** published *Hereditary Genius*, which discussed the pedigrees of distinguished families and suggested that ability and **intelligence** were inherited. Based upon **social Darwinism**, the survival of the fittest socioeconomic classes and systems, the British middle and upper classes proposed that they had achieved high social positions and wealth because they had inherited superior intellectual and physical gifts. However, the birthrate was decreasing among this group and increasing among **paupers** and "**degenerates**," leading to fear of the nation's decline. As in other Western cultures, the movement was fostered by the educated middle class; was considered a vital social program to prevent **racial degeneracy** and the decay of Western civilization; and was led by men, although many women belonged to its societies. Unlike in the **United States, Germany**, and other nations, however, few eugenics laws were passed.

In Britain two schools of thought concerning heredity and evolutionary change had emerged by 1901. The proponent of one school was British biologist William Bateson (1861–1926). He believed that mutations, or change in inherited material, were necessary for the evolution of the species, based upon the rediscovery of monk **Gregor Mendel**'s rules of inheritance. The other school, championed by Galton, and his protegee, statistician **Karl Pearson**, proposed that biometrics (application of statistics to biological phenomena) and Darwinian **natural selection** could explain the laws of heredity. Two organizations, each supporting one of these philosophies, became the leading institutions of the British eugenics movement. In 1904 Galton established the Eugenics Records Office in London, which merged with Pearson's Laboratory in 1907 and was called the **Galton Laboratory**. Its purpose was to research the relative importance of heredity and environment to human characteristics and diseases through the use of statistics. Pearson soon became its director and aimed to establish a scientific basis for eugenics through biometrics. Pearson and his coresearchers, including **Ethel Elderton**, produced many research reports and founded a scientific journal, *Annals of Eugenics*, in 1925, to publish their findings.

The second organization, the **Eugenics Society**, was established as the Eugenics Education Society in 1907 with the goal of promoting eugenics and lobbying for the implementation of eugenics laws. The society founded *The Eugenics Review* in 1909 to publish articles promoting its cause. Most of its members, including president **Leonard Darwin**, the driving force of Britain's eugenics movement, and physician **C. W. Saleeby**, who popularized eugenics through his numerous writings, embraced the theory of **Mendelian inheritance** and viewed statistics as unimportant. The differences in opinion concerning the mechanism of inheritance resulted in a bitter dispute that lasted until 1930 within the developing field of **genetics** and eugenics, as well as conflict between the Galton Laboratory and the Eugenics Society. Both theories, however, refuted **Lamarckian inheritance** regarding acquired characteristics. In 1930 statistician and eugenics supporter **R. A. Fisher** demonstrated that Mendel's work provided a foundation for **Darwinism** and natural selection, thus uniting the two schools of thought. By the 1950s biometrics and eugenics in Britain had evolved into the fields of population genetics, human genetics, and **social biology**.

The most influential years of the British movement were between 1901 and 1914. During this period numerous books, pamphlets, and articles were published by British writers to educate the middle class about eugenics concepts. For example, Liverpool physician Robert Rentoul (1855–1925) in *Race Culture or Race Suicide?* (1906) advocated **negative eugenics** methods including **eugenic sterilization** and **eugenic segregation** of degenerates. Saleeby's many writings, including *Parenthood and Race Culture* (1909), were popular on both sides of the Atlantic. Psychologist Havelock Ellis's (1859–1939) *Problem of Race Regeneration* (1911) was part of a series on eugenics and sexuality topics. In spite of this advocacy the only major eugenics legislation passed was the Mental Deficiency Act of 1913, which allowed for the segregation of mentally disabled and ill individuals in state-run institutions.

During **World War I** the eugenics movement stagnated, and after 1918 it lost its political power as support from the professional middle class waned. In the postwar era, many eugenicists were alarmed by the elimination of the fittest young men in the **war**, leading a few to support pacifist ideology. Some British intellectuals began to satirize eugenics, including G. K. Chesterton (1874–1936), who wrote *Eugenics and Other Evils* (1922) and Aldous Huxley (1894–1963), who penned the classic *Brave New World* (1932). By the mid-1930s a few eugenicists, including Fisher, ceased to publicly advocate eugenics programs. Several reports, including the *Wood Report* (1929) on the incidence of mental deficiency, the *Brock Report* (1934) on sterilization, and the *Colchester Survey* (1938) on the genetics of **feeblemindedness**,suggested that many health and social problems were caused by both heredity and environment (**nature-nurture**).

In the last two decades of the twentieth century eugenics began to surface again in Britain. Using research from many cultures, British psychologist Richard Lynn (b. 1930) suggested that differences in intelligence among racial and ethnic groups could not be accounted for by cultural differences. He expressed concern that intelligent and creative individuals were not reproducing at the rate of individuals with less ability, leading to the potential decline of Western cultures. A resurgence of positive eugenics emerged, sometimes termed the *new eugenics*, in which couples seek a healthy child through **assisted reproduction**. The United Kingdom became a leader in this field and was the first nation to produce a healthy child, **Louise Brown**, through **in vitro fertilization**. It also became the first nation to successfully **clone** a mammal, "**Dolly**" the **sheep**. Britain continued to be a leader into the first decade of the twenty-first century in **genetic engineering** techniques that, for political reasons, were difficult to carry out in other Western nations.

FURTHER READING: Farrall, Lyndsay Andrew, *The Origins and Growth of the English Eugenics Movement, 1865–1925* (1985); Mazumdar, Pauline M. H., *Eugenics, Human Genetics, and Human Failings* (1992); Searle, G. R., *Eugenics and Politics in Britain, 1900–1914* (1976); Soloway, Richard A., *Demography and Degeneration* (1990); Stone, Dan, *Breeding Superman* (2002).

BROWN, LOUISE JOY (b. July 25, 1978)

Born in Oldham, England, Brown is the first human to have been born successfully as a result of **in vitro fertilization** (IVF)—fertilization of an egg with sperm in a petri dish. An offspring produced by this procedure is popularly known as a "**test-tube baby**." It is a **positive eugenics** technique as only the healthiest embryos are selected for implantation. Brown's parents had attempted to conceive by natural means for about a decade, but blocked fallopian tubes prevented conception. The Browns' gynecologist, Patrick Steptoe (1913–1988), in residence at the Oldham General Hospital, along with Robert Edwards (b. 1925), a Cambridge University physiologist, had attempted IVF in humans for over ten years. Although they had successfully fertilized eggs in petri dishes, all pregnancies from their procedure had lasted only a few weeks. In Louise's case, the embryo was deposited into the uterus of her mother in two, rather than four, days. This resulted in a successful implantation. Due to the mothers's high blood pressure, Brown was delivered a week early by cesarean section. Few details of Brown's life are known because she avoids publicity. She studied nursing, worked as a nursery school aide, and by age twenty-five worked for the post office in the Bristol area. A thousand other people born via in vitro fertilization attended her twenty-fifth birthday party in 2003. However, controversy concerning the ethics and morality of IVF continued into the twenty-first century. By 2004 over a million babies had been born following the procedure worldwide, which became a major facet of **assisted reproduction** and the **new eugenics**.

FURTHER READING: Edwards, Robert G., and Patrick C. Steptoe, *A Matter of Life* (1980); Henig, Robin Marantz, *Pandora's Baby* (2004).

BUCK v. BELL (274 U.S. 200 [1927])

A United States Supreme Court decision, *Buck v. Bell*, upheld a state's right to legally sterilize individuals considered "mentally defective." The case supported involuntary **eugenic sterilization** into the late 1970s and became a model for law in many other states. A white teenager named Carrie Buck (1906–1983) bore a child out of wedlock following rape by a relative of her adopted parents. Because of the social stigma of bearing an illegitimate child, to prevent further embarrassment to the parents, and to protect the relative, Buck was institutionalized at age eighteen by her parents in the Virginia State Colony for Epileptics and Feeble-Minded in 1924. When she was committed she was ruled "epileptic and feebleminded." In later years she and her child were found to be normal. At the same time as Buck's commitment, the state of Virginia enacted a sterilization law and sought a case to test the legislation. With the help of eugenicists **Harry H. Laughlin** and his associate **Arthur H. Estabrook** at the **Eugenics Record Office**, a test case was prepared using Buck as the subject. Because the potential existed that Buck could produce a "less than normal child," inasmuch as her mother was also considered **feebleminded**, she was ordered to be sterilized. Buck's guardian, R. G. Shelton, brought suit against the superintendent of the institution to prevent the surgery, but the circuit court ordered the procedure performed in accordance with the new Virginia state law. When the superintendent died, his replacement, J. H. Bell, continued the case. The Virginia Supreme Court affirmed the ruling of the lower court and held that Buck was the "potential parent of socially inadequate offspring."

The case went to the U.S. Supreme Court with the argument in Buck's defense that the Virginia law was void under the Fourteenth Amendment of the U.S. Constitution on the grounds that it denied her due process and equal protection under the law. The Supreme Court, based upon the principle of compulsory vaccination, upheld Virginia's law on May 2, 1927. It deemed that the health of the patient and the welfare of society may be protected in certain cases by the sterilization of "mental defectives." Noted Supreme Court justice Oliver Wendell Holmes delivered the opinion of the Court and suggested that "three generations of imbeciles are enough." Buck was subsequently sterilized against her will in 1927. Almost fifty years later, in 1981, *Poe v. Lynchburg Training School and Hospital* (the same institution but with a different name) overturned the Virginia law. Thousands of women had been forcibly sterilized for "legal" reasons including **alcoholism**, **prostitution**, **insanity**, and general **criminal** behavior from 1924 through 1979 under Virginia's **negative eugenics** sterilization program. In 2002 Virginia's governor apologized for the state's forced sterilization of about 7,500 people during the time the law was in force.

FURTHER READING: Carlson, Elof Axle, *The Unfit* (2001); Reilly, Philip R., *The Surgical Solution* (1991); Smith, J. David, *The Sterilization of Carrie Buck* (1989).

BURBANK, LUTHER (March 7, 1849–April 11, 1926)

A world-renowned plant breeder, Burbank was an early proponent of eugenics. Born in Lancaster, Massachusetts, he was the thirteenth child of a farmer and brick maker from a family proud of its old New England **Anglo-Saxon** stock. Educated in local schools, he attended Lancaster Academy four years but gained his scientific knowledge from the public library. In 1870, two years after the death of his father, he used his inheritance to buy property near his hometown, became a nurseryman and started experimenting with plant breeding. In 1875 he sold his land and moved to Santa Rose, California, where he developed an experimental garden and created new varieties of fruits, vegetables, and flowers. By 1890 Burbank's income from a variety of new hybrids was sufficient to enable him to devote full time to plant development. In 1905 the **Carnegie Institution of Washington** arranged with him to collate his scientific data as Burbank generally destroyed his field notes; however, the arrangement was terminated after five years.

In Burbank's genetic experiments, he crossbred the "fittest" plants. This research brought him into the **American Breeders Association**. In 1906 he became an original member of the association's Committee on Eugenics. Burbank's work with plants convinced him that the key to good breeding was a combination of **nature-nurture**, or **genetic** selection and the environment. He was one of few eugenicists who retained a belief in **Lamarckian inheritance** of acquired characteristics, long after it had been discredited by the scientific world. When the **Eugenics Registry** was established in 1914 at the **Race Betterment Foundation**, he was a member of its governing committee. Burbank was opposed to open immigration from **eastern and southern European** cultures due to "a large proportion of inferior representatives," and was concerned about **race suicide** among "the better classes of the community." He wrote little over his lifetime and did not seek publicity. In 1907 he published a booklet that carried the eugenics message for child rearing, *Training the Human Plant*. Burbank continued to experiment with plants for the rest of his life, and he achieved worldwide recognition and honors for his plant breeding. He was married twice, first to Helen A. Coleman (1890), whom he divorced in 1896, then to Elizabeth Waters (1916), his secretary, but he fathered no children with either wife. He died of complications from a heart attack at his home and was buried among his flowers.

FURTHER READING: Dreyer, Peter, *A Gardener Touched with Genius* (1975); Gould, Stephen Jay, "Does the Stoneless Plum Instruct the Thinking Reed?" (1992).

C

Sex selection and, soon, cloning and genetic engineering will alter the very idea of parenthood. What happens to the intrinsic value of human life when choosing a baby becomes the ultimate "shopping experience"?

Jeremy Rifkin, "Avoid this dangerous path," *USA Today*
(September 11, 1998)

CARNEGIE INSTITUTION OF WASHINGTON (1902–present)

Philanthropist Andrew Carnegie (1835–1919) founded the not-for-profit Carnegie Institution of Washington (CIW) during the **Progressive Era** to conduct and fund basic scientific research. Some of the institution's studies in the early twentieth century focused upon eugenics. Initially organized in January 1902, it was subsequently reincorporated by an act of Congress in 1904. In its first decade it supported **Luther Burbank**, whose research with plants expanded basic **genetics** information. In 1904 the institution established the Carnegie Institution of Washington's Station for Experimental Evolution at **Cold Spring Harbor**, New York, for genetics research. The first director was **Charles Davenport**, a geneticist and pivotal leader of the eugenics movement. During the second two decades of the century, the institution gave support to studies conducted by the **Eugenics Record Office** (ERO), which became the knowledge base of the eugenics movement. It published **Arthur H. Estabrook's** *The Jukes in 1915* (1916), which helped popularize eugenics. In 1918 **Mary Harriman**, the primary benefactor of the ERO, transferred the office to the CIW. In 1921 the Station for Experimental Evolution absorbed the ERO to form the institution's Department of Genetics, which remained under Davenport's direction. The ERO was maintained as a subsection of the department. During the late 1920s and into the 1930s **Harry H. Laughlin** and others at the ERO became more stridently **nativistic** and supported **Nazi Germany's** **eugenic sterilization** policies. These activities concerned leaders of the CIW, which in 1937 investigated Laughlin's research and deemed it unscientific. The institution discontinued funding, forced Laughlin to retire, closed down the ERO, and disassociated itself from eugenics. In the latter decades of the twentieth century, the CIW continued to support research in many areas of basic science, including developmental and molecular biology.

FURTHER READING: Allen, Garland E., "The Eugenics Record Office at Cold Spring Harbor, 1910–1940" (1986); Trefil, James, and Margaret Hindle Hazen, *Good Seeing* (2002).

CASE FOR STERILIZATION, THE (1934)

Written by **Leon Whitney**, the executive secretary of the **American Eugenics Society**, *The Case for Sterilization* presents arguments for voluntary **eugenic sterilization** and offers support for the *Buck v. Bell* decision, which upheld a state's right to preform legal sterilization of the **unfit**, including the **feebleminded**, the **insane**, and **criminals**. Aimed at the popular market, the book was written to educate the public about the benefits of eugenic sterilization. Its author considers the pros and cons regarding sterilization and recommends that the "fit" have more children. The 309-page book, published by Frederick A. Stokes Company, New York, contains eighteen chapters. Chapter titles include "Sterilization a Burning Issue To-day," "What Is Sterilization?" "The Relation of Mendelism to Sterilization," "Importing Trouble," "Degeneracy in the Making," "Voluntary or Compulsory, and "A Planned Society." Whitney presented material concerning the rules of **Mendelian inheritance** and **genetic diseases**. He discussed "degenerate" families such as the **Kallikaks** and the **Jukes** and argued that to prevent **racial degeneracy** and the survival of the nation, it was important to "cut off the useless classes by preventing their reproduction, and increase the better—that is, the useful and self-sustaining, not necessarily the more brilliant." The author supported **immigration restriction laws** and argued that unrestricted immigration had led to increased numbers of **degenerates** and a rise in social welfare costs. He praised **Germany's** mandatory sterilization law but suggested, "we in this country need have no fear lest any similar wholesale measure be adopted since we are not living under a dictatorship." The book received poor reviews from the *Journal of Heredity* but was popular.

FURTHER READING: Reilly, Philip R., *The Surgical Solution* (1991).

CATHOLIC CHURCH (ROMAN) OR CATHOLICS

The world's largest Christian organization, the Roman Catholic Church has opposed most eugenic and fertility control practices for decades. In many countries the church has exerted strong influence to prevent **birth control**, **abortion**, and **eugenic sterilization** legislation. However, during the early twentieth century the church took an ambivalent position on eugenics. Because it had not given a decision, eugenic issues, including eugenic sterilization, were open to discussion, especially in Europe. This changed when the Papal Encyclical on Christian Marriage (*Casti Connubi*) of December 31, 1930, and two subsequent decrees of March 18 and 21, 1931, were issued that condemned eugenics, birth control, companionate marriage, and divorce. Institutionalization for severe mental retardation and illness was acceptable.

Influential Catholics played various roles in the eugenics movement in Germany, Britain, and the United States. Catholic clergy held leadership positions in the **German eugenics** movement of the 1920s. Some theologians argued that eugenic sterilization, in principle, could be approved in suitable cases. Hermann Muckermann (1877–1962), a former Jesuit and family welfare reformer, became head of the Eugenics Department of the **Kaiser Wilhelm Institute**. He helped lobby for a eugenics sterilization bill in Prussia in 1932; however, it was defeated. In **Britain** Catholic clergy opposed not only artificial birth control but also potential eugenic sterilization legislation. As a result some British eugenicists were reluctant to advocate the use of contraceptives and sterilization out of fear of religious condemnation. British Catholic author G. K. Chesterton (1874–1936) satirized eugenics in *Eugenics and Other Evils*.

During the **Progressive Era** (1890–1920) large numbers of **eastern and southern European** Catholics migrated to the **United States**. Their sheer numbers and contrasting way of life triggered anti-Catholic feelings among "old-stock" **Protestants**, who formed the leadership cadre of the birth control, eugenics, **immigration restriction,** prohibition, and other progressive reform crusades. A contributing factor to these movements was the desire to control the behavior of immigrants, who were often seen as **degenerate.** The Catholic hierarchy and lay members fought these progressive reform efforts. Catholic writers, sanctioned by the church, produced tracts and books condemning artificial birth control as immoral and campaigned against the legalization of birth control devices or information. Crusades against leaders of the birth control movement, especially **Margaret Sanger,** who was born a Catholic, and public discussion of birth control were launched. Political pressure by the church prevented several states from passing eugenic sterilization laws. Because most eugenicists opposed abortions, this was not an important issue during the first half of the twentieth century.

In the last half of the twentieth century, the church continued to oppose artificial birth control and compulsory state sterilization laws. On July 25, 1968, the Papal Encyclical *Humanae Vitae* (Of Human Life) disapproved the birth control pill as a method of family planning. As a movement to legalize abortion emerged in the 1960s, the church opposed all legislation that allowed the procedure. Church members, particularly in the United States, formed a "pro-life" movement to rally support against the U.S. Supreme Court decision that legalized first-trimester abortions. By the last decades of the century the church opposed many aspects of the emerging **new eugenics.** The Vatican considered the birth in 1978 of **Louise Brown,** the first child produced by **in vitro fertilization** (IVF), "an event that can have very grave consequences for humanity" on the grounds that it separated the sexual act from procreation. On March 10, 1987, *in Donum Vitae* (Respect for Human Life in Its Origin and on the Dignity of Procreation), the Holy See declared IVF a mortal sin. It advised that surgery on live embryos could be done only if it did not harm the fetus or mother, and it prohibited using "stem cells" from human embryos for research. By the turn of the twenty-first century, however, many practicing Catholics in Western cultures approved of and used artificial birth control methods to plan their families and took advantage of assisted reproductive techniques to ensure a healthy child, even though these practices contradicted church doctrine.

FURTHER READING: Greeley, Andrew M., *The American Catholic* (1977); Leon, Sharon M, " 'Hopelessly Entangled in Nordic Pre-suppositions' " (2004); Reilly, Philip, *The Surgical Solution* (1991); Rosen, Christine, *Preaching Eugenics* (2004).

CAUCASOID

See Anglo *or* Anglo-Saxon Americans; Euro *or* European Americans; Nordics *or* Nordic Race

CHURCH OF JESUS CHRIST OF LATTER-DAY SAINTS

This Christian denomination, commonly known as the *Mormon* or *LDS Church,* was founded in the **Clean Living Movement** of the **Jacksonian Era** in the United States. Since the church's early years, eugenic concepts have underlain some of its practices, such as mandates for a healthy lifestyle, plural marriages, and reproductive guidelines. Joseph Smith (1805–1844), founder and prophet of the church, formed the religion in 1830 after a series of revelations and visions. The church expanded in the 1830s and established communities in Missouri and Ohio. Because of the Mormons' belief system and their practice of polygamy, they were considered un-American and persecuted. Smith and his brother were killed by a mob in Illinois in 1844. Forced out of the state in 1846, church members under the leadership of Brigham Young (1801–1877) migrated west and settled in the valley of the Great Salt Lake in Utah. In the 1880s polygamy was made a federal offense, and as a condition for statehood the church outlawed plural marriages in 1890. Utah became a state in 1896.

From its early years the LDS Church promoted a morally and physically healthy lifestyle. This health advice is addressed in church scripture, including the "Word of Wisdom" in the *Doctrine and Covenants*. In the nineteenth and very early twentieth centuries a widely accepted belief was the theory of **Lamarckian inheritance** of acquired characteristics. Because "unclean living" was believed to result in damage that could be passed on to offspring, the church frowned upon the use of **alcohol**, **tobacco**, and hot drinks and mandated chastity until marriage. A major tenet of the LDS Church is for "righteous"—and thereby healthy—

Brigham Young, an early leader of the Church of Jesus Christ of Latter-day Saints, and other church fathers promoted positive eugenics by encouraging "worthy" and healthy church members to have numerous children. They encouraged negative eugenics by discouraging mentally or physically unhealthy and "unworthy" members from marrying or having children. (From: *Journal of Heredity* [1916]. Main Library, Indiana University, Bloomington, Indiana. Image courtesy of the Digital Library Program.)

members to produce many children in order to give "pre-existing spirits" a chance to be born in "superior tabernacles"—healthy bodies and minds. Giving pre-existing spirits the best chance to be born healthy resulted in official guidance on reproduction and marriage. From the earliest years "marital abstinence" was encouraged when either partner had "impurities" that could be passed to offspring. Church fathers spoke against the marriage of "undesirables" like the **insane**, the **feebleminded**, **alcoholics**, or **paupers**. In 1925, at the peak of the eugenics movement, the predominantly Mormon state of Utah allowed **eugenic sterilization** of institutionalized "defectives." In practice, however, few sterilizations were performed compared with other states. By the end of the twentieth century **abortion**, **birth control**, and sterilization were frowned upon unless used for life-threatening or serious medical conditions. **Assisted reproduction** using nonspousal sperm or eggs was left to the judgment of husband and wife.

From their beginnings Mormons were both **race** and lineage conscious. Because they believed that they were descended from the ancient "ideal race" of the biblical Adam, they encouraged or allowed marriage only within their faith, to keep the race pure. In the nineteenth century the church sanctioned plural marriages that reflected this philosophy. In general, only the "better class" of "worthy men"—those who were healthy, wealthy, or intellectually sound—were allowed, or could even afford, a plural marriage. Some Mormon writers at the turn of the twentieth century suggested that the "stock was improved" if desirable men produced a large number of children and that plural marriage produced "superior offspring" because morally and physically worthy women generally married superior men. However, by the second half of the twentieth century possible **dysgenic** effects of these early marriages emerged. Some LDS couples with common ancestors from pioneer consanguineous marriages were found to have offspring with a higher incidence of **genetic diseases** than nonrelated couples.

FURTHER READING: Bush, Lester E., Jr., *Health and Medicine among the Latter-day Saints* (1993); Mauss, Armand L., *All Abraham's Children* (2003); Ostling, Richard N., and Joan K. Ostling, *Mormon America* (1999).

CLEAN LIVING MOVEMENTS

Periods of time when a surge of health-reform crusades, many with moral overtones, erupts into the popular consciousness were termed *Clean Living Movements* by the author in 1990. Reform activities during Clean Living Movements include temperance, sexual purity, diet, **physical culture**, **public health**, anti–tobacco and drug, and race-improvement campaigns. Interest in these issues rises and falls more or less simultaneously, and generally follows a religious awakening in which both evangelical sentiments and the development of new sects emerge. These movements manifest themselves in approximate eighty-year cycles, ranging from about seventy to ninety years for specific issues or causes. During Clean Living Movements reformers attempt to "clean up" society. **Germany**, and to some extent **Britain**, have also experienced these health and vitality movements bringing changes in public policies. In the **United States** widespread health agitation and subsequent reforms have, within a decade or so, coincided with the **Jacksonian** (1830–1860), **Progressive** (1890–1920), and **Millennium** (1970–   ) reform eras. Individual health crusades, as part of an overall Clean Living Movement, often experience a cycle of *moral suasion* (education and social pressure), *coercion* (public policies), *backlash*, and *complacency*. The reforming phase lasts from twenty to forty years, during which reformers attempt to change "bad habits" or behaviors that are perceived as harming the individual or society. As the cycle ebbs popular changes or reforms that make sense, on one hand, such as personal hygiene or sanitation, become institutionalized. On the other hand, a backlash often emerges against unpopular or restrictive reforms, such as the prohibition of **alcohol**. When reformers have failed to change behaviors even by legislation, a hereditarian or eugenics movement reaches its prime and the root cause of the social problems is then seen as inherited.

FURTHER READING: Engs, Ruth Clifford, *Clean Living Movements* (2000); Green, Harvey, *Fit for America* (1986); McLoughlin, William G., *Revivals, Awakenings, and Reform* (1978); Whorton, James C., *Crusaders for Fitness* (1982).

CLONES OR CLONING

The process of creating genetically identical cells or organisms is called *cloning*. This process is common in nature, for example, when plants reproduce asexually or when identical twins are produced from the same fertilized egg in mammals. The term *clone* was coined by British biologist J.B.S. Haldane (1892–1964) in 1963, although the concept had previously been discussed by several individuals including American geneticist **Hermann Muller** as "parthenogenesis." Cloning with genetic material removed from a somatic (body) cell and placed in an egg cell was first suggested by German researcher Hans Spemann (1869–1941) in 1938 in **Nazi Germany**. However, the lack of technology prevented the possibility of cloning at that time. In the post–**World War II** years cloning techniques were researched as part of the emerging field of **genetic engineering**. Clones of simple organisms using various techniques had been produced for decades, but the first successful cloning of a complex adult mammal from somatic cells and resulting in a live birth was accomplished by Ian Wilmut and Keith Campbell in Scotland. On July 5, 1996, a lamb named **"Dolly" the sheep** was born. A technique called *somatic cell nuclear transfer* was used. The nucleus from a body cell of an adult sheep was removed and injected into the egg cell of another sheep from which the genetic material had been removed. Then, as in the process of **in vitro fertilization**, after several cell divisions of the fertilized egg the resulting pre-embryo was transplanted into the uterus of yet another ewe and allowed to grow to term. In subsequent years dogs, cats, horses, and other animals have been cloned. However, high rates of failed pregnancies, genetic defects, and health problems have been observed among cloned animals.

The popular perception of cloning at the beginning of the twenty-first century was that clones would exhibit identical personality traits and characteristics to the adult original. However, clones are not complete genetic duplications of the donor inasmuch as the host egg has differences in its mitochondrial genetic material. After the cloning of Dolly, U.S. president Bill Clinton (b. 1946) in 1997 banned federal funding of human cloning. The following year the U.S. Food and Drug Administration took jurisdiction over human cloning, considering it a medical treatment that requires an extensive review process. In 1997 nineteen European states approved a ban on human cloning, and California banned cloning to initiate pregnancy. By 2003 Arkansas, Iowa, Louisiana, Michigan, North Dakota, Rhode Island, and Virginia had human cloning–restriction laws. However, "therapeutic cloning"—creating a human embryo to harvest and culture "stem cells" (cells that can develop into any tissue) from a donor to replace damaged or diseased tissues is being researched in some countries. As of 2005 no living human baby has resulted from the cloning of adult somatic cells. The potential of cloning humans is one aspect of the **new eugenics** and involves considerable controversy.

FURTHER READING: Harris, John, *Clones, Genes, and Immortality* (1998); Kolata, Gina Bari, *Clone* (1998); Winters, Paul A., (ed.), *Cloning*, (1998).

COLD SPRING HARBOR

The epicenter for eugenics research and the eugenics movement during the early twentieth century, Cold Spring Harbor, a community on Long Island, New York, is known for its biological, genetic, and molecular biology research. The first research institution at Cold Spring Harbor was established in 1889 when a land holding from a whaling company was bestowed to the Brooklyn Institute of Arts and Science for a marine biology laboratory. In 1904 the **Carnegie Institution of Washington** (CIW) established the Station for Experimental Evolution under the direction of **Charles Davenport**, a pioneer geneticist, to carry out genetic research based upon **Gregor Mendel's** rediscovered laws of inheritance. In 1910 **Mary Harriman**, a rich widow, donated property next to the station to establish the **Eugenics Record Office** (ERO) to engage in eugenics research, with Davenport as director and **Harry H. Laughlin** as superintendent. In 1918 Harriman transferred the ERO to the CIW. The station was renamed the Department of Genetics and the ERO became a subsection of this department. The ERO undertook **family history and pedigree studies** until 1939, after which the CIW withdrew its support out of concern that the ERO's research lacked a scientific basis and some of its members had publically admired the **eugenic sterilization** program of **Nazi Germany**. In 1941 geneticist Milislav Demerec was appointed director of both the biological laboratory and the Department of Genetics. Research focused upon the nature of the gene. By the mid-twentieth century, Cold Spring Harbor had become an internationally renowned genetics research and educational center. In 1962 the CIW gave up support of the Department of Genetics, which merged with the biological laboratory. The institution was renamed the Cold Spring Harbor Laboratory. **James Watson**, a discoverer of the structure of **DNA**, became director in 1968 and the laboratory shifted to virology and cancer research. In 1988 Watson was named director of the **Human Genome Project**, an initiative to identify the sequence of DNA in the human **genome**, or "human blueprint." The Cold Spring Harbor Laboratory's DNA Learning Center opened in 1988 and the *Image Archive on the American Eugenics Movement* Web site, which features material from the early eugenics movement, went on line in 2000. In 1999 a doctoral program was instituted. At the turn of the twenty-first century, the laboratory conducted research and education in molecular biology, neurobiology, plant genetics, **genetic engineering**, and genomics.

FURTHER READING: Allen, Garland E., "The Eugenics Record Office at Cold Spring Harbor, 1910–1940" (1986); Haller, Mark H., *Eugenics* (1984); Witkowski, Jan, *Illuminating Life* (2000). http://www.eugenicsarchive.org/eugenics (downloaded September 24, 2004).

CONSUMPTION
See Tuberculosis

CRIME AND CRIMINALS
Throughout history *crime* and *criminals* have generally been associated with the underclass and poverty. Because criminality was often viewed as running in families, it was assumed to be inherited. From the mid-nineteenth century, however, opinions as to whether criminality was caused by heredity or environment (**nature-nurture**) have waxed and waned. Crime, however, is a social construct. A behavior that is considered a crime in one generation or culture may be acceptable in another. For example, until the late 1880s the legal age of consent for sexual intercourse for girls was ten in several American states. Today sexual activity between an adult and a child of that age would be considered a serious crime. Based upon French naturalist **Jean-Baptiste Lamarck**'s theory of the inheritance of acquired characteristics, French physician Benedict Morel (1809–1873) suggested that criminality along with **alcoholism**, epilepsy, **insanity**, and **feeblemindedness**, were different manifestations of hereditary **degeneration**. Cesare Lombroso (1836–1909), a founder of criminology, investigated the physical traits of numerous criminals and claimed that a "criminal type" of human

existed. However, others during this era believed that poor environments caused crime and criminality.

During the **Progressive Era** (1890–1920) in the United States crime was associated with immigrants, poverty, **prostitution**, and alcoholism. U.S. crime statistics in 1890 suggested that immigrants were more likely to be prisoners, compared with the white native-born population. Crime syndicates, such as the Mafia among Italian populations and tongs in Chinese communities, were seen as a major force behind violent crime in large cities. The reported high crime rate among immigrants helped ignite the **nativism** surge at the end of the nineteenth century, culminating in the **Johnson-Reed Immigration Restriction Act of 1924**. Because crime was considered inherited, **eugenic sterilization** was recommended as a preventive measure to reduce further crime and **racial degeneracy** (decline of the health and fitness of humankind). The inheritance of criminal tendencies appeared to be reinforced by **family history and pedigree studies** such as *The Jukes in 1915*, in which multigenerational criminality was documented. In the 1930s opinions as to the origins of crime, however, began to shift.

By the 1960s the causes of crime were claimed to be poor childhood environments, joblessness, and lack of educational and other opportunities. Suggestions that inherited traits for aggression lead to criminal and antisocial behaviors were rejected. In 1992 an academic forum was organized to discuss a genetic basis of criminal behavior and the potential of identifying by **genetic markers** individuals who might be predisposed to violent behaviors. The conference, "Genetic Factors in Crime," was cosponsored by the **Human Genome Project** of the National Institutes of Health (NIH). Because 25 percent of young **African American** males were then involved in the criminal justice system, some black leaders feared that this type of conference and its research could promote the **genetic screening and testing** of blacks to determine those biologically predisposed to violence, possibly leading to discrimination. Protests were lodged against holding the meeting, and the NIH withdrew its funding. Although that forum was canceled, one was held in 1995 but it skirted the issues. A number of studies in other countries and a few in the United States have found a correlation between aggressive behavior and abnormal brain chemistry or structure that is likely inherited. Criminality has also been found to be associated with low **intelligence**, poverty, joblessness, inability to stay in school, poor impulse control, and psychopathic personality. The precise role of heredity and environment in the manifestation of criminal and antisocial behavior continues to provoke debate and controversy.

FURTHER READING: Carlson, Elof Axle, *The Unfit* (2001); CIBA Foundation Symposium 1994, *Genetics of Criminal and Antisocial Behaviour* (1996); Rafter, Nicole Hahn, *Creating Born Criminals* (1997); Reilly, Philip, *Abraham Lincoln's DNA* (2000).

D

The general program of the eugenist is clear—it is to improve the race by inducing young people to make a more reasonable selection of marriage mates; to fall in love intelligently.

Charles Benedict Davenport, *Heredity in Relation to Eugenics* (1911)

DACK FAMILY, THE (1916)

One of several **family history and pedigree studies** conducted during the second decade of the twentieth century, *The Dack Family: A Study in Hereditary Lack of Emotional Control* was published by the **Eugenics Record Office** (ERO) at **Cold Spring Harbor**, New York. It contained fifty-two pages and was Bulletin No. 15 in the ERO's series of reports. This "socially troublesome Pennsylvania family," given the name "the Dack family," unlike many other family groups studied by eugenicists was not considered **feebleminded**. Instead, it was reported to have had a "high percentage of insanity, **criminality**, lack of emotional control, sexual immorality and shiftlessness." Anna Wendt Finlayson, a eugenics **field worker** at Warren State Hospital in Pennsylvania, was the author of the study. **Charles Davenport**, director of the ERO, wrote the preface and suggested that the study supported the theory that behaviors related to emotional control were inherited. This line of thinking led to support for passage of **eugenic sterilization** and other laws to prevent the **unfit** from reproducing.

The Dack kinship lived in west-central Pennsylvania near several small mining towns. They were uneducated, worked in mining, and were considered "undesirable" by middle-class members of the community. The Dacks were reported to have descended from two Irish immigrants, "William" and "Mary," who may have been first cousins. Information for the study was furnished by family members, neighbors, and physicians. The study examined three generations of the family and reported on a kinship of 754 individuals, of which 153 were direct descendants of William and Mary. Of those over the age of twenty, 72 were considered **degenerates**, causing many social problems; 25 were **insane**; 20 were "lazy or shiftless"; and 39 were "below average in intelligence." Thirty-four had "bad tempers," 30 were **alcoholic**, and 24 had "thieving tendencies." Of the women, about 27 were **prostitutes** or had engaged in sexual activity outside marriage. Davenport proposed that the Dacks reacted to their environment by being lazy, mentally dull, and having "a monkey-like instinct to steal and hide." It was estimated that about half may have been infected with syphilis, which the author suggested as a possible cause of degeneracy in one branch of the family. The study concluded that social reform needed to take "bad heredity" into consideration.

FURTHER READING: Carlson, Elof Axle, *The Unfit* (2001); Haller, Mark H., *Eugenics* (1984); Kevles, Daniel J., *In the Name of Eugenics* (1985); Rafter, Nicole Hahn, (ed.), *White Trash* (1988).

DARWIN, CHARLES ROBERT (February 12, 1809–April 19, 1882)

A British naturalist, Darwin proposed a theory for the gradual evolution of organisms through **natural selection**. This theory became an underlying theme of the early-twentieth-century eugenics movement, particularly in Britain. Darwin was born in Shrewsbury, England, the fifth child of a physician. His parents were both from distinguished and wealthy British intellectual and manufacturing families. **Francis Galton** was his cousin. Darwin's early education was at the local grammar school. At age sixteen he entered Edinburgh University for two years, then went on to Christ College, Cambridge, where he received a B.A. (1831). Soon after graduation he served as a naturalist on the British exploration ship the *HMS Beagle*, whose scientific mission was to explore South America and surrounding islands and circumnavigate the world. After the five-year voyage, Darwin returned home to live in London and, later, at his country home in Down, outside London (1842). Over the next twenty years he wrote many short monographs and used his observations of wildlife in the Galapagos Islands in an attempt to unify and explain various findings of the trip. In 1858 naturalist Alfred R. Wallace (1823–1913) sent him a paper containing the same explanation of evolution through the survival-of-the-fittest organisms that Darwin had deduced in the early 1840s but had not published.

Both Darwin's and Wallace's papers were read at the Linnaean Society of London. The following year Darwin expanded and elaborated upon his material in *On the Origin of Species by Means of Natural Selection; or, the Preservation of Favoured Races in the Struggle for Life* (1859). The work was so popular that its first edition, published November 24, 1859, sold out the first day. However, it was highly controversial, particularly in the United States, because it negated a biblical creation of humankind.

The evolution of higher forms of life out of more primitive organisms though natural selection and the survival of the fittest became known as **Darwinism**. When applied to other areas, including economic and political systems, ability and **intelligence**, or socioeconomic classes, it became known as **social Darwinism**. This concept was an underlying ideology of the early-twentieth-century eugenics movement as reformers attempted to improve the human race through pro-

British naturalist Charles Darwin proposed a theory of evolution of species based on natural selection that became an underlying theme of eugenics, especially in Britain. (From: *American Breeders Magazine* [1910], Main Library, Indiana University, Bloomington, Indiana. Image courtesy of the Digital Library Program.)

moting better marriage selection (**positive eugenics**) and preventing the **unfit** from reproducing (**negative eugenics**). In 1871 Darwin published a companion to his original work, *The Descent of Man and Selection in Relation to Sex*. In *Origin* he had focused upon the struggle for food, or means of survival, while the second book addressed the struggle for procreation. Darwin was elected to the prestigious Royal Society (1839). He married his cousin Emma Wedgwood (1839), with whom he had ten children. Many of them became well known, including **Leonard Darwin**, the primary leader of the **British eugenics movement**. Charles Darwin suffered increasing ill health over the years and died at his country home, of apparent heart disease, perhaps contributed by the Chagas parasite, which he may have contracted in South America.

FURTHER READING: Darwin, Charles, *On the Origin of Species by Means of Natural Selection* (1985); Dover, Gabriel A., *Dear Mr. Darwin* (2000); Jones, Steve, *Darwin's Ghost* (2000); Moorehead, Alan, *Darwin and the* Beagle (1982).

DARWIN, LEONARD (January 15, 1850–March 26, 1943)

The pivotal leader of the **British eugenics movement**, Darwin occupied leadership positions in many eugenics organizations. Like other British eugenicists he was concerned with the differential birthrate and the decline in population of the upper and middle classes and supported both **positive** and **negative eugenics**. Born in Down, Kent, north of London, Darwin was the seventh of ten children of **Charles Darwin**. He attended the Royal Military Academy, Woolwich, and joined the Royal Engineers of the British Army (1871). He was an instructor at the School of Military Engineering at Chatham (1877–1882), worked for the Intelligence Division of the Ministry of War (1885–1890), and was promoted to the rank of major (1889). He was a member of scientific expeditions including the "transit of Venus" expeditions of 1874 and 1892. After retiring in 1890 at age forty, Darwin became involved in politics and was president of the Royal Geographical Society (1908–1911).

At age sixty Darwin found a new career as a eugenicist. After the death of **Francis Galton** he became president of the **Eugenics Society** (1911–1928) and served during its most active years. His tenure saw bitter conflict in the British eugenics movement between two schools of thought regarding the mechanism of inheritance. Statistician **Karl Pearson**, who used biometrics (applying statistics to biology) at the **Galton Laboratory** and adhered to **Darwinism**, clashed with biologists, who subscribed to the rules of **Mendelian inheritance** and emerging **genetic** theories. Darwin saw both theoretical schools as credible, and under his leadership the society at-

Leonard Darwin, a son of Charles Darwin, was a pivotal leader of the British eugenics movement. (From: *Eugenics in Race and State* [1923], Main Library, Indiana University, Bloomington, Indiana. Image courtesy of the Digital Library Program.)

tempted to remain neutral. Darwin took leadership positions in international eugenics events. He was chairman of the **First International Eugenics Congress** (1912). For the **Second International Congress of Eugenics** (1921) he gave the lead address, "Aims and Methods of Eugenical Societies." Although he was not able to attend the **Third International Congress of Eugenics** he sent a message, printed in *A Decade of Progress in Eugenics* (1934), in which he asserted, "My firm conviction is that if wide-spread Eugenic reforms are not adopted during the next hundred years or so, our Western Civilization is inevitably destined to such a slow and gradual decay as that which has been experienced in the past by every great ancient civilization."

Darwin became the early mentor to geneticist and statistician **R. A. Fisher** and encouraged him to bring together the Darwinian and Mendelian schools of thought concerning the mechanism of inheritance through the use of mathematics. Darwin wrote many eugenics articles, particularly for *The Eugenics Review*, the organ of the Eugenics Society. His books included *The Need for Eugenic Reform* (1926) and *What Is Eugenics?* (1928). During **World War I**, as a war effort, he recommended the control of **prostitution** to prevent **venereal disease**. He married Elizabeth Fraser (1882), who died in 1898, and in 1900 he married Charlotte Mildred Langton. Darwin had no children. He died at his home in Cripps Corner, Sussex, in southern England.

FURTHER READING: Bennett, J. H., (ed.), *Natural Selection, Heredity, and Eugenics* (1983); Mazumdar, Pauline M. H., *Eugenics, Human Genetics, and Human Failings* (1992); Searle, G. R., *Eugenics and Politics in Britain, 1900–1914* (1976).

DARWINISM

The process by which organisms survive to pass on desirable traits to their offspring through **natural selection** (survival of the fittest) and which over time leads to evolutionary changes in the species is called *Darwinism*. Darwinian thought became an underlying theme of the turn-of-the-twentieth-century eugenics movement, particularly in **Britain**, as reformers attempted to improve the human race through promoting better marriage selection (**positive eugenics**) and preventing the **unfit** from reproducing (**negative eugenics**). Darwinism grew out of the biological theory of evolution first given scientific support by British naturalist **Charles Darwin** and, independently, Alfred R. Wallace (1823–1913) in 1858. Although organic evolution of plants and animals had been discussed in intellectual circles since the beginning of the nineteenth century, Darwin and Wallace were the first to offer scientific support for the theory. During the last decades of the nineteenth century, the concept became credited primarily to Darwin via the popularity of his books *On the Origin of Species* (1859), focusing on plants and animals, and *The Descent of Man and Selection in Relation to Sex* (1871), applying the theory to humans.

Darwin called his theory on the mechanism of inheritance "pangenesis." It supported the prevailing theory of **Lamarckian inheritance** of acquired characteristics. Darwinian evolution through natural selection hypothesized that in any particular environment a divergent group of organisms arises. As a result of many factors including overpopulation, predators, climate, disease, and limited food supplies, organisms in the environment engage in a constant struggle for existence. Because the struggle is so intense, organisms with variations that are advantageous in a particular environment are more likely to produce their own kind. For example, a grey moth resting on a tree trunk of similar color is not as easily sighted by a hungry bird and is less likely to devoured compared with an all-white moth of the same species. The grey moth survives to reproduce and pass this trait on to its offspring. Since fewer all-white moths of the species survive, the species over time evolves into one that is colored grey. This struggle for existence was applied to human socioeconomic, ethnic, and cultural survival in the late nineteenth century, becoming known as **social Darwinism**, an underlying tenet of eugenics.

FURTHER READING: Darwin, Charles, *On the Origin of Species by Means of Natural Selection* (1985); Dover, Gabriel A., *Dear Mr. Darwin* (2000); Gould, Stephen Jay, *The Structure of Evolutionary Theory* (2002); Hooper, Judith, *Of Moths and Men* (2002); Jones, Steve, *Darwin's Ghost* (2000); Ruse, Michael, *The Darwinian Revolution* (1999).

DAVENPORT, CHARLES BENEDICT
(June 1, 1866–February 18, 1944)

A noted biologist, Davenport was the pivotal leader of the eugenics movement in the United States. He also introduced biometrics (the use of statistics with biology) into U.S. science. In his later years he embraced **nativist** eugenic ideology and favored **immigration restriction** and **eugenic sterilization** legislation. Davenport was one of eleven children raised on a farm near Stamford, Connecticut, where the family spent about half of each year. His father was an ardent crusader against **alcohol** and a strict Congregationalist from old Puritan stock that traced their ancestry to the Norman conquest of England. During the cold months the family lived in Brooklyn Heights, where his father was a successful real estate and insurance broker. As a child Davenport received private tutoring from his father and also worked in his business. In 1879 he entered the Polytechnic Institute of Brooklyn and graduated with a B.S. in civil engineering (1886). He later received a B.A. (1889) and Ph.D. (1892) in biology from Harvard College. Subsequently he became a biology instructor at Harvard and then taught at the University of Chicago (1899–1904). From 1898 to 1923 Davenport was head of the summer biological laboratory at the Brooklyn Institute of Arts and Science at **Cold Spring Harbor**, New York. In 1904 he persuaded the new **Carnegie Institution of Washington**, a philanthropic foundation, to establish a Station for Experimental Evolution near this laboratory. He was named director and remained in this position until his retirement in 1934.

Charles Davenport, a geneticist and pivotal leader of the U.S. eugenics movement, was director of the Carnegie Institution of Washington's Station for Experimental Evolution and the Eugenics Record Office at Cold Spring Harbor, Long Island, New York. (Image courtesy of the Cold Spring Harbor Laboratory Archives and the CSHL Eugenics Website.)

Influenced by British statistician and eugenicist **Karl Pearson**, Davenport introduced statistics to biology in a popular textbook, *Statistical Methods in Biological Investigation* (1899). He also served as an associate editor of Pearson's new statistics journal, *Biometrika* (1901–1908). At the station Davenport initiated biometry and **genetics** research with animals that led to an interest in human heredity. He developed **family history and pedigree studies**, for which he created a "family records" form to gather data on normal and abnormal human characteristics. These forms were distributed to a wide assortment of social groups. Results from the collected data became the basis for *Heredity in Relation to Eugenics* (1911), a widely quoted work that helped accelerate the eugenics movement in the **United States**.

In 1906 the Committee on Eugenics was formed within the **American Breeders Association**; Davenport became its secretary. Under his influence the committee became a section of the association in 1910. That same year the **Eugenics Record Office** (ERO), at Cold Spring Harbor, was founded. This office became a center for eugenics activities, both national and international. Davenport trained **field workers** to collect family data that resulted in many articles and monographs. He organized the **Eugenics Research Association** (1913) to facilitate communication between field workers and those active in the study of human heredity; he was elected its first president. As an army major (1918–1919) during **World War I**, Davenport supervised the testing and measurements of recruits. This resulted in the publication of several monographs concerning defects and diseases in recruits of different **races**.

In 1921 the ERO and the station were consolidated into the Department of Genetics, funded by the Carnegie Institution, with Davenport as director. He exerted leadership positions in all the important eugenic organizations over the course of the movement, including the **First**, **Second**, and **Third International Eugenics Congresses** and the **First**, **Second**, and **Third National Conferences on Race Betterment**. In the 1920s he corresponded with **German eugenics** leaders, among them **Fritz Lenz** and **Eugen Fischer**, the director of the **Kaiser Wilhelm Institute**. He was on German editorial boards and published in German journals. Davenport was a founder of the **Galton Society** and the **American Eugenics Society**. Like many other eugenicists, he feared that unrestricted immigration would lead to racial degeneracy and therefore encouraged **immigration restriction laws** and the screening of immigrants for inherited conditions. He opposed **Margaret Sanger** and her **birth control** movement on the grounds that contraception use among the "better stocks" would lead to **race suicide**.

Throughout his career Davenport thought in terms of single **Mendelian inheritance** for most human traits, including such tendencies as "nomadism" and **pauperism**. He ignored the force of environment, which in his later years brought criticism to his works. A prolific writer, Davenport published some 450 articles and books. He coauthored, with **Arthur H. Estabrook**, *The Nam Family* (1912), based upon fieldwork done at the ERO. Davenport married Gertrude Crotty (1913), with whom he had two daughters. He was active to the end of his life and died from pneumonia after boiling a whale carcass outdoors that he was preparing for an exhibition at a new maritime museum where he was curator.

FURTHER READING: Haller, Mark H., *Eugenics* (1984); Kevles, Daniel J., *In the Name of Eugenics* (1985); Pickens, Donald K., *Eugenics and the Progressives* (1968).

DE VILBISS, LYDIA ALLEN

(September 3, 1882–1964)

A pioneer infant and maternal health physician, **public health** educator, and birth control advocate, De Vilbiss championed the eugenics cause. Like many eugenicists of the time she was concerned with **racial degeneracy** from reproduction among the **unfit**. She developed and advocated simple **birth control** devices to prevent unwanted pregnancies in addition to supporting eugenic sterilization. Born in Allen County, Indiana, De Vilbiss graduated from Indiana Medical College, Indianapolis (1907), associated with Purdue University. She did postgraduate and surgical training at New York University and the University of Pennsylvania, was licensed as an M.D. in 1910, and practiced in Ohio for five years. She was director of the Division of Child Hygiene at the Kansas State Board of Health (1915–1919), where along with two nurses she organized a "hygiene train" that focused upon maternal and child health and took health education around the state. In 1920 she became professor of public health administration at Women's Medical College in Philadelphia. She also was a reserve officer of the U.S. Public Health Service. For most of her career she directed the Mothers Health Clinic in Miami, Florida.

De Vilbiss was involved with several programs with eugenic aims. In 1913 she became medical director of the Better Babies Bureau established by the *Woman's Home Companion* magazine. The bureau sponsored **better babies** contests, based on a program launched in Louisiana in 1908, in which infants were judged on health and **intelligence**. This eugenics and public health education program encouraged mothers to gather information to improve their own health in order to produce healthy offspring and improve the health of their babies. The **Fitter Families** program evolved out of this concept in the early 1920s. In 1921 De Vilbiss presented methods of pregnancy prevention at the first Birth Control Conference in New York City, organized by birth control crusader **Margaret Sanger**. De Vilbiss instructed physicians on methods of birth control, at that time considered radical, even illegal. She and Sanger clashed for a number of years after De Vilbiss backed out on an offer she had accepted in 1923 to head up Sanger's clinic, which did not have legal status. However, Sanger supported De Vilbiss' research in the 1930s.

De Vilbiss was director of the Mother's Health Clinic, Miami, Florida, and worked with other professionals to find a simple and inexpensive method of birth control for poor women. She developed a sponge attached to a string and permeated with a contraceptive powder. Her preliminary research results of this method in 1935 came to the attention of Sanger and the **American Birth Control League**. In 1936–1937 clinical trials of the technique were

Lydia De Vilbiss, a physician, health educator, and developer of birth control methods, was one of the few women eugenics leaders during the early-twentieth-century eugenics movement in the United States. (From: *Physical Culture* [1915]. Image courtesy of Lilly Library, Indiana University, Bloomington, Indiana.)

conducted by De Vilbiss in Miami and at Sanger's Birth Control Clinical Research Bureau in New York. The method was found to be reasonably effective.

De Vilbiss was active in eugenics organizations. She presented a paper on the Better Baby program at the **Race Betterment Conference** (1914) and was a member of the **American Eugenics Society**. She wrote several articles and a book with eugenics themes, including *Birth Control: What Is it?* (1923) in which she discussed **positive** and **negative eugenics**. She advocated reduction of the maternal and infant mortality rate through better medical care and reduction of the **feebleminded** and **insane** through **eugenic sterilization** or **segregation**. She promoted reproduction among the "healthy, intelligent stock." She also suggested that the institutionalized unfit should be sterilized and then released from custodial care if they had no "anti-social tendencies" so that they could have a normal sex life. She published in ***Birth Control Review*** and wrote a series of health articles for *Physical Culture* magazine. De Vilbiss married George Henry Bradford (1920) and lived in Miami, Florida, until her death.

FURTHER READING: Chesler, Ellen, *Woman of Valor* (1992); Holt, Marilyn Irvin, *Linoleum, Better Babies, and the Modern Farm Woman, 1890–1930* (1995); Reed, James, *From Private Vice to Public Virtue* (1978).

DEFECTIVES
See Unfit

DEGENERACY, DEGENERATES, AND DEGENERATION

From the latter part of the nineteenth through the early twentieth century, *degeneracy* and *degenerate* were used for the psychiatric diagnosis of **paupers**, **alcoholics**, **criminals**, the **feebleminded**, the **insane**, and persons who engaged in sexual activities outside marriage. It expanded beyond the individual to include cultures, nations, and **races**. Degeneration was thought to have physical, moral, mental, and economic causes and to be inherited through **Lamarckian inheritance** of acquired characteristics. Because the population of the "fit" was decreasing and the population of the **unfit**, or degenerate, was increasing, physicians and welfare reformers feared that this differential birthrate would result in the decline of national vitality and strength, leading to **racial degeneracy** and the decline of civilization. A primary purpose of eugenics was to halt this degeneration.

The concept of degeneration began to be formalized in eighteenth-century Europe, although the idea can be traced back to antiquity. It was discussed among reformers in the United States as "inherited realities" and in Europe as "degeneracy theory." By 1857 French physician Benedict Morel (1809–1873) proposed that **alcoholism** and other conditions were different manifestations of hereditary degeneration. Degeneracy might originate in the first generation through **alcohol** abuse, poor working conditions, **masturbation**, **venereal disease**, bad morals, toxic substances, or violations of the **laws of health**. Once established, degeneracy passed from generation to generation until it ended in insanity or "idiocy." Alcoholism, in particular, was seen as both a cause and result of inherited degeneration. In the United States during the first **Clean Living Movement** (1830s–1850s), **inherited realities**, which embodied both degeneracy and eugenic concepts, were discussed by several health reformers including Sylvester Graham, **Elizabeth Blackwell**, and **William Alcott**.

By the end of the nineteenth century, degeneracy theory led **public health** and social welfare reformers in several countries to champion the elimination of slums and poor environments in order to increase national vitality and reduce degeneracy. In 1914 **Charles Davenport**, pivotal leader of the eugenics movement in the **United States**, categorized individuals and their resulting offspring according to two types. The *aristogenic* were genetically fit and produced long lines of renowned men (outstanding women were rarely mentioned). The *cacogenic* produced many generations of both male and female degenerates. These generations of degenerates were identified in family history studies such as *The Dack Family* (1916), *The Jukes in 1915* (1916), *The Nam Family* (1912), and *The Kallikak Family* (1912). To prevent degenerates from reproducing in the early twentieth century, many eugenicists advocated **negative eugenics** programs such as **eugenic sterilization** or **segregation** in state-run institutions.

FURTHER READING: Carlson, Elof Axel, *The Unfit* (2001); Money, John, *The Destroying Angel* (1985); Rafter, Nicole Hahn, (ed.), *White Trash* (1988).

"DESIGNER BABY"

An infant produced by the process of selection of genetic material from individuals with specific characteristics (**germinal choice**) and/or genetic testing to develop a healthy embryo has been termed a "designer," "better," or "perfect baby." The term *designer baby* first appeared around 1985. Producing a genetically healthy infant, or one with desired characteristics, is a main feature of the "**new eugenics**" and is generally considered an aspect of **positive eugenics**. The procedures for a better baby, driven by consumer demand and used by both single women and couples when one partner is infertile, became common in the 1990s. In the process of germinal choice, sperm or eggs are selected by potential parents from **sperm banks**, relatives, or others based upon occupation, abilities, **intelligence**, lack of family diseases, and other characteristics. Fertilization is accomplished by **artificial insemination** or **in vitro fertilization**. In artificial insemination, prenatal testing of the embryo is often undertaken to screen for potential genetic or congenical conditions. When *in vitro* fertilization is used, embryos are generally tested for **genetic diseases** and the most healthy implanted. Another aspect of a perfect or healthy baby philosophy is the emphasis on good prenatal care and a healthy prenatal lifestyle on the part of the mother, including good nutrition, exercise, and avoidance of potentially toxic substances such as **tobacco**. The selection of gender (**family balancing**), **gene therapy**, or the potential of human **cloning** and other **genetic engineering** techniques could also be used to produce a designer baby. However, some of the techniques are fraught with ethical considerations and controversy.

FURTHER READING: McGee, Glenn, *The Perfect Baby* (1997); Rifkin Jeremy, *The Biotech Century* (1998); Stacy, Meg, (ed.), *Changing Human Reproduction* (1992); Stock, Gregory, *Redesigning Humans* (2002).

DICKINSON, ROBERT LATOU (February 21, 1861–November 29, 1950)

A gynecologist and sexologist, Dickinson was an early medical advocate of **birth control** and supported eugenics reform. Like many physicians and eugenicists of the early twentieth century, he was concerned with **racial degeneracy** through reproduction among so-called **degenerates**. Dickinson was born in Jersey City, New Jersey, one of five children. His father was a wealthy hat manufacturer from old-stock New England families. Dickinson enjoyed a privileged childhood in Brooklyn Heights, New York, and spent summers on an uncle's farm in Connecticut. He was privately tutored and attended schools in Switzerland and Germany. He graduated from Brooklyn Poly-technical Institute (1879). Although Dickinson was a talented artist, he entered the Medical School of Long Island Hospital and completed course work for an M.D. In 1882 he began to build a successful private practice. In 1886 he was hired by Long Island College Hospital as a lecturer in obstetrics and remained an active teacher until he retired from clinical practice. In the 1920s and 1930s he worked to change the New York State law regarding the dissemination of contraception. Initially at odds with **Margaret Sanger** and her lay birth control clinic, in 1923 Dickinson and medical colleagues founded the Committee on Maternal Health as a contraceptive research center. The committee became a clearinghouse for information on human fertility and marriage counseling and played a key role supporting contraceptive research in Europe. He was its secretary until 1937 and afterward chairman until 1950.

In December 1925 Dickinson and Sanger formed the Maternity Research Council to oversee Sanger's clinic if a dispensary license for it could be obtained. However, the license was refused due to New York State politics and opposition from the Roman **Catholic Church**. In 1920 Dickinson was elected president of the American Gyneco-logical Society. He served on the advisory council of the **American Eugenics Society**, edited the "birth regulation" department of *Eugenics*, and contributed to *Birth Control Review*. Dickinson advocated both **positive** and **negative eugenics** and recommended that the "fit" reproduce. He encouraged the curtailment of reproduction among the **unfit** and advocated **eugenic sterilization** for the **feebleminded**. For the AMA in 1929 he evaluated the sterilization program sponsored by the **Human Betterment Foundation** in California. Dickinson wrote more than 200 research papers. One of his most noted works was an atlas, *Human Sex Anatomy* (1933), used into the late twentieth century. Dickinson married Sarah Truslow, a founder of the Young Women's Christian Association, and fathered three daughters, one who died in infancy. He died at the home of a daughter in Amherst, Massachusetts.

FURTHER READING: Pivar, David J., *Purity Crusade* (1973); Reed, James, *From Private Vice to Public Virtue* (1978).

DNA (Deoxyribonucleic Acid)

The double helix molecule that encodes genetic information for most living material is termed *DNA*. It forms a spiral of two helical chains each coiled around the same axis and held together by pair bonds of bases, or nucleotides. Bonds that make up the pairs form only between the nucleotides adenine and thymine, and guanine and cytosine (A-T, G-C). An ordered sequence of base pairs found on a DNA strand is a **gene**. The procedure to determine the four-letter sequence is called "genetic sequencing." Genes are located on a particular part of a chromosome. Chromosomes are individual threads within the cell nucleus that self-replicate. Humans have twenty-three pairs of chromosomes. Swiss biochemist Friedrich Miescher (1844–1895) in 1870 isolated "nuclein" from the nuclei of white blood cells, which were later found to be the chemicals that make up DNA. In 1943 an American microbiologist, Oswald Avery (1877–1955), and colleagues demonstrated that DNA carried genetic information and suggested that it could be the gene. The three-dimensional structure of DNA was first identified in 1953 by **James Watson** and Francis Crick (1916–2004) after they viewed X-ray diffraction images taken by Rosalind Franklin (1920–1958), which had been shown to them by her colleague Maurice Wilkins (1916–2004) without her knowledge.

In the 1990s researchers mapped the human **genome**, the sequence of about 30,000 genes in human DNA in an international effort known as the **Human Genome Project**. Knowing the location and function of genetic material allows for **genetic engineering**, which includes procedures such as manipulating genes to eliminate **genetic diseases**, or even going as far as potentially reshaping the genetic makeup of the human species. A common procedure for changing genes is to create artificial, or *recombinant*, DNA from various base pairs from two or more sources. This was first accomplished in 1972 by American Paul Berg (b. 1926). The ability to identify DNA permits **genetic screening and testing** for specific inherited conditions and the creation of **genetic profiles**, or "fingerprints," that are unique for each individual. The identification and manipulation of human genetic material is an aspect of the **new eugenics**.

FURTHER READING: Clayton, Julie, and Carina Dennis, (eds.), *50 Years of DNA* (2003); Drlica, Karl, *Understanding DNA and Gene Cloning* (2004); Watson, James D., *DNA* (2003); Wilkins, Maurice, *The Third Man of the Double Helix* (2003).

"DOLLY" THE SHEEP

The results of the first successful cloning from somatic, or body, cells of an adult animal was the lamb Dolly, named after the popular American singer Dolly Parton. Once cloning techniques were perfected for mammals, the likelihood of cloning an adult human became possible. The first success in **cloning** an adult mammal was achieved by British researchers Ian Wilmut and Keith Campbell at the Roslin Institute in Edinburgh, Scotland, on July 5, 1996. Announcement of the lamb's birth appeared in the February 27, 1997, issue of *Nature*. Success came after trials with about 280 embryos. Before the creation of Dolly, cloning of mammals had been accomplished only by transplanting **DNA** (genetic material) from embryonic cells of one animal into embryonic cells from another in which the nucleus containing the DNA had been removed. After fusion of the material from both cells, the developing embryo was then transferred to the uterus of a third animal using the process of **in vitro fertilization**. Dolly's uniqueness is that this procedure was accomplished with the cells from the lining of an udder from an adult sheep, not an embryonic cell. After Dolly other animals were subsequently cloned using DNA from adult somatic cells. However, subsequent clones often developed health problems, including early aging. Dolly developed early-onset arthritis and was euthanized in 2003 due to a serious lung infection. Her cloning led to debate concerning the ethics and morality of human cloning. Wilmut and Campbell's cloning technique and the live birth of Dolly helped further the perfection of procedures that could be applied to humans. The potentiality of cloning humans, along with **genetic engineering** and **assisted reproduction**, lays the groundwork for the **new eugenics**.

FURTHER READING: Kolata, Gina Bari, *Clone* (1998); Yount, Lisa, (ed.), *Cloning* (2000).

DUGDALE, RICHARD LOUIS (c. 1841–July 23, 1883)

Social reformer and sociologist Dugdale was one of the first to study **feeblemindedness** and **criminality** in family kinships. He is most noted for *The Jukes*, which illustrated an example of the importance of **negative eugenics** programs to prevent the **unfit** from reproducing. Dugdale was born in Paris, France, one of three children, to English parents. His father was a manufacturer and journalist. Due to political problems and business reversals, the family returned in 1848 to England, where Dugdale attended the Somerset School for a few years. In 1851 he went to New York City with his family and attended public schools until he was fourteen. After he showed symptoms of heart problems the family moved to an Indiana farm. The situation did not work out, and the family returned to New York in 1860. Dugdale found employment as a stenographer, and upon his father's death he received a small inheritance, which he invested in a business. Dugdale also attended night classes at Cooper Union College, where he became interested in sociology and the problems of social welfare. When he had become financially successful, he involved himself with many social reform organizations in the New York City area.

In 1868 Dugdale became a member of the executive committee of the Prison Association of New York. His colleague Elisha Harris (1824–1884), a public health physician, had noticed that six generations of one family had produced numerous criminals and **paupers**. Around 1874 Dugdale was assigned to investigate surrounding county jails for the presence of members of this family. He noticed that many of the other inmates were also related to one another. Using private funds he made a detailed study of one large kinship in Ulster County, which he named the "Jukes." The study was published by the Prison Association as **The Jukes: A Study in Crime, Pauperism, Disease, and Heredity** (1877). This highly popular work became a model for **family history and pedigree studies** in the early-twentieth-century eugenics movement. Dugdale suggested that the **alcoholism**, **crime**, and **pauperism** of the Jukes were caused by both heredity and environment (**nature-nurture**). He also suggested that if children were removed from bad environments and placed into good ones, their hereditary tendencies for various social problems might be reduced. Dugdale subscribed to the inheritance of acquired characteristics as suggested by **Jean-Baptiste Lamarck**, but may not have been familiar with the actual theory of **Lamarckian inheritance**. In 1911 **Arthur H. Estabrook** of the **Eugenics Research Office** found Dugdale's original notes and manuscript and was able to identify the actual names of the family cited. This result was a follow-up study of the kinship and publication of **The Jukes in 1915** (1916), which helped popularize the eugenics movement. Dugdale never married and died at an early age, of heart disease, in New York City.

FURTHER READING: Carlson, Elof Axel, *The Unfit* (2001); Dugdale, Richard Louis, *The Jukes* (1970); Rafter, Nicole Hahn, (ed.), *White Trash* (1988).

DYSGENIC OR DYSGENICS

A late-nineteenth and early-twentieth-century term, *dysgenic*, or "bad heredity," is the opposite of *eugenic*, or "good heredity." The concept includes the transmission to offspring of negative or defective traits that could lead to **racial degeneracy** and the decline of the human race. Forces such as **war**, modern medicine, and public charities were considered dysgenic as they prevented the "fit" from reproducing and allowed the **unfit** to reproduce. The terms *cacogenic, unfit,* and ***degenerate*** were also used to describe those people with negative traits thought to be inherited. Such characteristics included **insanity, feeblemindedness, alcoholism, criminality,** and **tuberculosis**. The term *dysgenic* went out of general use by the mid-twentieth century.

FURTHER READING: Carlson, Elof Axel, *The Unfit* (2001).

E

Eugenics . . . is the science and the art of being well born.

David Starr Jordan, *The Heredity of Richard Row* (1911)

The Eugenic ideal—the belief that we owe a paramount duty to posterity dependent on the laws of heredity.

Leonard Darwin, "Preface," *Problems in Eugenics*, Vol. 2 (1912)

EASTERN AND SOUTHERN EUROPEANS

Immigrants from eastern Europe, including Poland and Russia, and from southern Europe, including Italy and Greece, were classified as *eastern and southern Europeans*. They emigrated to the New World in search of better economic conditions and, in some cases, religious freedoms. However, nativists considered these newcomers as threats to U.S. democracy and the country's **Anglo-Saxon**, Protestant heritage. Before the American Civil War most immigrants had hailed from Britain or Ireland. In the postwar era Scandinavians and Germans, plus more Irish, dominated the flow. Other than the Irish, until 1881 most immigrants were **Protestant**. They arrived as families, settled on farms, and soon assimilated. In the 1890s unskilled peasants from southern and eastern Europe poured into the already crowded eastern slums. Because these new immigrants appeared to resist becoming Americanized, increased animosity toward them arose among the Anglo American establishment.

By the turn of the twentieth century eugenic and **nativist** reformers, most of whom were of northern European heritage, feared that eastern and southern European immigrants lacked the biological wherewithal to adapt to U.S. culture and climate. They were concerned that the new arrivals would dilute the genetic makeup of the nation and lead to **racial degeneracy** and **race suicide** of the old American stock. They also feared a Bolshevik political takeover by Russian Jewish immigrants or a papal conquest by Irish, Italian, and Polish **Catholic** immigrants. During the second decade of the century some surveys showed that eastern and southern European immigrants were more likely institutionalized for **insanity, tuberculosis,** and **crime. Intelligence testing** of recruits for **World War I** showed that they had lower **IQ** scores compared with northern Europeans. Fears of the new immigrants resulted in the **Johnson-Reed Immigration Restriction Act of 1924**, which dramatically decreased immigration from these "undesirable" regions. Over the course of the twentieth century, however, the descendants of eastern and southern European immigrants became educated and economically stable, moved to the suburbs, intermarried with other ethnic groups, and assimilated into the dominant culture. Few differences were found in their **intelligence** compared to other Europeans. By the third generation few could speak

Eastern and southern European immigrants were considered "inferior racial stock," prompting an immigration restriction movement. (From: *Journal of Heredity* [1917], Main Library, Indiana University, Bloomington, Indiana. Image courtesy of the Digital Library Program.)

the language of their ancestors. Only their surnames gave a hint of their ethnic origins, and they were considered **Euro Americans**, or whites.

FURTHER READING: Jacobson, Matthew Frye, *Special Sorrows* (1995); Kraut, Alan M., *Silent Travelers* (1994); Marchione, Margherita, *Americans of Italian Heritage* (1995).

ELDERTON, ETHEL M. (December 31, 1878–May 8, 1954)

A researcher at the Galton Eugenics Laboratory at University College, London, Elderton undertook many eugenics research studies concerning the contribution of heredity and environment to human characteristics. She was the third of eight children, was educated near her home at Streatham High School, and attended Bedford College, where she exhibited mathematical ability and became an eugenics enthusiast. However, upon the death of her father in 1890 she returned home and taught at a school run by her mother to help support her younger siblings. Her older brother became a distinguished actuary and a colleague of eugenicist and pioneer statistician **Karl Pearson**. As a result of a recommendation from Bedford College, in 1904 Elderton became **Francis Galton**'s assistant and analyzed fingerprints and **family pedigrees**. In 1906 she was appointed a Galton Research Scholar in the newly established Eugenics Records Office (later called the **Galton Laboratory**) and secretary to the laboratory. She successively became a Galton Fellow and assistant professor at the laboratory and later received a doctor of science degree. Elderton was awarded the Weldon Prize from Oxford University (1919) for her contributions to the science of biometrics (applying statistics to biology) and made a fellow of University College, London (1921). From 1925 to 1933 she was assistant editor of the *Annals of Eugenics*, founded by Pearson. Over her career Elderton analyzed the data and wrote many of the reports published by the Galton Laboratory.

Elderton coauthored with Pearson the Eugenics Laboratory Lecture Series, which was published at intervals. In an important work of this series, *Relative Strength of Nurture and Nature* (1909), of which she was sole author, she contended that health, physical, and mental traits in children were associated with inheritance, not environment, and that heredity was the primary mode of transmission of human behaviors and abilities. She argued that "improvement in social conditions will not compensate for a bad hereditary influence" and that the only way to keep the nation strong was to have the fittest members reproduce. A series of reports concerning alcoholic families also suggested that **alcoholism** was inherited. These reports were controversial, and a number of British scholars including physician and eugenics popularizer **C. W. Saleeby** challenged them. They also became a factor in the dispute between British geneticists and eugenicists who embraced the rules of **Mendelian inheritance** and opposed the Galton Laboratory's biometric school, which supported **Darwinism** and **natural selection** as the mechanism of inheritance. Many studies analyzed by Elderton over the first three decades of the twentieth century refuted the theory of **Lamarckian inheritance** of acquired characteristics and illustrated the complex nature of human traits. She retired in 1933, the same year as Pearson, never married, and died in Watford, Hertfordshire, north of London.

FURTHER READING: Love, Rosaleen, " 'Alice in Eugenics-land': Feminism and Eugenics in the Scientific Careers of Alice Lee and Ethel Elderton" (1979).

ELIOT, CHARLES WILLIAM (March 20, 1834–August 22, 1926)

President of Harvard University and an educational reformer, Eliot was prominent in many health and social-reform movements of the early twentieth century, including **social hygiene**, **public health**, and temperance. He was an early advocate of eugenics although not a major leader of the movement. Born in Boston to a socially prominent old-stock New England Unitarian family, Eliot enjoyed a privileged childhood. He attended Boston Latin School before entering Harvard College at age fifteen; he graduated with high honors (1853). After graduation Eliot taught mathematics and chemistry at Harvard, studied abroad, and upon returning to the United States became a professor of chemistry at the Massachusetts Institute of Technology in 1865. Eliot's publication of his philosophy of educational reform led to his election as president of Harvard at age thirty-five, in 1869. During his forty-year tenure he was instrumental in transforming Harvard from a college to a prestigious university.

Eliot promoted several causes with underlying eugenics implications. He was the first president of the **American Social Hygiene Association**, organized in 1913 to combat **venereal disease** and **prostitution** and to promote sex education. He supported peace initiatives out of fear that **war** would lead to **race suicide** and the decline of Western civilization. He advocated segregation of **feebleminded** women in state institutions during their child-bearing years as an effort to prevent illegitimate children, prostitution, and venereal disease. Eliot was against the melting-pot concept of assimilation and interracial marriage on grounds that it would lead to social problems. In his later years he supported Prohibition. His major involvement in the eugenics movement was as a member of the planning committee for the **First International Eugenics Congress** in 1912. Eliot married Ellen Peabody in 1858. The couple had four children. She died the day after he became Harvard's president. In 1877 Eliot married Grace Mellen Hopkinson. He died at his summer home on the Maine coast.

FURTHER READING: Cotton, Edward H., *The Life of Charles W. Eliot* (1926); James, Henry, *Charles W. Eliot, President of Harvard University, 1869–1909* (1930); Saunderson, Henry Hallam, *Charles W. Eliot: Puritan Liberal* (1928).

ESTABROOK, ARTHUR HOWARD (May 9, 1885–December 6, 1973)

A biologist and eugenicist, Estabrook published several works that became part of the theoretical basis for, and helped popularize, the early-twentieth-century eugenics movement. Estabrook was born in Leicester, Massachusetts. He received his undergraduate degree from Clark College (1905) in Massachusetts and a doctorate in biology from Johns Hopkins University (1910) in Maryland. Estabrook was a **field worker** and researcher at the **Eugenics Record Office** (ERO) and the Station for Experimental Evolution of the **Carnegie Institution of Washington** (1910–1929) in **Cold Spring Harbor**, New York. During **World War I** he served as a captain in the Sanitary Corps of the U.S. Army. Between 1920 and 1936 he was involved with a campaign to eliminate **venereal disease** in New York City. After leaving the ERO he worked as a researcher (1929–1941) for the predecessor of the American Cancer Society. In the post–**World War II** era he held various administrative positions in mental hygiene and public health, including working in several capacities (1951–1963) for the U.S. State Department in Washington, D.C., until his retirement.

Estabrook was most noted for eugenics research and the promotion of **eugenic sterilization** and custodial care of the **unfit**. In 1911 he published *The Nam Family*, coauthored with **Charles Davenport**, director of the ERO, based upon his fieldwork. In

Arthur Estabrook, a eugenics researcher, published several studies, including *The Jukes in 1915* and *The Nam Family*, that became part of the theoretical basis for—and helped popularize—the early-twentieth-century eugenics movement. (Courtesy of the American Philosophical Society and the CSHL Eugenics Web site.)

1914 the field notes for **Richard Dugdale's** original study *The Jukes* (1877), a case study of a "degenerate family," were discovered. Estabrook traced the contemporary descendant of this family and published the results of his inquiry as *The Jukes in 1915* (1916). In this classic **family history and pedigree study**, he concluded that half the family was **feebleminded** and recommended permanent custodial care and sterilization to prevent their defective "**germ-plasm**" from being transmitted to future generations. Estabrook carried out or assisted state agencies with eugenic surveys. In 1915 he supervised surveys of two Indiana counties, the results were published as *Mental Defectives in Indiana*. Another survey led to *Mongrel Virginians* (1926), coauthored with Ivan E. McDougle. In 1927 Estabrook testified at the *Buck v. Bell* proceedings that resulted in the Supreme Court decision to uphold the right of the state of Virginia to sterilize "defective" individuals against their will. In 1917 Estabrook was named to the executive council of the **Eugenics Research Association**, then at the peak of its influence, and was elected its president in 1925. He married Jessie McCubbin (1911), and after her death in 1931 married Anne Ruth Medcalf the same year. Although well known in the early part of the century as a eugenicist, Estabrook was largely forgotten by the end of his life. In his later years Estabrook lived near Utica, New York.

FURTHER READING: Bix, Amy Sue, "Experiences and Voices of Eugenics Field-workers" (1997); Engs, Ruth Clifford, *The Progressive Era's Health Reform Movement* (2003).

EUGENIC MARRIAGE-RESTRICTION LAWS

A **negative eugenics** measure to prevent marriage among the mentally and physically unhealthy, eugenic marriage-restriction laws were enacted as part of the **Progressive Era**'s eugenics movement in the **United States** and other countries. The laws were primarily implemented to prevent **venereal diseases**, especially syphilis, that could be passed down to unborn children. Until 1900, when monk **Gregor Mendel**'s laws of inheritance were rediscovered, many people believed in **Lamarckian inheritance** of acquired characteristics and **social Darwinism**, which implied that reproduction among the **unfit** would lead to a proliferation of **degenerates** such as **paupers**, **alcoholics**, and the **feebleminded**. Reformers championed programs that they felt would benefit humankind on both humanitarian and economic grounds. **Social hygiene** reformers, in particular, pushed for laws that required a "marriage health certificate" indicating freedom from venereal disease. Some states in the 1890s began to forbid marriage if either party was "insane, an idiot, feeble-minded, epileptic or had not been cured of syphilis or gonorrhea." By 1912 physical examinations to rule out venereal disease were required in Connecticut, Washington, Utah, Michigan, and Colorado before a marriage license was granted. In nine states epilepsy was also grounds to deny marriage, and in two states, "drunkenness." By 1912 some type of marriage restriction had been enacted in thirty-four states or jurisdictions in the United States. The vast majority forbade marriages in which either party was **insane** or feebleminded. Marriage-restriction laws were not always strongly enforced.

FURTHER READING: Haller, Mark H., *Eugenics* (1984); Paul, Diane B., *Controlling Human Heredity* (1995).

EUGENIC SEGREGATION

The legal requirement that the **unfit**, including **feebleminded** and **insane** individuals, live in state-run institutions during their reproductive years to prevent their breeding was termed *eugenic segregation*. Confinement of **degenerates** in institutions away from the rest of society increased during the late nineteenth century on both sides of the Atlantic. In 1890 fourteen states in the **United States** maintained institutions for the mentally disabled. That year 40 percent of inmates in these institutions were immigrants or children of immigrants. Whether they were old-stock **Anglo Americans** or recent immigrants, most were **paupers**. Fear of feebleminded **prostitutes**, who not only bore "degenerate children" but also transmitted **venereal diseases**, became a particular concern. Social workers suggested that the state should construct more institutions to house these "defective" individuals under "permanent, maternal care by the good Mother State." However, custodial care was expensive, and **eugenic sterilization** began to be seen as an attractive alternative. **Charles Davenport**, the pivotal leader of the eugenics movement, argued that illegitimate reproduction among "imbeciles" must be prevented. If segregation in institutions was not possible, he maintained that sterilization should be required to "prevent the reproduction of their vicious germ plasm." Most **negative eugenics** supporters argued for both segregation in institutions and sterilization.

FURTHER READING: Haller, Mark H., *Eugenics* (1984); Paul, Diane B., *Controlling Human Heredity* (1995); Reilly, Philip R., *The Surgical Solution* (1991).

EUGENIC STERILIZATION AND LAWS

The sterilization of the **unfit**, including the **insane**, the **feebleminded**, **criminals**, epileptic, and those with **genetic diseases** was termed *eugenic sterilization*. The procedure was a **negative eugenic** measure to prevent people with "inherited abnormalities" from reproducing. Until the twentieth century the only way to sterilize males was by castration; sterilization of females was so risky it was rarely done. In 1899 vasectomy for males, a simple procedure done under local anesthesia, was first described by physician A. J. Ochsner. **Harry Sharp**, a physician at the Indiana State Reformatory, used the procedure that year to prevent **masturbation**, which was thought to lead to **degeneracy**, and to prevent "mental defectives" from reproducing. In 1905 the Pennsylvania legislature passed the first eugenic sterilization law, but the governor vetoed it. Two years later, under Sharp's guidance, Indiana passed and adopted the first compulsory sterilization law. By 1915 fifteen states had passed legislation modeled on the 1907 Indiana law.

Harry H. Laughlin, superintendent of the **Eugenics Record Office** in **Cold Spring Harbor**, New York, surveyed sterilization laws across the nation and published *Eugenical Sterilization in the United States* (1922), which included a "model eugenical sterilization law." This model was adapted by other states and countries. Outside the United States governments began enacting legislation for involuntary sterilization in the late 1920s: the Swiss canton of Vaud (1928), Denmark (1929), the Canadian provinces of Alberta (1928) and British Columbia (1933), Sweden and Norway (1934), Finland and Danzig (1935), and Estonia (1936). After *World War I* German eugenicists proposed eugenic sterilization legislation, but they were unsuccessful. However, in **Nazi Germany** a compulsory sterilization law somewhat based upon Laughlin's "American Model" was passed in 1933, and as a result up to 400,000 individuals, mostly ethnic Germans, were sterilized. Sterilization of the unfit was debated in the British parliament as part of the **British eugenics** movement, but it never passed.

States that instituted eugenic sterilization laws and states that subsequently repealed, revoked, or declared unconstitutional such laws, 1907–1931.

DATE	STATES THAT PASSED A STATUTE OR REFERENDUM	STATES THAT REPEALED, REVOKED, OR DECLARED LAW UNCONSTITUTIONAL
1907	Indiana	
1908		
1909	Connecticut, California, Washington	
1910		
1911	Nevada, New Jersey, Iowa	
1912	New York	
1913	Oregon, North Dakota, Kansas, Michigan, Wisconsin	Iowa, Oregon, New Jersey
1914		
1915	Nebraska, Iowa	New York
1916		
1917	South Dakota, Oregon, New Hampshire	
1918		Michigan, Nevada
1919		
1920		Indiana
1921		
1922		
1923	Alabama, Michigan, Montana, Delaware	
1924	Virginia	
1925	Idaho, Minnesota, Maine, Utah	
1928	Mississippi	
1929	Arizona, Delaware, Idaho, Nebraska, North Carolina, West Virginia	
1931	Oklahoma, Vermont	

Adapted from Laughlin (1922, 1–4), Whitney (1934, 302), and Gosney and Popenoe (1929, 160–174). Some states over this time period repealed and then reinstated the act or declared the act constitutional or unconstitutional.

In the United States several publications in the late 1920s, including ***Sterilization for Human Betterment*** (1929) by California businessman **E. S. Gosney** and eugenics supporter **Paul Popenoe**, helped fan the sterilization movement. By the 1930s more than thirty states had passed sterilization laws. Most allowed for sterilization of institutionalized "confirmed criminals, idiots, imbeciles, and rapists." However, the laws varied from state to state, and a combination of lax enforcement and court challenges left them largely unenforced. In the 1920s legal challenges were brought against them. In some states, such as Colorado, members of the **Catholic Church** were instrumental in preventing sterilization laws from being passed. European Catholic clergy were divided on the issue until 1930, when the papal encyclical *Casti Connubi* condemned the practice. An increased numbers of sterilizations were performed in the 1930s after the involuntary sterilization law of the state of Virginia was upheld by the 1927 U.S. Supreme Court decision ***Buck v. Bell***. In total, around 60,000 people were believed to have been sterilized by the 1960s. The sterilization laws were continually challenged, and although a few states still have them as statues, they are rarely enforced. By the 1970s vasectomies had become common among middle-class men who requested them because they desired no additional children.

FURTHER READING: Haller, Mark H., *Eugenics* (1984); Reilly, Philip R., *The Surgical Solution* (1991).

EUGENICAL NEWS (1916–1953; *Eugenics Quarterly*, 1954–1968; *Social Biology*, 1969–present)

The *Eugenical News*, subtitled *Current Record of Race Hygiene* until 1939, was founded in 1916 by American eugenicists **Charles Davenport** and **Harry H. Laughlin** as a clearinghouse for the activities of the **Eugenics Record Office** (ERO) at **Cold Spring Harbor**, New York. It published short, popular articles promoting eugenic concepts and reported research related to eugenics. Themes included the "Menace of the Feeble-minded," "Population Problems," and "Man in the Ice Age." It actively championed **immigration restriction**, **eugenic sterilization**, and other eugenic measures. The newsletter reprinted minutes of **Galton Society** meetings, profiled biographies of prominent leaders, and included reviews of eugenics books. Beginning in the late 1920s and lasting through the late 1930s *Eugenical News* contained articles in favor of **Nazi Germany**'s **race hygiene** program. The *News* was the official organ of the **Eugenics Research Association** (ERA), an offshoot of the ERO, from August 1920 to December 1938, and of the Galton Society from 1925 to December 1938. It also became the official voice of the **American Eugenics Society** (AES) when this organization terminated its own publication, ***Eugenics***. In June 1938, when the ERO was forced to close, the *Eugenical News* was transferred to the AES in New York City and changed from an advocacy to a more scientific-based journal. The *News* began as a quarterly with the 1939 issue and published studies of human populations and behaviors with titles such as "Population Growth and Fertility Trends in the United States." In 1954 the journal was replaced by *Eugenics Quarterly*, intended "to further knowledge of the biological and sociocultural forces affecting human populations." In 1969 it became *Social Biology*, a publication for the Society for the Study of Social Biology.

FURTHER READING: Engs, Ruth Clifford, *The Progressive Era's Health Reform Movement* (2003); Haller, Mark H., *Eugenics* (1984).

EUGENICAL STERILIZATION IN THE UNITED STATES (1922)

Written by eugenicist **Harry H. Laughlin**, superintendent of the **Eugenics Record Office** at **Cold Spring Harbor**, New York, *Eugenical Sterilization in the United States* details the sterilizations laws of each state and provides a model law that was later adapted by German eugenicists and implemented in **Nazi Germany**. However, because of the book's advocacy of sterilization the **Carnegie Institution of Washington** and the Rockefeller Foundation, which helped fund the ERO's research studies, refused to publish the work. The Chicago Psychopathic Laboratory of the Municipal Court of Chicago, a project of Judge **Harry Olson**, provided the financial support for its publication. Olson also wrote the introduction. The 502-page book was printed in Chicago by F. Klein Co. Its inside cover shows a photo of a statue of three generations of a family with the caption "Keep the Life Stream Pure." It contains twenty-six chapters and several tables detailing the states that had passed, vetoed, or annulled various sterilization laws. Chapter titles include "Analysis, by States, of Sterilization Laws Enacted Prior to January 1, 1922," "Litigation Growing Out of the Several Eugenical Sterilization Statutes," "The Right of the State to Limit Human Reproduction in the Interest of Race Betterment," "Eugenical Diagnosis," "The Anatomical and Surgical Aspects of Eugenical Sterilization," and "Model Eugenical Sterilization Law."

The volume was intended to help lawmakers form policy to implement and regulate **eugenic sterilization** and to aid judges in determining which individuals were **dysgenic** and needed to be sterilized. It was also aimed at state officials to help them locate potentially "defective" individuals, and at "individual citizens who, in the exercise of their civic rights and duties, desire to take the initiative in reporting . . . specific cases of obvious family degeneracy." The book includes tables and sample forms for judges and institutions to use for authorizing sterilization. It details state laws that allow eugenic sterilization and the status of fifteen states that had laws "unopposed to sterilization." Of these states, California by 1922 had performed the most sterilizations (2,558). None had been performed in Nevada, New Jersey, or South Dakota. The book reports that sterilizations had only been performed in state institutions for the **insane**, **feebleminded**, and **criminal**. The publication received mixed reviews. **Leonard Darwin**, head of the British **Eugenics Society**, suggested that although the book was a valuable resource on eugenics sterilization, constitutional questions of compulsory sterilizations were of great concern.

FURTHER READING: Reilly, Philip R., *The Surgical Solution* (1991).

EUGENICS (1928–1931; *People*, 1931)

The **American Eugenics Society** (AES) launched its official journal, *Eugenics: A Journal of Race Betterment*, in October 1928 and published it until February 1931. Most issues of *Eugenics* were devoted to a single theme, such as immigration, psychology, medicine, or crime. Typical article titles included "Immigrant Birth Rates" and "Marriage and Birth Rates at Bryn Mawr." The title page featured the silhouette of British naturalist Francis Galton, the father of eugenics, and a likeness of the **Fitter Families** contest medal used by the **Race Betterment Foundation** and the AES. During its brief life the journal's editorial board included promoters of the eugenics movement such as **Harry H. Laughlin**, **Samuel J. Holmes**, **Roswell H. Johnson**, **Leon Whitney**, and

Albert E. Wiggam. In 1931 *Eugenics* was superseded by *People*, "A Magazine for All the People, Official Organ of the American Eugenics Society, and Devoted to All Phases of the Eugenics Movement." Only one issue was published, in April 1931, by the Galton Publishing Company. Following the discontinuance of these publications the AES contracted with the **Eugenics Research Association**, publishers of *Eugenical News*, to make that periodical the official organ of the society. This relationship continued until the end of 1938.

FURTHER READING: Engs, Ruth Clifford, *The Progressive Era's Health Reform Movement* (2003); Haller, Mark H., *Eugenics* (1984).

EUGENICS EDUCATION SOCIETY
See Eugenics Society

EUGENICS RECORD OFFICE (1910–1939)

The heart of the American eugenics movement in its three decades of reform activity was the Eugenics Record Office (ERO) at **Cold Spring Harbor**, Long Island, New York. This organization helped to facilitate and coordinate all aspects of the movement in the **United States** and was the only eugenics institution with its own headquarters, research facilities, and paid staff. Most noted eugenicists of the era were associated with it. In 1902 Andrew Carnegie established the **Carnegie Institution of Washington** (CIW), which two years later set up the Station for Experimental Evolution at Cold Spring Harbor under the direction of **Charles Davenport**, a geneticist. Davenport enlisted **Mary Harriman**, widow of a railroad magnate, in founding and supporting an institution for eugenics research and education. Harriman donated a tract of land adjacent to the experimental station, which was established as the Eugenics Record Office on October 1, 1910, under the direction of the **American Breeders Association**'s (ABA) **Eugenics Section**. Davenport, who was secretary of the section, became resident director of the ERO and **Harry H. Laughlin**, superintendent, or assistant director.

The major purpose of the ERO was to "serve eugenical interests in the capacity of repository and clearinghouse." To this end the office, in conjunction with the **American Breeders Association**'s **Eugenics Section**, provided forms for recording health problems of families going back several generations. **Alcoholism**, **insanity**, **feeblemindedness**, **tuberculosis**, and Huntington's chorea were among the health issues investigated. The ERO trained **field workers** to collect data and maintained a permanent storage facility for the information. Other purposes of the office were to build up a "collection of traits of American families," to "study offspring in terms of inheritance of specific traits," to give advice to prospective marriage partners, and to encourage new eugenic research centers. It supported both **positive eugenics**, encouraging talented individuals to increase their reproduction, and **negative eugenics**, preventing "the propagation of defectives."

During its active years the ERO published numerous **family history and pedigree studies**. It established a **eugenics registry** and in 1916 launched a news bulletin, the *Eugenical News*, to provide information concerning activities of the office. The publication became the official journal of the **Eugenics Research Association** in 1920. Funding for the ERO came chiefly from Harriman, although in its first years philanthropist J. D. Rockefeller Jr. and the YMCA also helped to underwrite it. In 1918 Harriman

The Eugenics Record Office was the center for eugenics research in the United States during the early twentieth century. (From: *Eugenics Record Office Report No. 1* [1913], Main Library, Indiana University, Bloomington, Indiana. Image courtesy of the Digital Library Program.)

transferred the ERO to the CIW. The Station for Experimental Evolution in 1921 absorbed the ERO to form the Department of Genetics, under Davenport's directorship. The ERO was maintained as a subsection of the department.

In the early 1930s the focus on the cause of human behaviors and problems shifted from heredity to environmental factors. However, some eugenics advocates, including Laughlin, continued to champion simple **Mendelian inheritance** as the root of human problems and conditions. They also praised the **race hygiene** policies of **Nazi Germany**. Such activities concerned the leaders of the CIW, which in 1937 investigated Laughlin's research, found it unscientific, forced him to retire, and closed down the ERO. In 1939 it became the Genetics Record Office. The records from the family history studies were moved to the Dight Institute at the University of Minnesota, and, later, to the **American Philosophical Society** in Philadelphia.

FURTHER READING: Haller, Mark H., *Eugenics* (1984); Osborn, Frederick, "History of the American Eugenics Society" (1974).

EUGENICS REGISTRY (1915–c. 1939)

Established at the **Second National Conference on Race Betterment**, the Eugenics Registry was founded to collect information from individual families concerning "hereditary traits" and to classify families as eugenically fit or **unfit**. The registry, under the guidance of physician and eugenics supporter **John Harvey Kellogg**, was a "partnership" of the **Race Betterment Foundation**, Battle Creek, Michigan, and the **Eugenics Record Office** (ERO) in **Cold Spring Harbor**, New York. Members of the board included eugenics movement pioneers such as educator **David Starr Jordan**, economist **Irving Fisher**, plant breeder **Luther Burbank**, and ERO director **Charles Davenport**. Its purpose, as stated on its family-information survey forms, was, "(1) To make an inventory and record of the socially important hereditary traits and tendencies of the individual; (2) To point out, as far as possible, the conditions under which these traits and tendencies may express themselves in succeeding generations; (3) To contribute to the growth and spread of our knowledge of natural inheritance in man; (4) To assist in the maintenance and increase of natural endowments and to combat race decay." After data were collected, names of subjects who met certain standards were enrolled in a "human pedigree" book. Forms for data collection were given to willing participants through various clubs and organizations. The registry collected information on thousands of families during its years of operation.

FURTHER READING: Haller, Mark H., *Eugenics* (1984); Schwartz, Richard William, *John Harvey Kellogg, M.D.* (1981).

EUGENICS RESEARCH ASSOCIATION (1913–1938)

A professional organization, the Eugenics Research Association (ERA) emerged out of a 1913 conference for **field workers** at the **Eugenics Record Office** (ERO) in **Cold Spring Harbor**, New York. Its purpose was to facilitate communication among workers, study human heredity, and promote research. **Charles Davenport**, director of the ERO, was elected its first president. In 1916 the ERA, under Davenport and **Harry H. Laughlin,** superintendent of the ERO, launched a monthly newsletter, the *Eugenical News*. The ERA became an important group for the organization and for the coordination of eugenics research and, in turn, human **genetics** on the national and international level. Eminent scholars and reformers from many fields, united by their interest in heredity, became members of the ERA.

By 1928 the ERA had 300 members. It presented research papers and held annual meetings at Cold Spring Harbor, but did little else. With its limited funds, the association subsidized some minor research projects and judged a few essay contests on eugenics. In 1928 **Frederick Osborn**, a former businessman and nephew of anthropologist **Henry Fairfield Osborn**, began to finance some research programs for the ERA. Soon he became treasurer and a leading member of the association. Under his leadership the ERA published a series of monographs, the Eugenics Research Association Monograph Series, which began around 1929. Each monograph focused upon a eugenic issue, including topics like "Mental Tests and Heredity," "Some Biological Aspects of War," and "Comparative Birth-Rate Movements among European Nations." When genetics research in the late 1920s and 1930s did not support many of the presuppositions of eugenics, and when Laughlin and ERA president Clarence Campbell (1868–1956) openly supported **Nazi Germany's race hygiene** policies, the organization began to lose credibility. The last annual meeting of the ERA was held in June 1938. Members voted to transfer the association's property to the newly organized Association for Research in Human Heredity, which was never activated.

FURTHER READING: Allen, Garland E., "The Eugenics Record Office at Cold Spring Harbor, 1910–1940,"(1986); Haller, Mark H., *Eugenics* (1984).

EUGENICS REVIEW, THE (1909–1968; *Journal of Biosocial Science,* 1969–present)

The major British publication advocating eugenics over the course of the early-twentieth-century eugenics movement, *The Eugenics Review* was the voice of the Eugenics Education Society (later named the **Eugenics Society**). Founded in April 1909, the journal was intended to promote eugenics among the middle class and the educated. During the height of the eugenics movement well-known authors and eugenicists on both sides of the Atlantic, among them educator **David Starr Jordan** in the

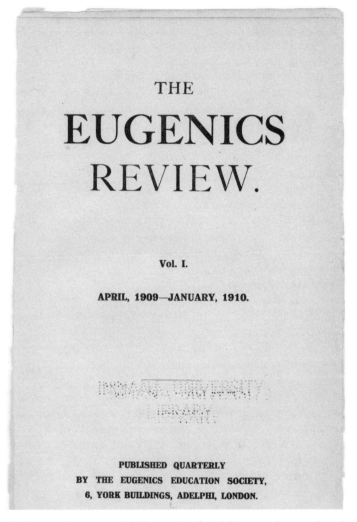

THE

EUGENICS REVIEW.

Vol. I.

APRIL, 1909—JANUARY, 1910.

PUBLISHED QUARTERLY
BY THE EUGENICS EDUCATION SOCIETY,
6, YORK BUILDINGS, ADELPHI, LONDON.

The Eugenics Review was a British magazine founded in 1909 that served to educate the middle class concerning eugenics and lobby for eugenic causes. In 1969 it became the *Journal of Biosocial Science*. (From: *Eugenics Review* [1909], Main Library, Indiana University, Bloomington, Indiana. Image courtesy of the Digital Library Program.)

United States and statistician **Karl Pearson** in **Britain**, contributed to the journal. Included were articles promoting eugenics, reports of eugenic programs, some data-based biometric works, book reviews, and philosophical theses. Subject areas embraced biology, anthropology, politics, ethics, religion, and sociology. Titles ranged from "Crime and Eugenics in America" to "The Eugenics of War." Although heredity was considered most important, the publication's early editors suggested that environmental factors also determined human behaviors. Over its years the journal published a range of articles including topics considered controversial by some groups such as **venereal disease**, **birth control**, the population explosion, **assisted reproduc-**

tion, and the genetic base for behavior, **intelligence**, and disease. Examples of titles included "Sociology, Biology and Population Control" and "An Inquiry into the Psychological Effects on Parents of Artificial Insemination and Donor Semen." The trend of the journal over time was toward more academic topics.

With shifts in the political and social climate regarding eugenics and advances in knowledge in the medical sciences and **genetics**, the journal was discontinued in December 1968. It was replaced the following year by the *Journal of Biosocial Science* which publishes research findings with a "common ground between the biological and the social sciences" and focuses upon social and biological aspects of reproduction, ecology, genetics, health and epidemiology, applied psychology, sociology, education, criminology, and demography. The sponsors of this new journal were the Galton Foundation (1969–1986), the Parkes Foundation (1987–1990), and the Biosocial Society (1991–present). In 1997 the journal began an electronic version. It is considered a respected international journal in biosocial science and publishes original research papers, reviews, short reports, and a debate section. Recent articles include "The Influence of Birth Order on Birth Weight" and "Pacific Island Suicide in Comparative Perspective."

FURTHER READING: Soloway, Richard A., *Demography and Degeneration* (1990).

EUGENICS SOCIETY, UK (1926–1989; Eugenics Education Society, 1907–1926; Galton Institute, 1989–present)

A major force of the **British eugenics** movement was the Eugenics Education Society. In 1907 a young **social hygiene** reformer, Sybil Gotto (1886–1955), organized members of the Moral Education League and other interested individuals to establish the society. Physician **C. W. Saleeby** also was instrumental in its creation. Its purpose was to promote public awareness of eugenics and of health, moral, and social problems like **venereal disease**, **alcoholism**, and **pauperism** thought to lead to **racial degeneracy**. Its aims were to educate the public for responsible parenthood following eugenic ideals; to "spread the laws of heredity"; and to "further eugenic teaching at home, in the schools, and elsewhere." Members included eminent individuals, university professors, social workers, biologists, and others of the educated middle class. By 1913 it had over 1,000 members. The first president of the society was a lawyer, Montague Crackanthorpe (1832–1913), a friend of **Francis Galton**. Gotto was secretary. Galton was made honorary president in 1908, a post he retained until his death in 1911. That same year **Leonard Darwin**, **Charles Darwin**'s son, became president, a position he retained through 1929. The society influenced the eugenics movements in Canada, the **United States**, and Australia. It was briefly associated with the **Galton Laboratory**, but the purpose of the society was to establish eugenics as a sociological and political ideology, not research, which was the mission of the Galton Laboratory. The two groups also clashed over the nature of inheritance. The society officially supported both **Mendelian inheritance** and **Darwinism** as the mechanism of inheritance. However, its more noted scientists championed **genetics** as defined by monk **Gregor Mendel** rather than biometrics, or statistics applied to biology as used by statistician **Karl Pearson** and his team at the Galton Laboratory. In 1926 the society was renamed the Eugenics Society, and in 1989, the Galton Institute.

The society was engaged in publication and during its early years produced educational pamphlets, journals, and books to enlighten the middle class about the importance of eugenic reform. In 1909 the organization established the *Eugenics Review* as its official publication, which it retained until 1968. In 1969 this publication was replaced by the *Journal of Biosocial Science*. It partially supported the *Annals of Eugenics*, a research publication, founded in 1925 by Pearson. In the latter third of the century the society published proceedings of its symposia, supplements to the journal, and reprints of classic eugenics works.

In its early years the society engaged in social action and lobbying. Soon after its formation it protested the closing of homes for "chronic inebriate women" in the London area. It argued for the study, treatment, and prevention of venereal disease, which was seen as the primary "social evil" that led to **dysgenic** births. It lobbied for the care of the **feebleminded** and influenced the passage of the British Mental Deficiency Act of 1913. The society facilitated the organization of the **First International Eugenics Congress**, held in London in July 1912. Since 1914 the society has sponsored an annual Galton Lecture. During **World War I** it supported a maternity home for wives of professional men who had been drafted or adversely affected by the war. In the 1930s it advocated contraceptive research, and intermittently over forty years fought for the legalization of voluntary **eugenic sterilization**.

During **World War II** the organization arranged for a few British children of "good stock" to be sent to Canada. In the postwar years the society endorsed early experimentation with human **artificial insemination**. In 1959 it was left a substantial portion of the estate of British **birth control** leader **Marie Stopes**. The legacy included her birth control clinic, library, and other assets. Due to decreasing membership and increasing negative perception concerning eugenics, the society changed its scope and focus. In 1963 it became a tax-exempt charity that proposed to bring together the biological and sociological sciences for the mutual exchange of ideas concerning heredity and environment (**nature-nurture**). Beginning in 1963 the organization began to sponsor an annual symposium to discuss biosocial issues. In 1967 it formed the Galton Foundation, a grant-giving agency. In 1989 the society changed its name to the Galton Institute and its purpose to "the interdisciplinary study of the biological, genetic, economic and cultural factors relating to human reproduction, development and health." Still active, the institute sponsors scholarly symposia and publications.

FURTHER READING: Keynes, Milo, (ed.), *Sir Francis Galton, FRS* (1993); Mazumdar, Pauline, M. H., *Eugenics, Human Genetics, and Human Failings* (1992); Searle, G. R., *Eugenics and Politics in Britain, 1900–1914* (1976); Soloway, Richard A., *Demography and Degeneration* (1990).

EUGENICS SURVEY OF VERMONT (1925–1936)

A series of studies, the Eugenics Survey of Vermont was affiliated with the Department of Zoology of the University of Vermont and directed by biologist and eugenics supporter **Henry F. Perkins**. The initial aim of the survey was to study so-called "**degenerate**" families in Vermont and to determine the causes of degeneration in order to support a **eugenic sterilization law** for the state. To carry out its studies, the survey used records of families registered with both private and public charity and welfare organizations. The Vermont Children's Aid Society and the Social Service Exchange were the two primary organizations from which data were collected. The survey's advisory board was composed of directors of correctional facilities, mental health hospitals, educators, public welfare officers, and private charity groups. It was financed by two prominent women of the Vermont Children's Aid Society. Results of the studies were reported in a series of annual reports and two books. The survey was closed in 1936, due to lack of funding.

The survey went through three phases. In its early and most influential years, 1925–1928, it studied degenerate families. The results were used to launch a crusade for legalized sterilization. Social worker Harriett Abbott, who had been trained as a eugenic **field worker** at the **Eugenics Record Office**, conducted the majority of the investigations. She created a series of **family history** charts based upon records and interviews for each family studied. The first report, published in 1927, discussed 4,620 individuals from sixty-two "**dysgenic** families." Almost a fourth were listed as **paupers**, and many were classified as **feebleminded**, **criminals**, **alcoholics**, sex offenders, and bearers of "illegitimate children." The **genetic disease** Huntington's chorea was found in one family. Perkins concluded that these families owed their "degeneracy" to heredity and launched a crusade for legal sterilization. In 1928–1929 the survey focused on hereditary factors that led to progress or decline in rural communities and investigated social conditions that affected family size. During the final years of the survey, 1929–1931, social forces influencing population trends, migration patterns, and types of employment were studied. Results from the Eugenics Survey of Vermont helped persuade the state's legislators to implement a "voluntary" sterilization law in 1931 to keep Vermonters genetically fit.

FURTHER READING: Gallagher, Nancy L., *Breeding Better Vermonters* (1999).

EURO OR EUROPEAN AMERICANS

Americans with European ancestry are termed *Euro* or *European American*, although the term did not become common until the late twentieth century. It was coined by anthropologist Margaret Mead (1901–1978) in 1949. In the 1970s, as the terms *black* and *African American* replaced *colored* or *Negro* for Americans with ancestry from sub-Saharan Africa, *European American* rose as a parallel term for *white*, or Americans with European pedigrees. From the early nineteenth century immigrants from Ireland, and in the midcentury immigrants from northern Europe, emigrated to the United States. Over a couple of generations they had more or less assimilated into the American culture. During the decade on either side of 1900 **eastern and southern European** Roman **Catholics** and eastern European **Jews** entered the nation in increasing numbers. "Old-stock" **Protestant Anglo-Saxon Americans** worried that these new immigrants were not assimilating and were of "inferior stock," compared with earlier immigrants from Britain and northern Europe. **Nativist** reformers encouraged **immigration restriction** laws to curtail immigration from eastern and southern Europe.

Over the course of the twentieth century, however, most eastern and southern European immigrants assimilated into the dominant culture and were identified as whites or Euro Americans, not by their ethnic ancestry, such as Italian or Polish American. By the turn of the twenty-first century some "white pride" groups used the term *European American* as a political statement. Similar to nativists of the earlier **Jacksonian** and **Progressive Eras**, these groups feared the disappearance of old Anglo-

Saxon Protestant values, violent **crime**, and large social welfare programs, along with other political and social problems due to unregulated immigration from non-European cultures. The 2000 U.S. census reports that 75.1 percent of the population of the United States is white. The most prevalent European ancestry includes German (15.2%); British (12.5%); Irish (10.9%); Italian (5.6%); approximately 3 percent each Polish, French, and Scandinavian; and 1 percent or less for other European countries.

FURTHER READING: Howe, Louise Kapp, (ed.), *The White Majority* (1971); Swain, Carol M., *The New White Nationalism in America* (2002).

F

. . . and let the unmarried choose healthy companions or none.

Orson Fowler, *Love and Parentage* (1847), 43.

Woman . . . is the natural conservator of the race, the guardian of its blood.

Albert Wiggam, *Fruit of the Family Tree* (1924), 280.

FAIRCHILD, HENRY PRATT (August 18, 1880–October 2, 1956)

A social scientist, Fairchild was active in various aspects of the early-twentieth-century eugenics movement. He helped refocus the eugenics field into population studies as the movement waned in the middle decades of the century. Born in Dundee, Illinois, Fairchild was the son of an educator from a Congregationalist family with **Anglo-Saxon** colonial roots. He spent his childhood in a rural area prominent with non-English-speaking immigrants, which led to his interest in the "immigrant problem." After attending public schools in Crete, Nebraska, he graduated with an A.B. (1900) from Doane College in that community. He taught at several institutions, including in Turkey, before earning a Ph.D. from Yale University (1909) with a dissertation on Greek immigration; he then taught at Yale (1910–1918). During **World War I**, Fairchild was director of personnel at the War Camp Community Service in New Haven, Connecticut. In 1923 Fairchild served as special immigration agent for the United States Department of Labor. He taught at New York University until his retirement in 1945. He was involved with many organizations embracing eugenics, population studies, family planning, and immigration problems.

Fairchild was a charter member of the **American Eugenics Society** and active in the society's **immigration restriction** activities. He served on the advisory council (1923–1927), was secretary-treasurer (1926), president (1929–1931), and a member of the board of directors (1932–1940). He supported immigration restriction as a measure to ensure old American values and argued that the "the problem of immigration is but a part of the great conservation movement. It has to do with the conservation of the American people, and all that it stands for." Unlike many early eugenicists Fairchild favored **birth control** and saw it as an essential aspect of eugenics. In the 1920s he became involved with the birth control movement and suggested that limiting population was necessary to securing international peace. He served on the advisory council of birth control leader **Margaret Sanger**'s Birth Control and Clinical Research Bureau and as vice president of the Planned Parenthood Federation (1939–1948).

Out of his concern over worldwide population pressures Fairchild participated in the first World Population Conference in Geneva, Switzerland, in 1926 and helped found the Population Association of America; he served as its first president (1931–1935). By 1932 he argued that eugenics, inasmuch as it now focused on social and economic problems, appropriately belonged among the social sciences, not biology. Fairchild wrote numerous articles and books on eugenics, birth control, population, and immigration, including *Immigration* (1913), *The Melting-Pot Mistake* (1925), and *Race and Nationality* (1947). He married Mary Eleanor Townsend (1909), with whom he had one daughter. He died at the home of his daughter.

FURTHER READING: Kline, Wendy, *Building a Better Race* (2001).

FAMILY BALANCING

The term *family balancing* came into use during the mid-1990s to describe the selection of a child's gender for nonmedical reasons. Methods emerged in the mid-1990s to provide married couples with at least one child the opportunity to increase the chance of having another child of the opposite sex, thus "balancing" their family. Two techniques in this regard have proven successful. "Pre-conception gender selection" (selecting embryos with the desired sex) had been used since the late 1970s to help couples avoid passing on a gender-linked **genetic disease**. With this technique, **in vitro fertilization** was used. Although expensive, it has proven successful in selecting the desired gender. However, it is now more common to separate sperm through the "MicroSort" method, which uses fluorescent dye to identify the Y (male) or X (female) chromosome in the sperm. After the sperm are separated, those for the desired gender are then inserted into the uterus via **artificial insemination**. Family balancing is legal in the **United States** and many other countries. However, in **Britain** preconception gender selection is allowed only for medical reasons. Family balancing is a **positive eugenics** technique.

FURTHER READING: Stock, Gregory, *Redesigning Humans* (2002).

FAMILY HISTORY AND PEDIGREE STUDIES

The formal study of kinship groups and their inherent characteristics was termed *family history* or *pedigree studies* in the late nineteenth and early twentieth centuries. Information for these studies was collected through field investigations or through "pedigree" (genealogical) charts. Studies were done for both "outstanding" and **unfit** families. British naturalist **Francis Galton**, the father of eugenics, pioneered the method by using genealogical charts of notable British families. He reported his findings in *Hereditary Genius* (1869) and suggested that high achievement and **intelligence** ran in certain families, including his own, the Darwin-Wedgewood-Galton family. In **Germany** the pedigrees of various intellectual and noble families were also studied in the early twentieth century. For the studies in **Britain** and Germany family pedigree charts were primarily used. However, in the **United States** family history studies based upon field investigations were more common and focused on **degenerates** rather than prominent families.

Richard Dugdale produced the first family history in the United States after investigating a rural New York family known for its **crime** and **pauperism**, whom he called *The Jukes* (1877). **Arthur H. Estabrook**, who followed up on living descendants of this family and published *The Jukes in 1915* (1916), concluded that defective **genes** had caused continuing social and health problems in the family. **Alexander Graham Bell**, inventor of the telephone, was one of the first to use eugenic **field workers** to collect

29 —EXTERNAL EAR—PINNA
291 —Helix
2911 —Descending helix
2912 —Transverse helix
2913 —Inner helix
2914 —Darwinian tubercle
2915 —Navicular fossa
292 —Antihelix
2922 —Upper ramus
2924 —Lower ramus
2926 —Oval fossa, elevation, relative to helix
293 —Lobule
2932 —Adherency
2934 —Lobular fold
2936 —Antitragus
2938 —Intertragial fossa
294 —Tragus
295 —Upper concha
297 —Lower concha
299 —General position

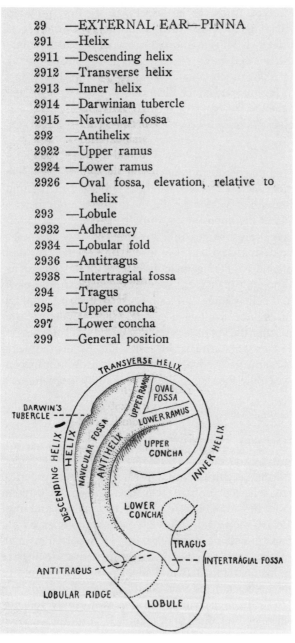

The Trait Book was used for family history and pedigree studies. Eugenic field workers trained by Charles Davenport at the Eugenics Record Office at Cold Spring Harbor, New York, used the codes within this book to standardize their genealogical reports. (From: The Trait Book, Main Library, Indiana University, Bloomington, Indiana. Image courtesy of the Digital Library Program.)

information for family studies. He investigated the inheritance of deafness among families on Martha's Vineyard in the early 1880s. Another early family history study was **The Tribe of Ishmael** (1888), conducted by Oscar McCulloch, an Indianapolis, Indiana, minister, who reported on a traveling clan to determine methods of reducing poverty.

Around 1910 geneticist and eugenics leader **Charles Davenport**, director of the **Eugenics Record Office**, designed a family record form to be filled out by eugenic field workers, whom he trained, to collect family history information. To construct a family pedigree the field worker interviewed neighbors, friends, relatives, charity workers, physicians, teachers, and others and examined public records. Pedigree charts were constructed based upon the Trait Book (1912), an indexed list of codes for physical, mental, and social characteristics and diseases developed by Davenport to standardize entries. If a trait such as laziness was found for several generations among blood relatives, the characteristic was assumed to have been inherited. These studies, conducted mostly between 1910 and 1930, included the **Dacks, Kallikats, Nams,**

Jukes in 1915, and *Hill Folk*. Several states also conducted series of studies, including *Mental Defectives in Indiana* (1916–1922) and the **Eugenics Survey of Vermont** (1924–1936).

The families investigated in these studies were mostly impoverished whites, although the *Mongrel Virginians* were of mixed **races**. In general these families had a reputation for being **alcoholics**, **criminals**, **feebleminded**, **insane**, lazy, **paupers** (inherited poverty), **prostitutes**, or having many out-of-wedlock children. Family pedigree information was also collected from middle-class individuals through clubs and conferences. Individuals submitting their family history could request the advisability of marriage based upon eugenic concerns. Many forms were collected at **Fitter Family** contests at state fairs. Health reformer **John Harvey Kellogg** also established a **Eugenics Registry Office** to collect family pedigrees in order to encourage marriages among the fit. Although family pedigree studies were considered scientific in their day, by the mid-twentieth century, they were deemed faulty because observations of traits were often subjective and it had been found that many human characteristics could not be explained by simple **Mendelian inheritance**. Impoverished environments and lack of educational and cultural opportunities began to be seen as major factors in generational poverty and other social problems found in kinship groups.

FURTHER READING: Bix, Amy Sue, "Experiences and Voices of Eugenics Fieldworkers: 'Women's' Work in Biology" (1997); Paul, Diane B., *Controlling Human Heredity* (1995); Rafter, Nicole Hahn, (ed.), *White Trash* (1988).

FEEBLEMINDED OR FEEBLEMINDEDNESS

The generic scientific term for mental disability from the mid-nineteenth through the mid-twentieth centuries was *feebleminded* or *feeblemindedness*. The term emerged in the 1850s when "idiot schools," or state asylums, were built for the mentally disabled. By the early twentieth century, "feebleminded" became a catchall category for anyone exhibiting unacceptable social behavior. It was believed that the feebleminded exhibited "immoral behaviors," did not have adequate reasoning powers, and lacked self-restraint. Although not all criminals were feebleminded, all feebleminded were seen as potential criminals and every feebleminded woman as a potential prostitute and transmitter of **venereal disease**. **Family history studies** such as the *Dacks, Kallikaks, Nams, Jukes*, and *Hill Folk* showed a high percentage of feebleminded in family lineages, thus supporting claims for the **genetic** transmission of mental and moral traits.

The "menace of the feebleminded" became a major concern of eugenicists. The "high-grade feebleminded," or **morons**, who looked normal, unlike more seriously mentally impaired individuals, who were generally institutionalized, often survived independently and reproduced. Many **criminals**, **paupers**, and **prostitutes** were found among their number. Since **intelligence** was considered inherited, it was feared that they would "reproduce their own kind," thus leading to **racial degeneracy** and the downfall of civilization. Eugenic popularizers including **Lothrop Stoddard** and **Albert E. Wiggam** argued that human intervention such as charity and **public health** had increased the number of feebleminded by not allowing **natural selection** to run its course. **Henry Goddard**, who headed a state asylum in Vineland, New Jersey, in 1906 began a study of the cause of the condition. Believing it was based on simple **Mendelian inheritance**, he recommended identification of the feebleminded by using the new **IQ** or **intelligence tests** and prevention of reproduction by custodial care.

However, by the mid-1910s **eugenic segregation** in institutions ceased to be cost-effective, so **eugenic sterilization** became the new solution. Sterilization laws were enacted in many states, and thousands of mentally ill or disabled individuals were subsequently sterilized. By the 1970s the term *feebleminded* fell out of favor, to be replaced by phrases such as *mentally* or *developmentally retarded, handicapped, disabled, impaired* or *challenged*.

FURTHER READING: Carlson, Elof Axle, *The Unfit* (2001); Ingalls, Robert P., *Mental Retardation* (1986); Kline, Wendy, *Building a Better Race* (2001).

Field workers, who were mostly women, were trained at the Eugenics Record Office to collect data for family pedigree and genealogical studies. (From: *Eugenics Record Office Report No. 1* [1913], Main Library, Indiana University, Bloomington, Indiana. Image courtesy of the Digital Library Program.)

FIELD WORKERS

Individuals who collected information for genealogical **family history and pedigree studies** in the second and third decades of the twentieth century were termed eugenic *field workers* or *field psychologists*. Eugenic field work was considered an applied science, and the resulting reports were used to support the **eugenic sterilization** movement in the **United States**. Inventor **Alexander Graham Bell** was among the first to use field workers in a study of deaf-mutism in the late nineteenth century. Most field workers of the early-twentieth-century eugenics movement were trained at a summer school run by eugenicists **Charles Davenport** and **Harry H. Laughlin** at the **Eugenics Record Office, Cold Spring Harbor**, New York. From 1910 until 1924 more than 250 field workers were instructed in human heredity, field research techniques, and the construction of human pedigree charts. The summer school was largely funded by **Mary Harriman**. Most trainees were young women—only twenty-five were men—with bachelor's degrees or higher in education, social work, or medicine who considered training in eugenics field work as a path toward professional advancement. However, few of these women actually gained opportunity for advancement.

Field workers performed a variety of tasks. They were often affiliated with an institution and investigated the heredity of an inmate by interviewing relatives as well as friends, employers, and professionals. They administered intelligence tests, acted as social workers, and looked for potential candidates to be institutionalized among family members of the inmate. They analyzed each individual in regard to mental, moral, and physical traits, developed pedigree charts; and gave expert witness testimony in legal sterilization cases. Several field workers became prominent in their own right. **Arthur H. Estabrook** became a noted eugenics researcher, publishing *The Nam Family*, *The Jukes in 1915*, and *Mongrel Virginians*, and biologist **Wilhelmine Key** conducted research at the **Race Betterment Foundation**. In the early decades of the twentieth century family pedigree studies were considered scientific. However, by the 1940s they were deemed faulty inasmuch as observed traits and characteristics of individuals were often subjective. However, methods used by the field workers influenced modern ethnographic data collection in addition to influencing research in the developing fields of psychology, sociology, and social work.

FURTHER READING: Bix, Amy Sue, "Experiences and Voices of Eugenics Field-Workers: 'Women's' Work in Biology" (1997); Paul, Diane B., *Controlling Human Heredity* (1995); Rafter, Nicole Hahn, (ed.), *White Trash* (1988).

FIRST INTERNATIONAL EUGENICS CONGRESS (1912)

As the early-twentieth-century eugenics movement spread on both sides of the Atlantic, the First International Eugenics Congress brought together eugenics leaders from the **United States**, **Britain**, **Germany**, and other countries. The congress, held at the University of London July 24–30, 1912, was hosted by the British **Eugenics (Education) Society**. Attendance included 324 individuals from around the world. The conference name tags bore the likeness of **Francis Galton**, the father of eugenics. This major race improvement, or eugenics, conference was aimed at all organizations and individuals studying eugenics, **race hygiene**, and heredity.

Britain's pivotal eugenics leader, **Leonard Darwin**, son of naturalist **Charles Darwin**, served as chairman. Other British delegates included eugenics author **C. W. Saleeby**, geneticist William Bateson (1861–1926), and Winston Churchill (1874–1965), then secretary of state for home affairs and later prime minister. Eugenics supporter and statistician **R. A. Fisher**, as a student, served as a steward; statistician **Karl Pearson** did not attend.

From the United States, "vice presidents," or those who helped organize the conferences, included inventor **Alexander Graham Bell**, eugenics leader **Charles Davenport**, university presidents **Charles Eliot** and **David Starr Jordan**, and conservationist Gifford Pinchot (1865–1946). Other notable American delegates were anthropologist **Henry Fairfield Osborn**; eugenics supporter **Madison Grant**; and geneticists William E. Castle, Frederick A. Woods, and Raymond Pearl. From Germany, eugenics leader **Alfred Ploetz** and Max von Gruber, the first German professor of hygiene in Munich, served as vice presidents; German eugenics researcher Agnes Bluhm presented a paper. Norwegian eugenics leader Jon Mjøen and several representatives from Italy and France also presented papers.

The conference sessions were divided into four categories, the relationship of eugenics to (1) biological research, (2) sociological and historical research, and (3) legislation and social customs; and (4) the practical applications of eugenic principles. The papers and proceedings from the conference program were published by the Eugenics Society in two volumes. Vol. 1, *Problems in Eugenics: Papers Communicated to the First International Eugenics Congress* (1912), contained 490 pages. Most of the papers presented at the conference were found in this volume. Specific titles included "Eugenics and Militarism," "The Influence of Race on History," and "Backward Children."

PROBLEMS IN EUGENICS.

VOL. II.

REPORT OF PROCEEDINGS

OF THE

First International

Eugenics Congress

HELD AT

THE UNIVERSITY OF LONDON,

July 24th to 30th, 1912.

Together with an Appendix containing those Papers communicated
to the Congress not included in Volume I.

PUBLISHED BY

THE EUGENICS EDUCATION SOCIETY,

KINGSWAY HOUSE, KINGSWAY, W.C.

1913.

The proceedings from the First International Eugenics Congress, held in London in 1912, contained the presentations and discussion of various eugenics topics. (From: *Problems in Eugenics* [1913], Main Library, Indiana University, Bloomington, Indiana. Image courtesy of the Digital Library Program.)

Vol. 2, *Problems in Eugenics: Report of Proceedings*, contained 189 pages. Although it included a few papers, its primary focus was discussion following presentations, grouped under session headings such as "Biology and Eugenics," "Practical Eugenics," "Education and Eugenics," "Sociology and Eugenics," and "Medicine and Eugenics." Discussion centered upon how to prevent the **unfit** from breeding through **eugenic segregation**, **eugenic sterilization**, and **eugenic marriage** certificates to certify health, and how to encourage the "fit" to reproduce. A planning committee was formed to plan the next congress, in 1913. However, due to **World War I** the committee did not meet until October 1919, in London, where it agreed to hold a **Second International Congress of Eugenics** in New York in 1921.

FURTHER READING: Haller, Mark H., *Eugenics* (1984); Kühl, Stefan, *The Nazi Connection* (1994).

FIRST NATIONAL CONFERENCE ON RACE BETTERMENT (1914)

Sponsored by the **Race Betterment Foundation** and organized by health reformer and eugenics supporter **John Harvey Kellogg**, the First National Conference on Race Betterment forged a link between Kellogg's program of healthy living and other health reforms of the **Progressive Era**. The conference, held January 8–12, 1914, at Battle Creek, Michigan, brought together notable eugenicists, **public health** professionals, and reformers focused on problems believed to cause **race degeneracy**. The stated purpose of the conference was "to assemble evidence as to the extent to which degenerative tendencies are actively at work in America, and to promote agencies for race betterment." These included both hereditarian and environmental factors (**nature-nurture**). Presentations and photographs of exhibits were published in a 635-page volume, *Proceedings of the First National Conference on Race Betterment* (1914). Topical sessions included "Statistical Studies," "General Individual Hygiene," "Alcohol and Tobacco," "Child Life," "Sex Questions," "School and Industrial Hygiene," "City, State, and National Hygiene," and "Eugenics and Immigration."

More than 400 delegates attended the conference, presided over by Stephen Smith, vice president of the State Board of Charities of New York City. Eugenics leaders including **Charles Davenport**, **Irving Fisher**, and **Harry H. Laughlin** helped plan the meeting or made presentations. Kellogg, who believed in **Lamarckian inheritance** of acquired characteristics even though it had been refuted by the second decade of the century, recommended healthy lifestyle changes in order to prevent inherited **degeneracy**. These recommendations, as addressed by conference presenters, reflected both **positive** and **negative eugenics**. To promote marriage between "fit" individuals, Kellogg suggested the establishment of a **Eugenics Registry** to collect **family history pedigrees**. "**Better Babies**" contests and mental and physical competitions for youth were encouraged. Negative eugenics including **immigration restrictions**, and **eugenic sterilization** of the **insane** and **feebleminded** were recommended. Environmental efforts such as the elimination of **tobacco**, **alcohol**, and **prostitution** through stricter laws, and methods to prevent **tuberculosi**s, **venereal disease**, and other illnesses; sanitation; personal hygiene; diet; and physical fitness were championed. The conference received much public attention, and as a result Kellogg and his associates sponsored the **Second National Conference on Race Betterment** the following year.

FURTHER READING: Money, John, *The Destroying Angel* (1985); Schwarz, Richard William, *John Harvey Kellogg, M.D.* (1981).

FISCHER, EUGEN (June 5, 1874–July 9, 1967)

An internationally known anthropologist, Fischer was one of the most influential researchers in the **German eugenics** movement of the post–**World War I** Weimar and **Nazi Germany** eras. He headed the **Kaiser Wilhelm Institute** (KWI) of Anthropology, Race Hygiene, and Eugenics, the major center for eugenics research in Germany, and

German anthropologist Eugen Fischer, head of the Kaiser Wilhelm Institute, was a major leader of the German eugenics movement after 1920. (Image courtesy Archiv zur Geschichte der Max-Planck-Gesellschaft, Berlin-Dahlem.)

was one of the first researchers to apply **Mendelian inheritance** to human characteristics. Born in Karlsruhe, the son of a businessman, he was raised a Roman **Catholic** but as an adult grew away from the church. When he was two his family moved to the Freiburg im Breisgau area. Fischer studied medicine and received a doctorate in both medicine and anatomy (1898) from the University of Freiburg. He studied under geneticist August Weismann (1834–1914), who had developed the **germ-plasm** theory of inheritance, which influenced Fischer's lifetime interest in **genetics**. In 1900 he completed an advanced dissertation in order to teach at the university level. He was appointed lecturer in both anatomy and physical anthropology at the University of Freiburg, and in 1904, assistant professor. Fischer lectured in Würzburg in 1912 and a year later returned to Freiburg im Breisgau. In 1918 he was appointed professor and director of the Institute for Anatomy, where he remained until 1927, when he was appointed professor of anthropology at the University of Berlin, director of the KWI, and head of the anthropology department at the institute. In 1937 he was elected rector of the University of Berlin, against the wishes of the Nazis, and retired from all these positions in 1942.

Around 1903 Fischer became interested in Mendelian inheritance and its application to human populations. He received a grant in 1908 from the Berlin Academy of Sciences to investigate a mixed-race population of Dutch settlers and native Hottentots in the village of Rehoboth in German Southwest Africa. The resulting book, *The Rehoboth Bastards and the Cross-breeding in Man* (*Die Rehobother Bastards und das Bastardierungsproblem beim Menschen*, 1913) was the first major study to support Mendelian inheritance of physical characteristics among human populations. With this publication he founded what became known as the "anthropobiological" school of anthropology, which focused on the genetics of heredity. After moving to Berlin to head the KWI, he studied the inheritance of **tuberculosis** and the origins of human **races**.

Fischer was involved with all the leading German and international eugenics organization. Encouraged by **Alfred Ploetz**, the father of German eugenics, whom he met in 1908, Fischer formed the Freiburg branch of the **German Society for Race Hygiene** in 1910. Fischer's student **Fritz Lenz**, leader of the post–World War I eugenics movement, became secretary of the organization. In 1922 Fischer became a member of the Prussian Council for Racial Hygiene. He was president of the German and the Berlin Race Hygiene Societies (1928–1933) until the Nazis asked him to step down. In the late 1920s he interacted with American eugenicists, including **Charles Davenport** and **Lothrop Stoddard**, and was sent a copy of **Harry H. Laughlin**'s *Eugenical Sterilization in the United States* (1922), which influenced eugenic sterilization laws in Nazi Germany. In 1921 Fischer, along with Lenz and botanical geneticist Erwin Baur (1875–1933), published *Human Heredity and Race Hygiene* (*Menschliche Erblehre und Rassenhygiene*), which became the leading race hygiene text of the German eugenics movement during both the Weimar Republic and Nazi eras. In 1929 Fisher was appointed associate editor for the leading German eugenics journal, *Archive for Racial and Social Biology*, and remained until its demise at the end of World War II. He wrote several books and numerous articles not only for German, but also for American and British journals.

As Nazi power grew in the post–World War I era, Fischer began to accept "racial purity" as desirable for eugenics programs, but he did not support National Socialism until the party's takeover in 1933. Fischer and others at KWI were called upon to support the implementation of the 1933 sterilization laws and, later, to determine the **racial classification** of individuals. In 1937 the Nazis made him a member of the Prussian Academy of Sciences. Fischer likely manipulated within the Nazi regime to keep his institute open and did not join the party until 1940. He helped to reorganize the field of anthropology, which by the late 1930s began to separate into two separate disciplines—physical anthropology and human genetics. In the post-**World War II** era he was made an honorary member of the reconstructed postwar German Anthropological Association. He married Else Walter around 1898, with whom he had three children, and died in Freiburg im Breisgau.

FURTHER READING: Gessler, Bernhard, *Eugen Fischer (1874–1967)* (2000); Kühl, Stefan, *The Nazi Connection* (1994); Weindling, Paul, *Health, Race and German Politics between National Unification and Nazism, 1870–1945* (1989); Weingart, Peter, "German Eugenics between Science and Politics" (1989).

FISHER, IRVING (February 27, 1867–April 29, 1947)

An economics professor, Fisher was a pivotal figure not only in the eugenics movement but of many other **Progressive Era** health reform movements. Through his leadership positions and publications he influenced campaigns embracing diet and nutrition, **physical culture**, **tuberculosis**, **public health**, and prohibition. Fisher, the third of four children, was born in Saugerties, New York, the son of a Congregational minister descended from early New York settlers. Soon after his birth his family moved to Rhode Island, then Connecticut, and finally St. Louis. The year his father died he graduated from Smith Academy in St. Louis (1884). He received a B.A. from Yale University, where he was class valedictorian. Under full scholarship he immediately began working on a doctorate in mathematics. Fisher joined the Yale faculty in 1890 as an instructor and completed his Ph.D. (1891). He then studied in Berlin and Paris (1893–1894). Upon returning to Connecticut in 1894 he switched from the mathematics department to the department of political economy, where he remained the rest of his career. Over his lifetime Fisher became one of the nation's leading economists.

Fisher integrated eugenics as an underlying theme in several health reform crusades. He campaigned against **racial poisons** such as **alcohol**, habit-forming drugs, **venereal disease**, **tobacco**, tea, and coffee and advocated teaching eugenics in schools and colleges. He favored **eugenic sterilizations laws** and eugenic **segregation** of "defectives" in institutions; he supported **positive eugenics** programs such as the **Fitter Families** campaign. Fisher held leadership positions in most eugenics organizations. At the **Second International Congress of Eugenics** (1921) he spearheaded a committee that evolved into the **American Eugenics Society**; he was the society's first president (1922–1926). He was vice president of the **Third International Congress** (1932), served on the executive committee of the **First, Second, and Third National Conferences on Race Betterment**, was president of the **Eugenics Research Association** (1920), and was a member of the **Eugenics Registry**'s governing committee. He supported the **immigration restriction movement** but opposed **birth control** unless it was extended to "inferior racial stocks" on the grounds that both birth control and **war** would lead to **race suicide** among the educated and wealthy. Fisher's fortune and reputation were marred by the 1929 stock market crash. Several days before the crash he had reassured investors that stock prices had achieved a new and permanent plateau and were not overpriced. In 1935 he retired from Yale. Over his lifetime he published more than 2,000 works and received many awards. He married Margaret Hazard (1893) and they had three children. Fisher died in New York City of colon cancer.

FURTHER READING: Fisher, Irving Norton, *My Father, Irving Fisher* (1956).

FISHER, R(ONALD) A(YLMER) (February 17, 1890–July 29, 1962)

As internationally renowned British pioneer in statistics, Fisher provided the scientific underpinnings for the developing fields of statistics, **genetics**, and evolutionary biology during the early decades of the twentieth century. Born in London, England, Fisher was the last of seven children; his father was a partner in a fine arts auction firm. As a child Fisher was gifted in mathematics and interested in the natural sciences. He attended the private school Harrow and received a scholarship to Gonville and Caius College at Cambridge University (1909–1912). Upon graduation as a "Wrangler" (high honors), he spent a year of graduate work studying mathematics, theoretical physics, biometry, and genetics. The next six years he worked in a variety of pursuits including as a farm laborer in Canada, statistician in London (1913–1915), and public school teacher (1915–1919) during **World War I** after being rejected for military service due to poor eyesight. Fisher gained a reputation as a brilliant mathematician. In 1919 he rejected an offer from noted statistician and eugenics supporter **Karl Pearson** to work at the **Galton Laboratory** in London. Determined to conduct his own research, he instead accepted a position at the Rothamsted Experimental Station, north of London. There he pursued his interest in genetics by breeding various animals. Fisher left this job in 1933 to occupy the Galton Chair of Eugenics at University College, London, when Pearson retired. At the laboratory he established in 1935 a blood-typing department that developed information on the inheritance of the Rh factor. Fisher left the laboratory in 1943 to become Balfour Professor of Genetics at Cambridge, where he remained until his retirement in 1957. He was president of Gonville and Caius College (1956–1959) and then moved to Adelaide, Australia, to work with several former students as a statistical researcher for the CSIRO, a government research organization.

Like many other eugenicists, Fisher accepted **social Darwinism** and was concerned about the decline in the birthrate among the middle and upper classes. He admired German philosopher Friedrich Wilhelm Nietzsche's (1844–1900) **superman** philosophy as a forerunner of a "new human race" and supported **positive eugenics** programs to encourage the "fitter classes" to reproduce. He was against **birth control** on the grounds that it prevented the rich from procreating. As a student Fisher formed the Cambridge Eugenics Society in 1911, and there he met **Leonard Darwin**, prime leader of the **British eugenics** movement. Darwin became a mentor to Fisher and influenced his thinking on eugenics, **genetics**, and **Darwinism**. While a student Fisher assisted at the **First International Eugenics Congress** (1912) in London, joined the **Eugenics Society** that same year, and later served as its secretary (1920–1930). In the 1920s Fisher's influence brought more scientists into the society. Beginning in 1914 he wrote many book reviews for *Eugenics Review*, the society's journal. He was also editor of *Annals of Eugenics* (1939–1941), the professional journal of the Galton Laboratory. Fisher was actively involved with the **Third International Congress of Eugenics** (1932), held in New York. Between 1920 and 1934 he campaigned for voluntary **eugenic sterilization** of the **unfit**. In 1928 he published a prototype legislation for parliament, to no avail. In 1932 he chaired a committee for the Eugenics Society to investigate family allowances.

Fisher demonstrated in 1918 that human inheritance was consistent with the principles of **Mendelian inheritance** and complementary to **Darwinism** and biometrics (using statistics with biology) developed by Pearson. Fisher's theory was not fully developed until publication of his most seminal work, *The Genetical Theory of Natural Selection* (1930), which brought together biometrics and the mechanics of genetics. It was dedicated to Leonard Darwin. A misunderstanding between Fisher and Pearson concerning a comment in one of Fisher's papers triggered a lifelong conflict between the two pioneer statisticians. Fisher was one of the first researchers (1923) to question the validity of **family pedigree studies**, first introduced by **Francis Galton** and used by Pearson and his lab. Over his lifetime Fisher published more than 300 papers and seven books and received numerous honors including a knighthood (1952). He married Ruth Eileen Guinness (1917), the daughter of a minister; unlike most middle-class professionals, who had few children, he practiced positive eugenics in his marriage and fathered two sons and six daughters. In 1943 he separated from his wife. Fisher died in Adelaide, Australia, after surgery.

FURTHER READING: Box, Joan Fisher, *R. A. Fisher* (1978); Fienberg, Stephen E., and D. V. Hinkley, (eds.), *R. A. Fisher* (1980); Fisher, Ronald Aylmer, Sir, *Natural Selection, Heredity, and Eugenics* (1983).

FITTER FAMILIES CAMPAIGN

A prime example of a **positive eugenics** program, which emerged near the peak of the eugenics movement in the **United States**, was the Fitter Families campaign, modeled after livestock judging at county fairs. Families were judged in terms of mental, emotional, physical, and intellectual fitness. Out of recognition "that humans and livestock are controlled by the same laws of heredity," it was hoped that young adults who knew their personal eugenical history would be more prudent in their selection of mates, thus leading to a "fitter human stock." The campaign evolved out of the **Better Babies Movement**, popular in the pre–**World War I** years, and the **First National Conference on Race Betterment**. At this 1914 conference, hosted by physician **John Harvey Kellogg**, "mental and physical perfection contests" were held. Several thousand schoolchildren and about 600 babies were tested for "mental and physical efficiency." Prizes were awarded to children scoring highest in each age group, from six months to nineteen years. In connection with the contests, educational literature was distributed to mothers concerning diet and hygiene. Children found to be sick were referred to their physicians. The conference and the contest attracted wide media attention throughout the United States and laid the groundwork for the Fitter Families contests.

Mary T. Watts (d. 1926) and Florence Brown Sherbon (1869–1944), two pioneers of the health examination for babies movement, organized the first Fitter Families contest around 1920 at the Kansas Free Fair. The aim of the program was to introduce the concept of positive eugenics and periodic health examinations for all family members. Healthy children would grow up to provide "Fitter families for the future." For the examination, a form was adopted from the Life Extension Institute, an organization that encouraged annual physical examinations, and the **Eugenics Record Office**. These **family history** forms asked for medical, physical, dental, psychiatric, and other histories. The examination was free to contestants. In 1924 sponsorship of the contest was underwritten by the **American Eugenics Society** and extended to other states. It became so popular that most fairs had to limit entries. Families were classified according to size and given a letter grade; the winning family in each class received a trophy. All contestants with a B+ or better were given a bronze medal bearing the inscription, "Yea, I have a goodly heritage." A copy of the medal became the logo for the title page of the journal *Eugenics*. In 1928 the **Race Betterment Foundation**, sponsored by Kellogg, underwrote the project. The contests were held into the late 1930s.

FURTHER READING: Dorey, Annette K. Vance, *Better Baby Contests* (1999); Holt, Marilyn Irvin, *Linoleum, Better Babies, and the Modern Farm Woman, 1890–1930* (1995); Rydell, Robert, *The World of Fairs* (1993).

FOWLER, LYDIA FOLGER (May 5, 1822–January 26, 1879)

Considered the second female physician in the United States, Fowler taught female physiology to women and promoted hygiene, diet, physical exercise, and temperance to improve the health of offspring. Born on Nantucket Island, Massachusetts, one of seven children, she was from **Anglo-Saxon** colonial stock. Her father worked in a number of trades. Fowler spent her childhood and gained her early education in Nantucket, where she became interested in mathematics and astronomy. She went to Wheaton Seminary (1838–1839) and returned home to teach (1842–1844). She married Lorenzo Niles Fowler, a phrenologist, who with his brother **Orson Fowler** popularized **phrenology** and hereditarian ideals in the **Jacksonian Era**. She traveled with her husband and brother-in-law on lecture tours, and in 1847 began her own lectures on phrenology, hygiene, "**laws of health**," and **physical culture** to women in an effort to improve the health of children. Her lectures were popular and were published as *Familiar Lectures on Physiology* (1847) and *Familiar Lectures on Phrenology* (1847), which were aimed at youth and went through several printings.

Her interest in hygiene motivated her to seek a medical degree, a difficult proposition for women of the era. She began medical studies in 1849 at the "eclectic" Central Medical College in Syracuse, New York, and graduated in 1850, the second woman to receive a medical degree. (The first was **Elizabeth Blackwell**, who also promoted hygiene and hereditarian ideals.) After graduation Fowler stayed at the college and demonstrated anatomy to female students, became head of the "Female Department," and then professor of midwifery and diseases of women and children (1851). She was the first woman to hold a professorship in a U.S. medical college. When the college closed in 1852, Fowler moved to New York City with her husband and practiced medicine and lectured until 1860. She was also involved with women's rights and promoted temperance reform. After 1860 she and her husband lectured in Europe and in 1863 settled in London, where she continued to lecture on hygiene, diseases of women, temperance, and the laws of health as methods of producing healthy babies and improving the human race. Her marriage to Lorenzo in 1844 resulted in one daughter sixteen years later. She died at her home in London, of pneumonia.

FURTHER READING: Stern, Madeline B., *Heads and Headlines* (1971).

FOWLER, ORSON S(QUIRE) (October 11, 1809–August 18, 1887)

A phrenologist who sought the perfectibility of the human race, Fowler and his publishing company brought a protoeugenics message, or "**inherited realities**," to the middle class in the pre–Civil War era. He suggested techniques for improving the heredity of offspring that in turn could lead to the improvement of the human race. Fowler was born in Cohocton, County, New York, the oldest of three children. His father was a Congregationalist deacon and farmer of colonial **Anglo-Saxon** stock. Deciding to become a minister, Orson took preparatory studies, attended Ashfield Academy, and graduated from Amherst College (1834) in Massachusetts, where he was introduced to **phrenology**, the pseudoscience of determining human characteristics and traits from the shape of the human head. Instead of entering a seminary, he began to lecture on phrenology. With his younger brother, Lorenzo Niles (1811–1896), Fowler moved to Washington, D.C., Philadelphia, and in 1842 New York City to proselytize phrenology through writings and lecture tours. He also developed a successful publishing business. He became acquainted with "Grahamism," the strict vegetarian diet and exercise regime advocated by health reformer Sylvester Graham. Around 1844 the brothers were joined by their brother-in-law Samuel Wells, a phrenologist and hydropath, and formed the Fowlers and Wells Company, which published numerous works discussing health and heredity during the **Clean Living Movement** of the **Jacksonian Era**. In 1860 the Fowler brothers dissolved their partnership. Lorenzo and his physician wife **Lydia Folger Fowler** went to Europe, where they lectured on health, phrenology, and physiology. In 1863 they permanently settled in Britain. Orson moved to Boston and lectured and wrote for the rest of his life.

The Fowler brothers' publishing ventures began in 1836 with *Phrenology Proved, Illustrated, and Applied* which went through thirty editions. They began the *American Phrenological Journal* (1838–1911), which focused on health-reform issues, gave "head readings" of notable persons (both good and bad), discussed the need to contract "hygenic" marriage and "urged every precaution during gestation and early childhood to ensure the development of healthy, moral, and intelligent children." In 1840 the brothers started the *Phrenological Almanac*, which rapidly spread interest in phrenology and hereditarian ideas to all segments of society. Orson wrote numerous tracts and books that touted protoeugenic ideas for improving the human race including *Fowler on Matrimony* (1841), *Marriage and Parenthood* (1847), and *Hereditary Descent, Its Laws and Factors, Applied to the Improvement of Mankind* (1848). Fowler continued writing into his old age. He was married three times: to Martha Brevoort Chevalier (1835), with whom he had two daughters; to Mary Aiken Poole (1865); and to Abbie L. Ayres (1882), with whom he had three children. He died near Sharon Station, Connecticut.

FURTHER READING: Nissenbaum, Stephen, *Sex, Diet, and Debility in Jacksonian America* (1988); Stern, Madeline B., *Heads and Headlines* (1971); Walters, Ronald C., *American Reformers 1815–1860* (1978).

FRUIT OF THE FAMILY TREE, THE (1924)

Written by science and eugenics writer Albert Wiggam, *The Fruit of the Family Tree* served to popularize eugenics among the middle class, especially women, in the early twentieth century. It also gave specific advice to various professionals on how they could advocate eugenics in their fields. Published in Indianapolis, Indiana, by the Bobbs Merrill Company, it contained 395 pages. The book was dedicated to "the health, intelligence and beauty of the unborn." In the preface Wiggam explained that "eugenics is the basis of the new sociology, and sociology is the cap sheaf of all the sciences. Eugenics is, therefore, the application of human intelligence to human evolution." The publication contained twenty-one chapters, with titles such as "What Heredity Tells and How It Tells It," "What Twins Tell about Heredity," "Is Brain-power Inherited?" "Can We Make the Human Race More Beautiful?" "Woman's Place in Race Improvement," "Does Heredity or Environment Make Men?" and "What You Can Do to Improve the Human Race."

Wiggam discusses what eugenics is, and what it is not. For example, he explains that eugenics is not "free love," "creating a super race," or "birth control." He gives background information on theories of **genetics**, and of **nature-nurture** (heredity and environment). The author expresses concern that **public health** and charity were saving the weak and **unfit**, leading to **racial degeneracy** and the deterioration of the quality of the human race. He suggests that eugenics needs to be practiced by both the state and the individual. Recommendations include the establishment of a state board of heredity to collect a eugenic **family pedigree** for every family and a minimum **intelligence** level for marriage. Eugenics as practiced by the individual includes **positive eugenics** such as marrying well and producing more children for the "future of the race." Women, in particular, were advised to "meet the great call of race improvement." **Negative eugenics** was suggested to decrease reproduction among the unfit. Although **birth control** for the unfit was advocated, the author expressed concern that the middle class and "fit" were limiting their number of children by its use. The use of "rational birth control" was endorsed, which Wiggam argued would result in a gradual decrease of the "badly born" and an increase of the "well born." This popular work went through three printings in the 1920s.

G

We have wonderful new races of horses, cows, and pigs. Why should we not have a new and improved race of men?

John Harvey Kellogg, *Proceedings of the First National Conference on Race Betterment*, 1914

GALTON, FRANCIS (February 16, 1822–January 17, 1911)

A brilliant British naturalist with a wide variety of interests, Francis Galton is recognized as the father of modern eugenics. He coined the phrases *nature versus nurture* in 1874 and *eugenics* in 1883. Born in Birmingham, England, the son of a successful banker and businessman, Galton received private tutoring, studied medicine as an apprentice in Birmingham (1938), attended Kings College (1839–1840), and pursued mathematics at Trinity College, Dublin, Ireland, where he earned a pass degree (1844). His father died in 1844 and left him a considerable amount of money. This inheritance enabled him to abandon his formal education. From late 1845 until 1852 he explored the Middle East and then southwestern Africa. The published account of his adventures brought him fame as an explorer. In 1854 Galton was awarded the Gold Medal of the Geographical Society and in 1856 he was elected a fellow of the Royal Society. Afterward he conducted research and established the scientific basis of several fields including fingerprinting and weather patterns.

Galton developed an interest in heredity after the publication of *Origin of Species* (1859), written by his cousin **Charles Darwin**. Galton proposed that **natural selection** also held true for humans. He suggested that talent, ability, **intelligence**, and distinction of any kind "ran in families" and declared that the human race could be improved by *controlled breeding*. In an effort to support these hypotheses he studied families who had achieved distinction over several generations. His findings were first published as essays in 1865, and later expanded into *Hereditary Genius* (1869) with the theme that human traits, in particular great abilities, are inherited. These works, along with *English Men of Science: Their Nature and Nurture* (1874), laid the foundation for eugenics. Galton was also one of the first to reject **Lamarckian inheritance of acquired characteristics**. He believed, although with no experimental evidence, that hereditary material was passed unchanged from generation to generation.

In 1883 Galton published *Inquiries into Human Faculty and Its Development*, a collection of essays discussing the results of his varied research studies; in this tract he first defined *eugenics*. Discovering that advances in the study of heredity were being hampered by the lack of quantitative findings, Galton founded anthropometric re-

British naturalist Francis Galton is considered the father of eugenics. (From: *Conclusions to Memories of My Life* [1908], Main Library, Indiana University, Bloomington, Indiana. Image courtesy of the Digital Library Program.)

search and devised instruments for exact measurements of human physical traits including the method for fingerprinting. He studied twins (**twin studies**) as a method to determine which characteristics and traits were inherited and which were environmental. He also attempted to develop **intelligence tests**. From his interest in statistics he founded the Biometric Laboratory at University College in London (1884) and with others established the journal *Biometrika* in 1901 to publish statistical research. In 1904 he founded the Eugenics Records Office, which soon became known as the **Galton Laboratory** at the University College, London, to further scientific studies in eugenics. Galton was named honorary president of the newly formed **Eugenics (Education) Society** (from 1908 until his death in 1911) and of the **German Society for Race Hygiene**.

Galton established an endowment to support his eugenics laboratory upon his death, along with a Galton Eugenics Professorship. With Galton's recommendation, his first protégé, statistician **Karl Pearson**, was granted the professorship. Over his lifetime Galton published more than 15 books and around 200 papers. For his many scientific contributions he was knighted in 1909. In 1853 he married Louise Butler, the daughter of a prominent clergyman; they had no children. He died at home in Surrey after several years of poor health. By the beginning of the twenty-first century, even though he had founded and pioneered many scientific fields, Galton became relatively unknown.

FURTHER READING: Galton, Francis, Sir, *Memories of My Life* (1974); Gillham, Nicholas W., *A Life of Sir Francis Galton* (2001); Keynes, Milo, (ed.), *Sir Francis Galton, FRS* (1993).

GALTON INSTITUTE

See Eugenics Society

GALTON LABORATORY, UK (Eugenics Records Office 1904–1907; Francis Galton Laboratory for National Eugenics, 1907–1945; Galton Laboratory, University College, London, 1946–present)

The Eugenics Records Office (not to be confused with **Charles Davenport**'s **Eugenics Record Office** in the United States) was founded in 1904 by **Francis Galton** at University College, London. It was one of the first centers established for the study of **eugenics** and human **genetics**. The laboratory's mission was to establish eugenics as a science, as opposed to the **Eugenics Society**, whose aim was public education. During the twentieth century the research focus of the Galton Laboratory evolved from eugenics to genetics and biotechnology. The laboratory published scientific reports and journals and was a major player in the early-twentieth-century **British eugenics movement**. The laboratory over the first third of the century and during the peak of the eugenics movement was intimately connected with its leader, **Karl Pearson**, a statistician, and protégé of Galton. Pearson headed the Biometric Laboratory at University College and in 1906 took over directorship of Galton's research office. A year later he merged the two units and called it the Francis Galton Laboratory for National Eugenics. Attempting to make it a scientific center, Pearson and his team focused on research, training of professionals, and publications. During its first year it introduced a research report, *Laboratory Memoirs*.

Upon Galton's death in 1911, an endowment was established to support the eugenics laboratory along with a Galton Eugenics Professorship. At Galton's recommendation, Pearson was given the first professorship and a new department was created, the Department of Applied Statistics, which included both the statistics and eugenics laboratories. Its research focused upon statistical analysis of data related to heredity and environmental factors considered related to **alcoholism, intelligence, insanity, feeblemindedness**, mental defects, **criminality**, albinism, and other physical and mental human traits. In 1913 the department was renamed the Department of Applied Statistics and Eugenics by the college. Support for the laboratory before **World War I** came from private donors and after the war from government grants. In order to publish both genetics and eugenics research, Pearson in 1925 launched the *Annals of Eugenics*.

Pearson retired in 1933 and the department was reorganized. Anthropology joined the eugenics division and **R. A. Fisher**, the new Galton Professor, headed the Department of Eugenics. That same year the laboratory was placed officially within University College, London. In 1944 biometry and eugenics were combined again into one department, headed by geneticist J.B.S Haldane (1892–1964). In 1965 the department was renamed the Department of Human Genetics and Biometry, and the following year the department became known as the Department of Biology. The laboratory was renamed the Galton Laboratory. In the last decades of the twentieth century the research focus of the laboratory included human molecular genetics, developmental and evolutionary genetics, the human **genome**, and population genetics.

Further Reading: Jones, J. S., "The Galton Laboratory, University College London," in Keynes, Milo, (ed.), *Sir Francis Galton, FRS* (1993); Soloway, Richard A., *Demography and Degeneration* (1990).

GALTON SOCIETY (1918–c. 1939)

In the United States the most elitist and **nativistic** eugenics group to emerge during the eugenics movement was the Galton Society. It was founded April 2, 1918, by **Charles Davenport**, its first president, along with naturalists and eugenics supporters **Henry Fairfield Osborn** and **Madison Grant**. The society was dedicated to "the promotion of study of racial anthropology, and the origin, migration, physical and mental characteristics, crossings and evolution of human races, living and extinct." Early members included nativist writer **Lothrop Stoddard**, Princeton University biologist Edwin G. Conklin (1863–1952), biologist Frederick A. Woods, and psychologist Robert M. Yerkes (1876–1956). These men (few women belonged to the society) were considered distinguished scholars by their contemporaries and served as leaders and guides within the eugenics movement.

Financed by **Mary Harriman**, widow of a well-known railroad magnate, the Galton Society was highly exclusive and admitted only elite native-born Americans. Its prestige arose from its noted members, who gathered at luncheons and dinners at the American Museum of Natural History to hear papers and discuss the problems of heredity and **race**. The group advocated eugenic **sterilization** programs, **immigration restriction laws**, legislation against racial intermarriage to promote "racial purity," and laws to prevent reproduction among the **unfit**. The society had formal ties with the **Eugenics Research Association** and the **Eugenics Record Office**, and was a financial sponsor of the **Third International Congress of Eugenics**, held in New York in 1932. In the late 1920s it began to avidly support **Nazi German** views through its official journal *Eugenical News*. These sentiments raised concern among some geneticists and eugenicists, many of whom disassociated themselves from the organization in the mid-1930s. By **World War II** the society had lost its leadership and credibility as a scientific body and ceased to be active.

Further Reading: Engs, Ruth C., *The Progressive Era's Health Reform Movement* (2003); Haller, Mark H., *Eugenics* (1984).

GENE OR GENES

The classic definition of a *gene* is a physical and functional "unit of heredity." The term was coined in 1909 by Wilhelm Ludwig Johannsen (1857–1927), a Danish botanist. It was adapted from **Charles Darwin**'s term *pangenesis*, for all units of reproduction in an organism. Genes contain instructions for building specific proteins that cause biochemical reactions in a cell. They act in concert with one another and with the cellular environment. Specific environmental factors are thought to switch genes "on" or "off," resulting in biochemical reactions that trigger certain behaviors. For example, the increase in daylight in spring may biochemically trigger birds to build nests and lay eggs. When the structure of **DNA** (the genetic code) was identified in 1953, genes were found to be made up of tightly coiled spiral strands of DNA in an ordered sequence, or pattern, of nucleotide "base pairs." Segments of DNA in the gene can be spliced, replicated (*cloned*), and recombined with other DNA segments to change the function of the gene. This technique of producing recombinant DNA, a major aspect of **genetic engineering**, was first developed in the 1970s and has led to the manipulation of genes in many areas of agriculture and other sciences. Genes contain information from both parents and come in pairs; each member of the pairs is an *allele*. If one, or both alleles, has a defect or is missing, a **genetic disease** can result. The potential for manipulating the allele, or even the whole gene, through twenty-first-century genetic engineering techniques in order to eliminate the condition and have a healthy child is an aspect of the **new eugenics**.

FURTHER READING: Carey, Gregory, *Human Genetics for the Social Sciences* (2003); Dawkins, Richard, *The Extended Phenotype* (1999).

GENETIC COUNSELING

The process of advising prospective parents that their children might be born with a particular condition or genetic disease is termed *genetic counseling*. The prototype of genetic counseling began in 1910 with establishment of the **Eugenics Record Office (ERO)** in New York State. The ERO collected information from individuals, including diseases and "peculiarities" of family members, and used the information to encourage those with a known family history of certain conditions, thought to be inherited, not to have children. These conditions included **insanity, feeblemindedness,** and **tuberculosis**. As the early-twentieth-century eugenics movement waned and environment, rather than heredity, came to be considered the prime determinant for many diseases and human behaviors, **family history and pedigree studies** were discredited and the office closed in 1939. Because of a lack of clear information about patterns of inheritance—let alone diagnostic tests and treatment for **genetic diseases**—genetic counseling was considered unimportant in the United States. In London in 1936, however, one of the first generic counseling centers, the Bureau of Human Heredity, was founded to advise families with histories of genetic diseases. Counseling generally encouraged couples to remain childless to avoid having a child with a condition.

In the post–**World War II** period of the 1940s and 1950s new technology uncovered more knowledge of the role **genes** played in inheritance. This led to the establishment of genetic counseling centers on both sides of the Atlantic by the 1960s. Newly developed **genetic screening and testing** techniques were able to determine with more certainty the probability of having a child with an inherited condition or genetic abnormality like Down syndrome. A pregnant woman now had the option of aborting a defective fetus. By the mid-1980s **genetic markers**—particular segments of DNA associated with a genetic disease—had been found. By the turn of the twenty-first century, over 800 genetic diseases could be identified, making genetic counseling integral to reproductive medicine. Genetic counseling is an aspect of the **new eugenics**.

FURTHER READING: Duster, Troy, *Backdoor to Eugenics* (2003); Hubbard, Ruth, and Elijah Wald, *Exploding the Gene Myth* (1993); Kerr, Anne, and Tom Shakespeare, *Genetic Politics* (2002).

GENETIC DISEASES OR DISORDERS

A condition caused by an abnormal or mutated **gene** is called a genetic disease or disorder. For centuries many conditions were known to run in families, but it was not until after the laws of **Mendelian inheritance** were rediscovered in 1900 that the knowledge was used to determine the probability of passing a genetic disease to offspring. The major groupings of genetic diseases include: *single-gene, chromosomal,* and

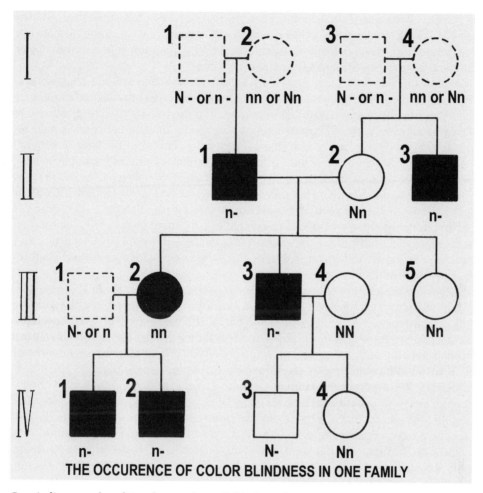

THE OCCURENCE OF COLOR BLINDNESS IN ONE FAMILY

Genetic diseases and conditions began to be studied in the early twentieth century. Color blindness, for example, is an X-linked trait. For the daughter in this diagram to have the condition, her father must be color blind and her mother must be a carrier of the trait giving her two X chromosomes with the defective gene. (From: *Journal of Heredity* [1919], Main Library, Indiana University, Bloomington, Indiana. Image courtesy of the Digital Library Program.)

multifactor disorders or conditions. Single-gene defects are caused by abnormalities of a single gene and follow simple Mendelian laws. Most single-gene defects are *autosomal recessive*. In order to exhibit the disease an individual must receive defective copies of the *allele*, or form of the gene, from both parents. A person who inherits only one copy of the defect from one parent is a *carrier*. If both parents are carriers, a child has a one in four chance of having the condition. Examples of recessive diseases include Tay-Sachs disease, predominant among those with eastern Europan **Jewish** ancestry; sickle-cell anemia, found among those with western **African** ancestry; and cystic fibrosis, mostly found among those with northern European ancestry. In *autosomal dominant* single-gene diseases, only one copy of the abnormal gene is needed to produce the disease; the probability of passing the condition to offspring is

50 percent. Examples include Huntington's disease and Marfan syndrome. *X-linked disorders* are due to abnormal genes on the X, or "female," chromosome. A male has a 50 percent probability of inheriting the condition if his mother is a carrier. Color blindness is a common X-linked inherited condition.

Chromosome disorders are caused by an excess or deficiency of chromosomes, which are found in the nucleus of cells. The most common chromosome disorder is Down syndrome (trisomy 21), called *mongolism* into the late 1970s. The syndrome is caused by an extra copy of chromosome 21 and results in mild to severe mental retardation, heart defects, and other physical problems. The most complex of genetic-related conditions are those resulting from interactions of one or more genes with one or more environmental factors. These multifactorial, or polygenic, traits are often found among family members but do not follow Mendelian laws of inheritance. Most likely chronic conditions such as **alcoholism**, heart disease, diabetes, and cancer fall into this category, along with about 50 percent of congenital birth defects such as the neural tube defect spina bifida. It is known that an allele of certain **DNA** (genetic code) sequences on a gene can increase the odds of breast and ovarian cancer, schizophrenia, and Alzheimer's disease.

Early-twentieth-century eugenicists were concerned that better medical care was allowing individuals with genetic dysfunctions to survive and to reproduce, thus adding their defective genes to the human gene pool, which would lead to **racial degeneration** and the weakening of the human species. By the end of the twentieth century **genetic screening and testing** and preimplantation diagnosis of embryos as a part of the **in vitro fertilization** process could identify numerous genetic diseases or congenital defects. Results from the **Human Genome Project** are helping to identify the location of genes, or **genetic markers**, that may be associated with various diseases. The possibility of eliminating a genetic disease from an offspring—or even the human race—by **genetic engineering** techniques is a major aspect of the **new eugenics**.

FURTHER READING: Buchanan, Allen, et al., *From Chance to Choice* (2000); Reilly, Philip R., *Abraham Lincoln's DNA* (2000).

GENETIC ENGINEERING

The artificial altering of genetic material of an organism resulting in the organism, or its cells, manufacturing new chemicals, performing different functions, and increasing the likelihood that its offspring will have new characteristics, is called *genetic engineering*. This scientific field and the potential of changing the genetic makeup of humans is a major aspect of the **new eugenics**. The phrase *genetic engineering* is often used interchangeably with *biotechnology*. It was coined in 1965 by geneticist Rollin D. Hotchkiss (b. 1911) at the **Wilhelmine Key** lecture (named for the pioneer geneticist and eugenicist) at the annual meeting of the American Institute of Biological Sciences. One of the first successful genetic engineering experiments was accomplished in 1973 when new **genes** were inserted into *E. coli* bacteria, which then reproduced. The technique used, termed *recombinant DNA* technology, involves the artificial addition, manipulation, deletion, or rearrangement of sequences of bases in **DNA** (the genetic code) in order to alter the phenotype (form) and/or function of an organism or population of organisms. By the mid-1970s the phrase *genetic engineering* began to be used to represent this recombinant DNA technology and the emerging field of molecular **genetics**. By the 1980s the technique was used in agriculture to develop disease-resistant plants, and by the early 1990s it was attempted in **genetic therapy** to treat specific diseases.

By the turn of the twenty-first century some genetic engineering procedures, such as research using human embryos, had raised controversy in the United States and resulted in many research projects using human tissue being undertaken in other countries. In *stem cell research*, for example, cells are extracted from early human embryos or created out of certain adult cells. They are manipulated to become blood, heart, liver, spinal cord, and other cells and have the potential of producing tissue to replace diseased or damaged organs. By 2001 *therapeutic cloning*, or the manufacture of **cloned** human embryos from an adult donor in order to extract stem cells to produce replacement tissue was achieved. Because the cloned embryo has the same DNA code as the donor, rejection of the manufactured tissue is less likely. Using genetic engineering tools it is potentially possible to genetically enhance a child's **intelligence** and personality, and to eliminate **genetic diseases** by inserting suitable genes into an embryo soon after conception. Genetic engineering tools when used to change human genetic material or to manipulate the human **genome** are major aspects of the **new eugenics**.

FURTHER READING: Buchanan, Allen, et al., *From Chance to Choice* (2000); McGee, Glenn, *The Perfect Baby* (1997); Rifkin, Jeremy, *The Biotech Century* (1998); Stacy, Meg, (ed.), *Changing Human Reproduction* (1992); Stock, Gregory, *Redesigning Humans* (2002).

GENETIC MARKER

A gene or segment of **DNA** found on a chromosome in a particular location is a *genetic marker*. Genetic markers are generally inherited and readily identifiable. A marker may, or may not, have a known function. Markers are used to determine **genetic profiles**, or "fingerprints," to identify individuals or populations and to test for a particular **genetic disease**. Known locations of genetic markers for a specific genetic disease are compared to samples collected from **genetic screening and testing** to determine if an individual has the marker for the condition.

FURTHER READING: Knight, Jeffrey A., (ed.), *Encyclopedia of Genetics* (1999).

GENETIC OR GENE THERAPY

Approaches to correct a defective **gene or genes** responsible for a **genetic disease** constitute *gene therapy*. Gene therapy attempts to replace, manipulate, or supplement nonfunctioning genes using recombinant DNA technology, a form of **genetic engineering**. Because genes tend to act in concert with one another and the environment, gene therapy often requires intervention with several genes. In direct gene therapy a normal, or desirable, gene is inserted into somatic (body) cells, germ-line cells (sperm or egg), or embryos. An abnormal, or undesirable, gene or genes can be "switched off" so that it no longer produces a negative effect. The most common technique for gene therapy is the insertion of a normal gene into the **genome** (the blueprint of an organism) to replace a nonfunctioning gene. The creation of synthetic genes to carry out the normal function is also possible. The new gene, or **DNA** sequence, is generally incorporated into the organism's cells by a vector such as a modified virus. An indirect form of gene therapy includes genetic pharmacology, in which certain drugs are designed to replace or increase the chemical products produced by the normal gene or genes. The first reported attempt at human gene therapy, in 1992, was for the "bubble boy syndrome," a condition in which the immune system does not function. This research, along with attempts for other conditions, has been only marginally successful. Gene therapy is a significant aspect of the **new eugenics** and has the potential of eliminating genetic diseases from the human population. However, some experts argue that eliminating genes in the reproductive cells, as opposed to somatic cells, could cause future problems for the human race. This is because some recessive traits for genetic diseases, such as sickle-cell anemia, which protects carriers from malaria, could be eliminated, thereby preventing evolutionary adaptation and survival in changing environmental conditions.

FURTHER READING: Buchanan, Allen, et al., *From Chance to Choice* (2000); Hubbard, Ruth, and Elijah Wald, *Exploding the Gene Myth* (1993); McGee, Glenn, *The Perfect Baby* (1997); Rifkin, Jeremy, *The Biotech Century* (1998).

GENETIC PROFILING OR PROFILE

The process of identifying an individual by his or her **DNA** pattern is *genetic profiling*. It is also called *DNA-based identification system* or *DNA fingerprinting*. British geneticist Alec J. Jeffreys (b. 1950) was the first to demonstrate that DNA could be used for identification in 1984. A DNA profile is unique for each individual, with the exception of identical twins, and has eugenic implications. A profile is created through **genetic testing or screening**. Blood, tissue, or bodily fluids are examined and DNA patterns (**genetic markers**) are identified from the DNA in the cell nucleus and, sometimes, the mitochondria, which are responsible for energy production. The resulting profiles are used for many purposes including research studies, the identification of inherited conditions, the identification of human remains, the determination of descendant or blood relatives, and in criminology to confirm or eliminate crime suspects. Genetic profiles are also used to test embryos for **genetic diseases** in the prenatal environment or before implantation during **in vitro fertilization** procedures. Individualized for specific persons they can be used for **gene therapy** or drug therapy. This unique identification system has caused concern about the possible use of DNA profiles for purposes beyond their original intent. The prospect of negative uses includes invasion of privacy; abridgement of civil rights; and discrimination in employment, health insurance coverage, medical treatment, or education.

FURTHER READING: Duster, Troy, *Backdoor to Eugenics* (2003); Hubbard, Ruth, and Elijah Wald, *Exploding the Gene Myth* (1993); Kerr, Anne, and Tom Shakespeare, *Genetic Politics* (2002); Stock, Gregory, *Redesigning Humans* (2002).

GENETIC SCREENING AND TESTING

The examination of blood, tissues, or bodily fluid samples to determine chromosomal or **DNA** configurations or to ascertain biochemical markers is the process of *genetic screening* or *testing*. Although the terms *screening* and *testing* are often used interchangeably they have a different focus. Genetic screening is a search of many people to determine those having a **genetic disease** that could be passed to offspring but unaware that they carry the condition. Genetic testing is generally aimed at specific individuals suspected of being at high risk for a particular inherited disorder, to confirm a suspected diagnosis of a genetic disease, to estimate the possibility of future illness, to predict the response to therapy, or to construct a **genetic profile** for identification purposes. Genetic screening began in the 1960s to identify newborns with the metabolic disease phenylketonuria (PKU), which causes mental retardation. Severe retardation can be prevented if the infant receives a diet low in protein. In the early 1970s screening for adult carriers of Tay-Sachs disease among European **Jews** and prenatal testing to determine if the fetus had inherited this condition were inaugurated. The program resulted in a decrease of this genetic disease, which causes death in infancy. Similarly, in the early 1970s screening of **African Americans** for sickle-cell anemia began. However,

Thoroughbred racehorses like Man O'War were developed over centuries of selective breeding. Eugenicists suggested that "human thoroughbreds" could also be developed by following the principles of genetics and encouraging the fittest "human stock" to marry each other and produce many children. (From: *Journal of Heredity* [1921], Main Library, Indiana University, Bloomington, Indiana. Image courtesy of the Digital Library Program.)

the fear of job discrimination led to the discontinuation of large-scale screening by the mid-1970s.

Genetic tests became more common during the last three decades of the twentieth century. In the 1970s prenatal diagnosis for genetic diseases or chromosomal abnormalities, such as Down syndrome, became regular medical practice. By the early 1990s embryos from parents using **in vitro fertilization** were tested so that only healthy ones were implanted. By the mid-1990s genetic testing to establish genetic profiles was used for the identification of remains, individuals, or **crime** suspects. However, the potential of data banks of DNA profiles led to concerns about privacy and discrimination. At the turn of the twenty-first century tests became available to identify more than 800 genetic diseases, leading to the possibility of individualized preventive medicine and treatment. All these uses of genetic testing and screening are an aspect of the **new eugenics**.

FURTHER READING: Duster, Troy, *Backdoor to Eugenics* (2003); Hubbard, Ruth, and Elijah Wald, *Exploding the Gene Myth* (1993); Kerr, Anne, and Tom Shakespeare, *Genetic Politics* (2002); Stacy, Meg, (ed.), *Changing Human Reproduction* (1992); Stock, Gregory, *Redesigning Humans* (2002).

GENETICS

"A branch of biology that addresses the physiology of heredity and its variations" is the classical definition of *genetics*. In the latter half of the twentieth century genetics became a complex field embracing many disciplines and focused on molecular research with **DNA**, the basic coding for heredity. British scientist William Bateson (1861–1926) proposed the term *genetics* in 1906 for a new branch of science, although the words *genetically* and *genetic relation* had been used since the mid-1860s. Genetics and **eugenics** developed together as one field, but began to diverge in the 1920s. The principle of heredity, "like begets like," had been observed since prehistoric times. Breeders had employed this principles to improve the quality of domestic animals, and farmers used it to produce better fruits and vegetables. **Gregor Mendel**, a Moravian monk, first discovered the mathematical laws of inheritance in the mid-nineteenth century, but his findings were not widely distributed. When his research was rediscovered in 1900 a surge of genetic research was performed on a variety of plants and animals in both Europe and the United States. The first professional organization for genetic research in the United States was the **American Breeders Association**, founded in 1903 to develop projects that led to "improved plants, animals, and men." The association was renamed the American Genetics Association in 1914 and its journal, the *American Breeders Magazine*, became *The Journal of Heredity* in 1915.

Classical genetics in the first decades of the twentieth century examined how genetic traits were transmitted and what influenced their changes or mutations. Chromosomes, colored threads found in the cell's nucleus, had been identified by the late 1880s and were suggested as a mechanism of inheritance. In 1909 the term *gene* was coined for this functional unit of heredity. The following year American zoologist Thomas Hunt Morgan (1866–1945) demonstrated that genes were located on chromosomes. By 1915 the modern chromosomal theory of heredity was established with publication of *The Mechanism of Mendelian Heredity* by Morgan and others, including his student **Hermann Muller.** Muller reported in 1926 that genes could be mutated, or changed, in fruit flies by X-ray. In the post–**World War II** years genetics began to overlap other disciplines including medicine, psychology, sociology, physics, and mathematics. In 1953 the structure of **DNA** that makes up genes was identified. A merging of technologies resulted in the emergence of new fields of study based upon genetic principles, including molecular biology, behavioral genetics, population genetics, **genetic engineering**, and **social biology**. Discoveries in medical genetics led to **genetic screening and testing** for inherited diseases and **assisted reproduction** techniques such as **artificial insemination** by the 1960s and **in vitro fertilization** by the late 1970s. During the last decade of the twentieth century the international **human genome project** mapped the **genome**, or "blueprint," for humans and research investigated the possibilities of **gene therapy**. The development of **genetic profiles**, or "fingerprints," for identification, genes patented for commercial use, and tailoring specific medications (pharmagenetics) and *stem-cell research* to make nondiseased tissues for specific individuals evolved. All these developments, which now make up the science of genetics, have the potential to eliminate **genetic diseases** to "improve the human race" or ensure healthy, or "**designer babies**." They have been labeled the **new eugenics**, thus bringing genetics back to its early association with eugenics.

FURTHER READING: Carey, Gregory, *Human Genetics for the Social Sciences* (2003); Dawkins, Richard, *The Extended Phenotype* (1999); Henig, Robin Marantz, *The Monk in the Garden* (2000); Sturtevant, Alfred H., *History of Genetics* (2001).

GENOME AND GENOMICS

All the **DNA** (deoxyribonucleic acid) found in an organism or in the nucleus and mitochondria of a cell that is required for development, growth, maintenance, and other processes throughout its lifetime is the *genome*. Genomics is the study of gene function and behavior as part of a system to determine how various genes interact and influence biological processes in an organism. The term *genome* as the complete set of genes for an organism was coined in 1965. The human genome has approximately 3 billion chemical base pairs that make up DNA. This DNA is found on 25,000 genes, which, in turn, are found on 23 pairs of chromosomes in the cell nucleus (until 1956 humans were thought to have 48 chromosomes). DNA is also found in the mitochondria of cells, which are involved with energy production. The **Human Genome Project**, an international research effort to sequence and map the genome of humans, began in 1990 and was completed in 2003. The development of the human genetic blueprint allows for the identification of specific genes that can cause **genetic diseases** and other characteristics. The complete knowledge of the human genetic code has eugenic implications. It is potentially possible to create human life from raw material; to identify, cure, or eliminate genetic diseases from the human population; to produce healthy or **"designer babies"** with certain characteristics, to manipulate **intelligence**, ability, and other human traits, and to create **genetic profiles** unique to each individual. Genomics is integral to the **new eugenics**.

FURTHER READING: Hawley, R. Scott, and Catherine A. Mori, *The Human Genome* (1999); Kerr, Anne, and Tom Shakespeare, *Genetic Politics* (2002); Kevles, Daniel J., *The Code of Codes* (1992); Ridley, Matt, *Nature via Nurture* (2003); Yudell, Michael, and Robert DeSalle, (eds.), *The Genomic Revolution* (2002).

GERMAN EUGENICS OR EUGENICS MOVEMENT (1890–1933) OR GERMANY

The German eugenics movement can be divided into the empire (1890–1918), republic (1918–1933), and **Nazi German** (1933–1945) eras. German eugenics was also called **race hygiene** (*Rassenhygiene*). This entry discusses the movement until 1933 when the National Socialists (Nazis) seized political power. Similar to the eugenics movements in other industrialized nations, the German movement had roots in **social Darwinism**, was supported by the educated middle class, and was considered a humanitarian effort. Based upon **Lamarckian theory** of acquired characteristics, it was argued that social welfare allowed the constitutionally weak and **unfit** (*minderwertig*) to survive and reproduce, leading to **racial degeneracy**, or the deterioration of the human race. This was coupled with the fear of **race suicide** among northern European peoples whose population was declining compared with other ethnic groups. In Germany these concerns developed out of rapid social, political, and industrial changes. To address these fears German physicians, who included both eugenicists and public health professionals, sought to increase the biological fitness, health, and "efficiency" of the nation. The German eugenics movement, as part of an overall health and **physical culture** movement, aimed to expand **public health** and **social hygiene**, eliminate "inherited" **degeneracy**, and increase the birthrate of the culturally and socially "fit."

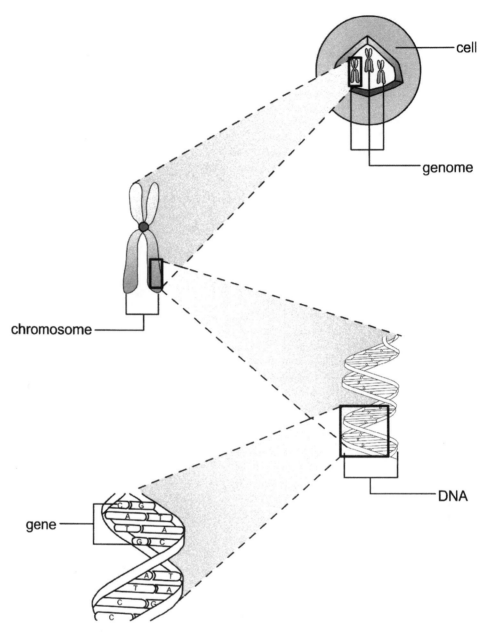

The genome is the "blueprint" for a making an organism. The chromosomes, which are found in the nucleus of a cell, are made up of strands of DNA. Sections of DNA are the genes that contain instructions for building specific proteins that cause biochemical reactions in a cell. (Source: Greenwood Publishing Group, Westport, CT.)

Eugenic concepts in Germany began to take shape during the rapid transformation of the country's agricultural states into an industrialized nation. This change resulted in numerous social welfare and health problems. In 1891 Munich physician **Wilhelm Schallmayer** first introduced the concepts of eugenics to address these problems in Germany in an essay discussing race degeneracy. Four years later another Munich physician, **Alfred Ploetz**, who had read the essay but was not familiar with British naturalist **Francis Galton**'s work on eugenics, defined procedures for improving the hereditary fitness of the human race as *race hygiene* (*Rassenhygiene*) in an book, which at the time did not generate much public interest. In 1900 Schallmayer entered a writing contest on evolution and its relationship to politics and won first prize for a work that promoted a "hygienic-sociological" approach to race degeneracy. The publicity surrounding this work, ***Heredity and Selection in the Life History of Nations*** (1903), appealed to liberal and left wing professionals and helped launch the German eugenics movement. The work was revised several times and became the theoretical basis and classic textbook of the early German movement.

In 1904 Ploetz, who became the pivotal leader of the movement in the pre–**World War I** years, founded the ***Archive for Racial and Social Biology***, which became Germany's primary academic eugenics journal. In 1905 he along with others formed a Eugenics Society that became known as the **German Society for Race Hygiene**. The society became the major voice of the eugenics movement and recommended methods of **positive eugenics** to increase the health and efficiency of all Germans. However, its suggestions, such as family allowances to encourage large healthy families, were not implemented. Although **negative eugenics** programs like sterilization of the unfit were considered desirable by some eugenicists, they were not seen as politically feasible in Germany. Like eugenicists in all other countries, German eugenicists rejected euthanasia of mentally ill and disabled individuals, and many of them were against **abortion**. Eugenics leaders participated in international conferences such as the **First International Eugenics Congress**. At the outbreak of the war the movement stagnated.

Paralleling the German eugenics movement was a "Nordic movement" that bore similarities to the **nativist** aspect of the eugenics movement in the **United States**. Nordic ideology (Nordicism) promoted the superiority of the **Aryan** or **Nordic race** as a "**master race**" and advocated "racial purity" based upon race theory introduced by **Arthur de Gobineau** and others. It also advocated the increase of reproduction among fit Nordics only and favored anti-Semitic views. A few race hygienists, including Ploetz, physician **Fritz Lenz**, and anthropologist **Eugen Fischer**, tolerated Nordicism in the developing German eugenics movement. However, most eugenicists criticized Nordic ideology. In the 1920s Nordicism was popularized by German race theorist **Hans F. K. Günther**. It gained many adherents in the post–World War I era due to fear of a Bolshevik (Communist) takeover, Germany's humiliating defeat in the war, and growing economic problems and social instability. National Socialism rose in this tumultuous decade, promising to solve the nation's problems and championing both Nordic superiority and race hygiene ideals.

During the aftermath of World War I, under the Weimar Republic of the 1920s, German eugenicists and other professionals became alarmed that so many of their fittest young men had been killed, while the unfit were left to reproduce at home.

They focused their efforts upon reducing the social cost of the unproductive and methods to reverse the declining population. In 1922 eugenicists, through the Society of Race Hygiene, strongly suggested implementation of their positive eugenics recommendations. In addition, they now advocated **eugenic sterilization** to decrease breeding among the unfit and degenerate. Eugenics began to be seen as the heart of all social reform and the source of health measures to increase the "efficiency" of the nation. In 1921 botanical geneticist Erwin Baur (1875–1933), along with anthropologist **Eugen Fischer** and Fritz Lenz, published *Human Heredity and Race Hygiene*, which became the standard text for German eugenics. That same year the German Genetics Association was founded, which included eugenics as an area of research. Human genetics research quickly gained governmental support, and Berlin became one of the leading genetic research centers in the world. Bavaria in 1923 became the first German state to establish a university chair of eugenics; Fritz Lenz was its first holder. By the mid-1920s health, biology, and medical books listed chapters on eugenics, and eugenics lectures were given in university hygiene classes. The **Kaiser Wilhelm Institute**, devoted to research in eugenics, genetics, social sciences, and other fields was established in 1927, with Fischer as director.

However, unlike other industrialized countries including the United States, no eugenics legislation was passed in Germany, due to concerns about individual rights. In the early 1920s friction developed between eugenicists in Munich with Nordic leanings, and those in Berlin who opposed this ideology. The German Federation for Population Betterment and Heredity was founded in Berlin during 1925–1926 to popularize eugenics among the general public. Because of its popularity, in 1931 the older Society for Race Hygiene changed its constitution, allowing the two national organizations to merge as the German Society for Race Hygiene (Eugenics). This new organization supported basic research and public health and welfare and stressed strategies to increase the population of the productive and to decrease reproduction among those unable to contribute to society. By 1930 many German eugenicists advocated mandatory sterilization laws for the unfit, like those enacted in the United States and other countries, in order to ease the social welfare burden accompanying the Depression. In 1932 a bill was introduced in Prussia by Berlin's eugenics leader and former Jesuit Hermann Muckermann (1877–1962), similar to the model suggested by American eugenicist **Harry H. Laughlin**; the proposal was defeated. After the Nazis took power in 1933 a sterilization law was quickly adopted and Nordicism merged with eugenics.

FURTHER READING: Weindling, Paul, *Health, Race, and German Politics between National Unification and Nazism, 1870–1945* (1989); Weingart, Peter, "German Eugenics between Science and Politics" (1989); Weiss, Sheila Faith, *Race Hygiene and National Efficiency* (1987); Weiss, "The Race Hygiene Movement in Germany" (1987).

GERMAN SOCIETY FOR RACE HYGIENE *(Deutsche Gesellschaft für Rassenhygiene, 1905–c. 1945)*

The first organization devoted to eugenics, the Society for Race Hygiene was established June 22, 1905, in Berlin. It was organized by physician **Alfred Ploetz**, who the previous year had founded the *Archive for Racial and Social Biology* in collaboration with his associate editors Anastasius Nordenholz (b. 1862) and ethnologist Richard Thurnwald (1869–1954). Psychiatrist and eugenicist Ernst Rüdin (1874–1952) was also a founding member of the organization. A goal of the society was to educate the middle class concerning eugenic ideals and to act as a model for **positive eugenics**. In addition to scientific discussions the society assembled for social gatherings and group outings such as hiking, aimed at the socially productive middle class. When several individuals from Norway and Sweden joined the group in 1907, the society renamed itself the International Society for Race Hygiene. By 1910 the organization had established chapters in several German cities, which united to form a national German subdivision. However, due to **World War I** and lack of interest in the society from other nations, the organization dropped "international" from its title and renamed itself the "German Society for Race Hygiene" in 1916.

The society began with twenty-four members and by 1913 had several hundred. The only requirements for membership were to be Caucasian and ethically, intellectually, physically, and economically fit. Potential members generally were required to be socially productive and to undergo a medical examination to prove their physical fitness. Most members were physicians, academics, civil servants, teachers, and intellectuals. **Catholics**, **Protestants**, and **Jews**, including people of various political persuasions, were members. The society planned to collect statistics from its members and their offspring and compare their productivity, "social usefulness," and fitness to the general public. The father of eugenics, **Francis Galton**, accepted the honorary presidency of the society in 1909. In 1911 it helped lay the groundwork for the **First International Eugenics Congress** held in London. In 1914 it expressed concerns about the decreasing birthrate, in particular among the upper and educated middle classes, and made a series of eugenic recommendations to reverse this trend. The group had become

Written by botanical geneticist Erwin Baur, anthropologist Eugen Fischer, and physician Fritz Lenz, *Human Heredity and Race Hygiene* became the basic German eugenics text throughout the post–World War I and Nazi German eras. (From: *Menschliche Auslese und Rassenhygiene* [1923], Main Library, Indiana University, Bloomington, Indiana. Image courtesy of the Digital Library Program.)

stagnant by the end of World War I, but was revived in 1922 when its headquarters was moved to Berlin and emphasis placed on **public health** and welfare. That year the society again promoted a series of eugenic recommendations built upon the 1914 aims. Proposed for the first time were **eugenic sterilization laws,** like those enacted in the United States, and **eugenic segregation** of "anti-social and other very defective individuals" in labor colonies.

Around 1911 some leaders of the Munich branch including Ploetz and his wife founded within the society a secret "Nordic ring," *Der Bogen* ("The Bow"), whose aim was the improvement of the **Nordic race** (Nordicism). By 1913 it had about fifteen members including Ploetz's protégé and future eugenics leader **Fritz Lenz.** This "archery club" continued to be active during the war years and in 1918–1919 transformed into a Nordic improvement group called the *Widar Bund.* In the Weimar Republic of the 1920s increased conflict arose between the social welfare–oriented Berlin chapter and the more pro-Nordic Munich chapter, now led by Lenz. Most society members outside of Munich were opposed to Nordicism and supported improving the efficiency and health of the entire population, not just the **Aryan race.** This conflict was a factor in the formation of the German Federation for Population Betterment and Heredity (*Deutscher Bund für Volksaufartung and Erbkunde*) in 1926 in Berlin by population registrars, public health officials, and medical professors—many who were Jewish—who wanted to popularize heredity and eugenics as a **public health** issue among all Germans. It successfully attracted governmental support. Loss of governmental funds to the society and increased tension within it between the Munich and the Berlin chapters prompted the Berlin chapter to add *Eugenik* to its name in 1930 and gravitate away from Nordic race-hygiene and purity ideology. In 1931 the society as a whole reformatted its constitution. This change allowed the federation and society to fuse as the "German Society for Race Hygiene (Eugenics)." This new organization embraced basic research along with public health and welfare interests. The society now had several thousand members and became more centralized. The Berlin chapter grew while the Munich one stagnated.

Upon the society's takeover by the Nazi government in November 1933, it was placed directly under the Reich Commission for National Health Service and became a government organization. Rüdin, now director of the **Kaiser Wilhelm Institute** for psychiatry in Munich, was appointed leader of the society, and its headquarters was transferred back to Munich. Rüdin eliminated *Eugenik* from the name of the society, and the older *Rassenhygiene* title was used. Liberal leaders and all Jews were asked to leave; membership was now restricted to Germans of Aryan ancestry. The society under **Nazi Germany** promoted Aryan supremacy and anti-Semitism and increased its focus on the prevention of **genetic diseases** and encouraging the fit to reproduce. It supported passage of sterilization and other eugenic laws. Over the twelve years of the Third Reich, many of the society's functions were taken over by other Nazi organizations, and as a result it lost its influence. At the end of the **World War II** few people attended meetings, and the society ceased to exist.

FURTHER READING: Kühl, Stefan, *The Nazi Connection* (1994); Weindling Paul, *Health, Race, and German Politics between National Unification and Nazism, 1870–1945* (1989); Weingart, Peter, "German Eugenics between Science and Politics" (1989); Weiss, Sheila Faith, *Race Hygiene and National Efficiency* (1987); Weiss, "The Race Hygiene Movement in Germany" (1987).

GERMINAL CHOICE

The voluntary selection of sperm or eggs from individuals with specific physical, mental, intellectual, or other characteristics to produce a pregnancy is *germinal choice*. The term was coined around 1960 by American geneticist **Hermann Muller** to describe a **positive eugenics** plan to promote a program of **artificial insemination** with the sperm of extraordinary men for couples in which the husband was sterile. In 1935 Muller had proposed the term *children of choice* for this plan. In 1979 Robert K. Graham established a **sperm bank**, the **Repository for Germinal Choice**, based upon Muller's philosophy. Its purpose was to store "superior sperm" from Noble laureates and other notable men for use by couples. By the early 1990s most large urban areas had centers for **assisted reproduction** along with storage facilities for screened human semen and, sometimes, eggs. Sperm is collected from healthy young men but not necessarily extraordinary individuals. Couples or single women wanting to become pregnant select semen from catalogs describing men with various characteristics. The semen is used for both artificial insemination and **in vitro fertilization**. Since those desiring pregnancy generally wish to have healthy infants free from **genetic diseases** and other conditions, or may want a child with a certain gender for **family balancing**, the concept of germinal choice, or choosing characteristics of sperm and eggs for pregnancy, is an aspect of the **new eugenics**.

FURTHER READING: Carlson, Elof Axel, *Genes, Radiation, and Society* (1981); Stock, Gregory, *Redesigning Humans* (2002).

GERM-PLASM, GERM CELLS, OR GERM-LINE CELLS

Coined by German physician August Weismann (1834–1914) in 1883, the term *germ-plasm* was defined as the "hereditary substance found within the germ-cells" or reproductive cells (egg and sperm), now called *gametes* or germ-line cells. The concept was also independently suggested by German biologist Wilhelm Roux (1850–1924) in 1888. It was proposed that both egg and sperm contribute chromosomes, or carriers of hereditary information, equally to the fertilized egg. It was initially speculated that the gametes were the only ones to carry the complete set of hereditary information and that each somatic, or body, cell contained only part of the hereditary information required for the specific type of cell—skin, brain, liver, and so forth. However, in the late twentieth century all cells were found to carry the complete **genome**, the blueprint for all hereditary instructions. The terms *germ-plasm* and *germ cell* were commonly used during the first part of the twentieth century in the newly developing fields of both **genetics** and **eugenics**. The terms changed meaning over the course of the twentieth century, and *germ-plasm* was generally only used in the agricultural sciences by the twenty-first century.

FURTHER READING: Carlson, Elof Axle, *The Unfit* (2001); Paul, Diane B., *Controlling Human Heredity* (1995); Weismann, August, *The Germ-Plasm* (2003)

GOBINEAU, JOSEPH ARTHUR DE (July 14, 1816–October 13, 1882)

An aristocratic French diplomat, author, and race theorist, de Gobineau was the first to propose a **"hierarchy of races"** in terms of ability and **intelligence**, and the superiority of the white, and especially the **Aryan**, race, as a scientific theory. His theories influenced eugenics laws in several countries during the early-twentieth-century eugenics movement. De Gobineau was born near Paris, the son of an army officer and descendant of a proud and aristocratic family who opposed the French Revolution and remained loyal to the royal family. As a child he was tutored at home and subsequently went to Switzerland with his mother and her lover. He attended the College of Bienne in Switzerland and in 1832 prepared unsuccessfully for the military. In 1835 de Gobineau went to Paris, where he attended classes, wrote, and was introduced to Royalists. His aristocratic connections enabled him to meet French statesman and writer Alexis de Tocqueville (1805–1859), which led to a diplomatic career. He also adopted his uncle's title, *comte* (count). Over four decades de Gobineau held positions in Berne, Hanover, Frankfurt, Teheran, Athens, Rio de Janeiro, and Stockholm. He was forced to retire from the diplomatic corps in 1876 and spent his remaining years in Italy.

De Gobineau was a prolific writer and published hundreds of manuscripts over his lifetime. His most influential work upon eugenics ideology was *Essai sur l'Inégalité des Races Humaines* (1853–1855). The book's first part was translated into English as *The Moral and Intellectual Diversity of the Races* (1856). In it he classified races on a hierarchy in terms of the advancement of civilization. Germanic and Teutonic **Nordic** Europeans were at the top. Gobineau did not support democratic ideals and viewed the French Revolution as the result of the bastardization of the northern Europeans that was leading to **racial degeneracy** and the decline of Western culture. In 1897 Gobineau met German musician Richard Wagner (1813–1883) in Rome and, later, his circle of literary friends in Bayreuth, Germany. De Gobineau's views impressed Wagner and, in turn, philosopher Friedrich Nietzsche (1844–1900), who popularized them in their creative efforts. Gobineau's work on race increased in popularity during the decades after his death in Turin, Italy.

FURTHER READING: Biddiss, Michael Denis, *Father of Racist Ideology* (1970); Gobineau, Arthur, Comte de *The Moral and Intellectual Diversity of Races* (1984).

GODDARD, HENRY HERBERT (August 14, 1866–June 19, 1957)

Goddard, an American psychologist and pioneer of mental assessments and testing, coined the term *moron* for higher-functioning **feebleminded**, or mentally disabled, individuals. His reports helped popularize eugenics in the 1920s and gave support for **eugenic sterilization, segregation,** and **immigration restriction laws** in the United States. Goddard was born in East Vassalboro, Maine, into a Society of Friends (Quaker) farming family descended from early **Anglo American** settlers. His father died when he was a child. He attended Quaker boarding schools and graduated with a B.A. (1887) from Haverford College (Pennsylvania), briefly taught, and returned to Haverford, where in he earned an M.S. in mathematics (1889). He taught in small colleges and earned a Ph.D. (1899) in psychology from Clark University, Worcester, Massachusetts. Goddard then taught pedagogy and psychology at the State Normal School in West Chester, Pennsylvania, until 1906. At this point he became director of psychological research at the Training School for Feeble-Minded Girls and Boys in Vineland, New Jersey. In 1918 Goddard left Vineland, and after brief employment with an Ohio governmental agency became professor of clinical and abnormal psychology at Ohio State University in 1922, a post he held until he retired in 1938.

As part of the evolving eugenics movement in the **United States,** Goddard pioneered intelligence testing as a method of identifying the **unfit.** In 1908 he translated into

English a standardized **intelligence test** developed three years earlier by French psychologists Alfred Binet (1857–1911) and Théodore Simon (1873–1961). Using this test with children at his institution, along with **family history** research methods developed by eugenicist **Charles Davenport** and information on the newly rediscovered theory of **Mendelian inheritance**, Goddard researched the hereditary nature of mental disabilities. He published a series of reports, including the internationally acclaimed *Kallikak Family* (1912), in which he concluded that feeblemindedness resulted from a simple recessive trait and that individuals with this condition were responsible for most social problems, including **pauperism**, **alcoholism**, and **venereal disease**. Prior to **World War I** Goddard worked at Ellis Island for the **Eugenics Record Office** (ERO). He tested for "mental defectives" among potential immigrants and asserted that a high percentage were morons. Based upon his research he championed eugenic sterilization and segregation for the feebleminded in state-run institutions and intelligence testing of immigrants.

Goddard was active in several eugenics organizations. In 1910 he was appointed chair of the Committee on Heredity of the Feeble-Minded for the **American Breeders Association**'s joint committee with the ERO. In the late 1920s he was a member of the advisory board of the **American Eugenics Society** and was on the central committee for the **Third National Race Betterment Conference** held in 1928. Goddard's early work, along with other eugenics studies, began to be challenged in the late 1920s as environmental theories for many human behaviors and characteristics emerged. Goddard defended his work, although he eventually repudiated some of his earlier claims. In 1947 he moved to Santa Barbara, California, where he died ten years later. He was married to Emma Florence but had no children.

FURTHER READING: Kline, Wendy, *Building a Better Race* (2001); Paul, Diane B., *Controlling Human Heredity* (1995); Richardson, Ken, *The Making of Intelligence* (2000); Ryan, Patrick J., "Unnatural Selection: Intelligence Testing, Eugenics, and American Political Cultures" (1997).

GOSNEY, EZRA SEYMOUR (November 6, 1855–September 14, 1942)

Gosney, a businessman, philanthropist, and eugenicist, became a financial supporter of eugenic and **social hygiene** causes in the United States. Born in Kenton, Kentucky, of French Huguenot descent, he received his preliminary education in local schools and at Caddo Grove, a Texas seminary. He worked his way through college and received a B.A. (1877) from Richmond (Missouri) College and a law degree from Washington University in St. Louis (1880). Gosney then established a successful law practice with the Missouri railroads. He moved to Flagstaff, Arizona, around 1888 due to illness, believed to have been **tuberculosis**. He continued his law practice and also became a successful banker. In 1905 he semiretired to Pasadena, California, where he became a leading civic figure and philanthropist and launched a successful citrus fruit–growing business. Through his agricultural interests Gosney became interested in eugenics and began to correspond with **Charles Davenport**, director of the **Eugenics Record Office**.

In 1926 Gosney founded the **Human Betterment Foundation**, incorporated in 1928, for the study of **eugenics sterilization**. **Paul Popenoe**, his neighbor and eugenicist, became head of research. Gosney believed that "sterilization is one of the few forms of philanthropy that tends to do away with its own need. Instead of being mere repair work, as so much charity is, it goes to the root of the evil and eradicates it." He favored voluntary **birth control** for responsible individuals and compulsory eugenic sterilization for the **insane** and **feebleminded**. Gosney was involved with many eugenics-related organizations including the **American Eugenics Society**, the **American Social Hygiene Association**, the **Eugenics Research Association**, and the American Genetics Association (formerly the **American Breeders Association**). He coauthored with Popenoe *Sterilization for Human Betterment* (1929) and was editor or author of other foundation publications. Gosney was married three times, widowed twice, and fathered two daughters. His wives were Tyrene Noyes (1886), Mae Hawkey (1893), and Sarah Hunt Dearborn (1924). After his death the Human Betterment Foundation was liquidated to establish a research fund at California Institute of Technology in Pasadena.

FURTHER READING: Engs, Ruth C., *The Progressive Era's Health Reform Movement* (2003); Kline, Wendy, *Building a Better Race* (2001).

GRANT, MADISON (November 19, 1865–May 30, 1937)

A leader of the early-twentieth-century eugenics, immigration-restriction, and conservation movements in the United States, Grant, through his publications, helped fan the **nativistic** branch of the eugenics movement on both sides of the Atlantic. Born in New York City the son of a socially prominent physician from an old New York Episcopalian family, Grant had a privileged childhood. He received a bachelor's degree from Yale University (1887) and a law degree from Columbia University (1890). Independently wealthy, he put his energies into **Progressive Era** causes. Grant became a noted naturalist. He helped found the New York Zoological Society (1895), was a trustee of the American Museum of Natural History, and collaborated with **Henry Fairfield Osborn**, its director, in forming the Save-the-Redwoods League (1919). Along with Osborn and **Charles Davenport**, director of the **Eugenics Record Office** (ERO), Grant cofounded the **Galton Society**. He was a charter member of the **American Eugenics Society**; a charter member and president (1919) of the **Eugenics Research Association**, and a major player in the **First**, **Second**, and **Third International Eugenics Congresses**.

Grant was most noted for his internationally acclaimed work *The Passing of the Great Race* (1916). This milestone work influenced the direction of the eugenics movement in regards to "social worth" among various **races**. Grant argued for the superiority of the **Nordic race**, which he suggested was a **master race**. He expressed concerns about **race suicide** among this group from **war**, **alcoholism**, **tuberculosis**, **venereal disease**, and a low birthrate. He contended that "the laws of nature require the obliteration of the unfit." To decrease the number of **unfit**, he recommended **eugenic sterilization** and **segregation laws**. He favored restricting immigrant groups whom he felt were "polluting the racial blood lines" of **Anglo Americans**. These groups included **eastern and southern European** immigrants, who were outbreeding the native-born American. German race-theorist **Hans F. K. Günther** lauded Grant for his works and theories as a model for Germany. Many of Grant's views and prescriptions were found in the racial policies of **Nazi Germany** in the 1930s. Among Grant's other major eugenics and immigration-restriction publications was *Conquest of a Continent* (1933), which was largely ignored due to a change in the intellectual climate by the early 1930s. Grant used his legal skills to help write and pass the **Johnson-Reed Immigration Re-**

striction Act. He was vice president of the influential Immigration Restriction League from 1922 until his death. Although he advocated a high birthrate for old-stock Americans, Grant never married or had children. He died in New York City of heart disease, after a long illness.

FURTHER READING: Grant, Madison, *The Passing of the Great Race* (1970); Paul, Diane B., *Controlling Human Heredity* (1995); Reilly, Philip R., *The Surgical Solution* (1991).

GREAT AWAKENINGS

Periods of time when religious, health, and social-reform movements surge are called *Great Awakenings*. Since colonial times, in the United States Great Awakenings have occurred in roughly 80- to 100-year cycles. They also have arisen to some extent in Europe, in particular, **Germany**. These movements in the United States have typically begun with religious revivals from which significant political, economic, educational, medical, health, and other social changes emerge. They generally extend from one to two generations, during which a reorientation in beliefs and values occurs. Awakenings attempt to bring society back to an imagined "golden age" free of **crime**, disruption, and immorality. Common themes in these massive social movements are women's rights; construction of a perfect or "millennial" society; a return to nature; concerns about the environment; emergence of new religious sects; and fear of immigrants, the **unfit**, and other "dangerous classes." Emphasis upon health issues with moral overtones can be described as the **Clean Living Movement** phase of a Great Awakening. The health and moral issues include ideals of recapturing "family values" and sexual purity; the elimination of **alcohol**, **tobacco**, or drugs from society; the promotion of vegetarianism and **physical culture** and exercise; and advocacy programs promoting **inherited realities**, eugenics, or **race hygiene**.

FURTHER READING: Engs, Ruth Clifford, *Clean Living Movements* (2000); McLoughlin, William G., *Revivals, Awakenings, and Reform* (1978); Strauss, William, and Neil Howe, *The Fourth Turning* (1997).

GÜNTHER, HANS F(RIEDRICH) K(ARL) (February 16, 1891–September 25, 1968)

A German social anthropologist and leading race-theorist of the interwar era (between **World Wars I** and **II**), Günther popularized the concept of "Nordicism," which considered the "superior" **Nordic race** as the creative force of Western civilization. He advocated eugenic measures to improve this **race**. Günther was born in Freiburg im Breisgau Germany, the son of a chamber musician. He studied languages in Paris and Freiburg and received a Ph.D. from the University of Freiburg (1914). He did not serve in World War I as he was declared unfit for service. He spent some time in Sweden and Norway as a freelance writer, passed the gymnasium teaching examination in 1919, and briefly taught. In 1920 he expounded his Nordic ideals with the publication of *Der Ritter, Tod und Teufel* ("The Knight, Death and the Devil"), a heroic folktale lauding the Nordic race. This publication brought him favorable attention from the growing National Socialist party. In 1930 when the Nazis gained power in Thuringia, they created a Chair of Social Anthropology for Günther at the University of Jena, against the wishes of other academics. In 1935 Günther was appointed to the Institute for Ethnology, Race, Biology, and Rural Sociology in Berlin and was awarded a scientific prize acknowledging his research as the philosophical foundation of Nazi Race Laws. From 1940 to 1945 he was professor of "racial science" at Freiburg University. He was awarded the Goethe Medal in 1941 for his work by Nazi leaders.

Günther was a prolific writer. He was commissioned to write the pro-Nordic *Rassenkunde des Deutschen Volkes* (*Ethnology of the German Nation*, 1922) by publisher J. F. Lehmann. This widely read work underwent fifteen editions through 1943. Of his numerous works, *Racial Elements of European History* (1927); was the only one published in English. Günther developed a theory on the connection between race and character and, thereby, culture. He advocated keeping the "Germanic blood pure" by preventing interbreeding with "inferior" individuals and races, including **Jews**. He favored eugenic programs that increased the population of Nordics and was concerned about their **race suicide** due to low birthrates. He supported **eugenic sterilization** and **marriage-restriction laws** to prevent **racial degeneracy** of Germanic peoples. The theoretical basis for **Nazi Germany**'s laws to keep the **Aryan race** pure was largely based upon Günther's writings. He did not, however, approve of using the term *Aryan*, meaning a "race," because he and other anthropologists considered it a language group. This was also true for the term *semitic*. Günther admired American eugenicists **Madison Grant**, and **Lothrop Stoddard** and their efforts in promoting **immigration restriction laws** to prevent inferior races and the **unfit** from entering the United States. In 1927 Günther coauthored a booklet with anthropologist **Eugen Fischer** with pictures of typical-looking Nordic Germans.

After **World War II** Günther claimed not to be a supporter of the Nazi political program. However, his claim was probably not true and he was removed from his academic position. To escape his reputation of being a Nazi, he published under the pseudonyms Ludwig Winter and Heinrich Ackermann. In 1958 Günther helped found the Northern League, a society that focused upon Nordic pride and aimed to foster friendship and solidarity among all Teutonic nations. With help from American geneticist **Hermann Muller** Günther was nominated as a "foreign member" of the American Society for Human Genetics and used this affiliation in an attempt to gain credibility for his research and writing, which he continued to the end of his life. He married Ida Faye (1923) and they had one daughter.

FURTHER READING: Kühl, Stefan, *The Nazi Connection* (1994); Field, Geoffrey G., "Nordic Racism" (1977); Weindling, Paul, *Health, Race, and German Politics between National Unification and Nazism, 1870–1945* (1989).

GUYER, MICHAEL FREDERIC (November 17, 1874–April 1, 1959)

Guyer, a zoologist, helped popularize eugenics in the United States through a widely used textbook. Unlike other eugenics adherents, he was not a leader in either the biological or social-reform aspects of the eugenics movement. For most of his career he was focused on medical education. Guyer was born on a farm near Plattsburg, Missouri. He attended local schools and the University of Missouri but transferred to the newly established University of Chicago, where he received his B.S. (1894). He subsequently did research at the University of Nebraska but returned to Chicago, where he received his Ph.D. (1900) in zoology. He chaired the Department of Biology at the University of Cincinnati (1901–1910) and then headed the Department of Zoology at the University of Wisconsin (1911–1945) until his retirement.

Biological research in pigeon spermatogenesis influenced Guyer to become interested in eugenics. At Wisconsin he began to offer a course in heredity and eugenics. From these lecture notes he published *Being Well-Born* in 1916, with a second edition in 1927. This popular work was used for over sixteen years as a high school and college textbook. It was designed to introduce the concepts of eugenics to the masses and to encourage the "fit" to have more children in order to prevent **race suicide** among **Anglo Americans**. He voiced concerns about unrestricted immigration and the higher prevalence of **insanity, feeblemindedness** and disease among the immigrant population, and noted that "the alien far outbreeds the native stock." However, unlike other eugenicists Guyer was ambivalent about **eugenic sterilization** and **segregation**. Instead, he encouraged education in the principles of eugenics and advised potential parents to "keep themselves in good physical condition by wholesome temperate living" and to protect their "immortal germ-plasm of which they are the trustees" from bad nutrition, **racial poisons**, or vice. He was concerned that **alcoholism** among pregnant women could cause "germinal or fetal poisoning." In his later years, Guyer published *Speaking of Man* (1942), in which he continued to maintain the hereditary predisposition of **crime**, disease, and mental illness, and disability even though environmental causes had become fashionable. He was on the advisory council of the **American Eugenics Society** from its early years and took part in the **Third International Congress of Eugenics**. He married Helen Stauffer (1898), who died in 1949, and fathered one son. Guyer died at his winter home in New Braunfels, Texas, in 1959.

FURTHER READING: Devlin, Dennis S., and Colleen L. Wickey, "Better Living Through Heredity: Michael F. Guyer and the American Eugenics Movement" (Winter 1984); Reilly, Philip, *The Surgical Solution* (1991).

H

Hygiene aims to improve the individual and eugenics to improve the race. The two must go hand in hand.

Albert Edward Wiggam, *The Fruit of the Family Tree* (1924)

Man is an organism—an animal; and the laws of improvement of corn and of race horses hold true for him also. Unless people accept this simple truth and let it influence marriage selection human progress will cease.

Charles Benedict Davenport, *Heredity in Relation to Eugenics* (1911)

HARRIMAN, MARY WILLIAMSON AVERELL (July 22, 1851–November 7, 1932)

The single largest individual donor to U.S. eugenics causes, Harriman, a wealthy widow, established the **Eugenics Record Office** (ERO), the primary research organization of the eugenics movement in the **United States** during the early twentieth century. Harriman, who descended from old stock **Anglo-Saxon** colonial families, was born in New York City, one of three children, to a prominent businessman. She spent most of her time in Ogdensburg, New York, and a few years at a New York City finishing school. At age fifty-seven her husband, Edward Henry Harriman (1848–1909), a former president of Union Pacific Railroad, died, leaving her an immense fortune. After his death she managed his vast properties and business interests and donated to a variety of charities and causes, including eugenics.

Harriman did not favor the establishment of large foundations and would not support a cause unless it personally interested her. She helped fund a variety of **public health** efforts including research laboratories, a **tuberculosis** sanatorium, and a home for the mentally retarded. Her daughter, Mary Munsey, who spent the summer of 1905 at the **Cold Spring Harbor**, New York, biological laboratory and studied with its director, geneticist and budding eugenicist **Charles Davenport**, introduced Davenport to her mother. Harriman was familiar with racehorse breeding and believed that the laws of heredity might also be used for improving the human race. Acquiring an interest in eugenics, she agreed to support Davenport in developing a research center. In 1910 she bought seventy-five acres next door to the laboratory to create the ERO. Its purpose was to study problems of human heredity and mental deficiency. The office was directed by Davenport, who in turn hired **Harry H. Laughlin** as superintendent. Pleased with the creation of the office, Davenport dedicated his text *Heredity in Relation to Eugenics* (1911) to Harriman. She funded the ERO on a yearly basis and in 1918 transferred it to the **Carnegie Institution of Washington**, along with an endowment. Harriman funded other eugenics efforts, including the **Second Interna-**

tional **Congress of Eugenics**, which was held in New York in 1921. In her later years she supported the arts. Her marriage to Harriman (1879) resulted in six children. Her son W. Averell Harriman (1891–1986) became a U.S. statesman. Mary Harriman died in New York after an operation for intestinal cancer.

FURTHER READING: Campbell, Persia, *Mary Williamson Harriman* (1960); Haller, Mark H., *Eugenics* (1984); Kevles, Daniel J., *In the Name of Eugenics* (1985).

HEREDITY AND SELECTION IN THE LIFE HISTORY OF NATIONS
(*Vererbung und Auslese im Lebenslauf der Völker*, 1903, 1910, 1918, 1920)

Written by pioneer German eugenicist **Wilhelm Schallmayer** and expanded from his prize-winning Krupp Essay of 1900, this classic work presented a "hygienic-socialist" approach to increasing the biological health and efficiency of the German nation. The first edition in 1903, published by Gustav Fischer in Jena, had 386 pages divided into two parts. The 1910 edition expanded to 464 pages. By 1919 it had gone through four editions and was the standard **German eugenics** textbook. Titles of chapters in the 1910 edition included "The Elements of the Darwinian Theory of Evolution," "The Death of a People in the Past and Present," and "Direct Improvements of Selective Reproduction." The book was not translated into English, although it was briefly discussed in the U.S. *Journal of Heredity*.

The social problems and eugenic proposals presented in the book, were originally outlined by Schallmayer in a brief essay written in 1891 in which he discussed **social Darwinism** and **degeneracy** theory. The book was also based upon these theories, which argued that modern medicine kept alive those who would have died through **natural selection**, and suggested that problems from urban living and industry were also having degenerative effects on the human race. Because the long-term power of a state was dependent upon the "biological vitality" of the nation, it was important to manage its "human resources" so as to improve the biological constitution of its people in order to increase the efficiency and productivity of the nation. Schallmayer recommended that Germany take an active part in reg-

VERERBUNG
UND AUSLESE

IN IHRER SOZIOLOGISCHEN
UND POLITISCHEN BEDEUTUNG

PREISGEKRÖNTE STUDIE
ÜBER
VOLKSENTARTUNG UND VOLKSEUGENIK
VON
DR. WILHELM SCHALLMAYER

MOTTO: FÜR DIE NATIONEN WIE FÜR DIE EINZELNEN IST
DAS HÖCHSTE GUT IHR ORGANISCHES ERBGUT

ZWEITE, DURCHWEGS UMGEARBEITETE UND VERMEHRTE AUFLAGE

VERLAG VON GUSTAV FISCHER IN JENA
1910

Written by pioneer German eugenicist and physician Wilhelm Schallmayer, *Heredity and Selection in the Life History of Nations* was a popular text for the early eugenics movement in Germany. (From: *Vererbung und Auslese* [1910], Main Library, Indiana University, Bloomington, Indiana. Image courtesy of the Digital Library Program.)

ulating the "hereditary efficiency" of its citizens by encouraging the biologically fit and socially productive to reproduce. These included **positive eugenics** programs to encourage middle-class civil servants, academics, military officers, and other professionals to marry young and produce many children. Other eugenic methods included tax breaks or bonuses for school-age children from these marriages, **family pedigree studies**, and "health passports."

The author suggested that marriage restrictions for the **unfit**, including the **insane**, the **feebleminded**, and people with chronic **alcoholism** and **tuberculosis**, were in the best interest of the state and the human race. However, he hedged on openly promoting **negative eugenic** methods such as **eugenic sterilization** or **segregation**. Schallmayer suggested that eugenics was an extension of **public health** that should be under the control of physicians. In order to halt population decline and **racial degeneracy**, he included suggestions similar to the health reform crusades of the U.S. **Clean Living Movement** that transpired during the **Progressive Era** (1890–1920). These included educational activities and legislation to curtail the consumption of **alcohol** and **tobacco**, the promotion of **social hygiene** to prevent and treat **venereal diseases**, and **physical culture** and exercise to improve physical fitness. The theme of improving the biological efficiency of the German people, as argued in this work, was the major thread running throughout the whole German eugenics movement until its decline in the 1940s.

FURTHER READING: Weiss, Sheila Faith, *Race Hygiene and National Efficiency* (1987).

HEREDITY IN RELATION TO EUGENICS (1911)

Written by geneticist and pivotal eugenics leader **Charles Benedict Davenport**, *Heredity in Relation to Eugenics* discussed the basis of **Mendelian inheritance**, the transmission of human traits, susceptibility to **genetic disease**, and eugenic recommendations. It was one of the first genetic books aimed at the educated middle class. Published by H. Holt and Company, New York City, it contained 298 pages. The book was dedicated to "Mrs. E. H. [**Mary**] **Harriman** in recognition of the generous assistance she has given to research in eugenics." It was divided into ten sections: "Eugenics: Its Nature, Importance and Aims," "The Method of Eugenics," "Inheritance of Family Traits," "Mutations and Their Eugenic Significance," "Geographic Distribution of Inheritable Traits," "Migrations and Their Eugenic Significance," "The Influence of the Individual on the Race," "The Study of American Families," "Eugenics and Euthenics," and "The Organization of Applied Eugenics."

Although many of the conditions the author described, such as hemophilia, are inherited, others, including complex behaviors such as **pauperism**, were later found not to be the result of simple Mendelian inheritance. The author discussed geographical distribution of inheritable traits, characteristics of various ethnic groups, and the eugenic significance of migration in carrying both "defective and valuable traits" into new regions. Like many other eugenicists, Davenport supported **social Darwinian** ideology of survival of the fittest and argued that the **unfit** were being allowed to survive through **public health** and modern medicine, which would lead to **racial degeneracy** and the decline of civilization. He advocated state eugenic surveys to collect **family history and pedigree** data and **eugenic sterilization**, **segregation**, and **marriage-restriction laws** to prevent the further decline of society.

FURTHER READING: Haller, Mark H., *Eugenics* (1984); Kevles, Daniel J., *In the Name of Eugenics* (1985).

HEREDITY OF RICHARD ROE, THE: A DISCUSSION OF THE PRINCIPLES OF EUGENICS (1911)

Aimed at the middle class, the easy-to-read short treatise on eugenic theory *The Heredity of Richard Roe* was written by educator and eugenics leader **David Starr Jordan**. It was published in Boston by the American Unitarian Association and contained 165 pages. The author used a fictitious individual, Richard Roe, and related both the "good" and the "bad" traits he inherited from various ancestors, along with possible outcomes in his life due to his genetic inheritance. The book was divided under approximately 100 subheadings, including "Race Characters," "The Thorough-bred," "Blood Will Tell," "Final Formula of Heredity," "Defects of the Mind," "Breeding of **Superman**," and "The Wholesome World." Principles of **genetics** and eugenic themes, including the rules of **Mendelian inheritance**; heredity and environmental factors in human characteristics (the **nature-nurture** debate); **family history studies** like *The Jukes*; and the concepts of **feeblemindedness**, **degeneracy**, and **pauperism** were covered. Jordan considered both nature and nurture important but suggested, "Nurture will do nothing unless Nature is first. Nature indicates possibilities. It is for Nurture to make them good."

HIERARCHY OF RACES

The classification of the "three races of man" into a hierarchy, based upon their perceived **intelligence** and ability was the *racial hierarchy*, or a *hierarchy of the races*, theory. French diplomat and author **Arthur de Gobineau** in 1853 constructed the racial hierarchy to explain the rise and fall of civilization throughout history based upon the social evolution of the different **races**. His theory was described in *Essai sur l'Inégalité des Races Humaines* ("Essay on the Inequality of Human Races; 1853–1855). The book's first part was translated into English in 1856 as *The Moral and Intellectual Diversity of the Races*. De Gobineau placed at the bottom of the hierarchy, or ladder, the "black race," which he described as innately endowed with animality and energy but severely limited in intellect. At the middle of the ladder was the "yellow race," who leaned toward apathy, lacked physical strength, liked mediocrity, and respected law. On top was the "white race," marked by energetic intelligence, perseverance in the face of obstacles, and an instinct for order and organization.

However, within the white race, "the noblest, and most highly gifted in intellect and personal beauty, the most active in the cause of civilization, [was] the Arian [*sic*] race." The most "superior" of the **Aryans**, in turn, were those of Teutonic or Germanic blood. De Gobineau contended that Aryans' love of migration and desire for conquest founded ten great civilizations. However, due to "race crossing" with "inferior races," the civilizations they established declined when the elite population decreased and lost its power due to democracy, which allowed inferior "mixed blood" people to come into power. When a nation declined, only a small quantity of "the blood of its founders" could be found, even though it had the same name. He proposed that Roman and Semitic cultures, including **Jewish** peoples, had degenerated because of race mixing with the black and yellow races.

The racial hierarchy theory and the purported superiority of northern Europeans influenced other theorists in the late nineteenth century. It was the underlying theory of the **immigration restriction movement** in the **United States**, supported by the

nativist branch of the eugenics movement. It became embodied in the **superman** and **master race** theories and superiority of the **Nordic race** ideology, particularly in **Nazi Germany**. Several eugenics supporters expanded de Gobineau's theory in their works, including **Lothrop Stoddard** and **Madison Grant** in the **United States** and **Hans F. K. Günther** in **Germany**.

FURTHER READING: Biddiss, Michael Denis, *Father of Racist Ideology* (1970); Gobineau, Arthur Comte de, *The Moral and Intellectual Diversity of Races* (1984).

HILL FOLK, THE: REPORT ON A RURAL COMMUNITY OF HEREDITARY DEFECTIVES (1912)

The first of a series of research reports from the **Eugenics Record Office** (ERO) exploring the hereditary nature of antisocial behavior, *The Hill Folk* was coauthored by Florence H. Danielson, a eugenics **field worker** trained by the ERO, and **Charles Davenport**, its director. This "Memoir No. 1," published by the ERO, was fifty-six pages long. The memoir, along with *The Nam Family*, *The Dacks*, *The Jukes in 1915*, and *The Kallikaks*, helped give a "scientific basis" to the eugenics movement and fostered support for **eugenic sterilization** and other laws to keep the **unfit** from reproducing. Danielson began to gather her sample in 1910 while employed as a field worker for Monson State Hospital, Palmer, Massachusetts. She traced five generations of a family through personal visits and interviews with family members, physicians, town officials, and neighbors, as well as review of existing court records. The "hill folk" lived in a fertile river valley amid the New England hills. They were known in their Massachusetts community for **feeblemindedness**, "immorality," and **alcoholism**.

The report claimed that of the 737 individuals traced about 25 percent had married cousins, 24 percent of married women had given birth to illegitimate offspring, and 10 percent were **prostitutes**. **Criminal** tendencies were found in 3.3 percent, 48 percent were feebleminded, and a large proportion were alcoholics. The authors compared this family with the "Jukes" and suggested that whereas the Jukes family had a high frequency of "criminal tendencies" among men and prostitution among women, the "hill folk" tended to be shiftless, had low-grade mentality associated with "sexual immorality," and showed a tendency toward minor criminal offenses. The authors estimated that the state, over sixty years, had spent a high proportion of its welfare fund on them. Based upon the report, Davenport proclaimed in the preface that "it is hoped that a presentation of the facts will hasten the so much desired control by society of the reproduction of the grossly defective." Some financial assistance for the study and the pamphlet's publication came from John D. Rockefeller, Jr. and **Mary Harriman**, the major sponsor of the ERO.

FURTHER READING: Carlson, Elof Axle, *The Unfit* (2001); Dugdale, Richard Louis, *"The Jukes"* (1970); Haller, Mark H., *Eugenics* (1984); Kevles, Daniel J., *In the Name of Eugenics* (1985).

HOLMES, S(AMUEL) J(ACKSON) (March 7, 1868–March 5, 1964)

A zoologist and prolific writer of eugenics articles and books, Holmes spent most of his career as an academic at the University of California. His publications helped popularize eugenics. Holmes was born in Henry, Illinois, and received a B.A. and M.S. in biology from the University of California and a Ph.D. in zoology from the University of Chicago (1897). After a year of public school teaching in California, he became a zoology instructor at the University of Michigan, and after six years transferred to the University of Wisconsin. In 1912 he went to the University of California, Berkeley, as a assistant professor in zoology, achieved the rank of professor in 1917, and remained at Berkeley for twenty-seven years before retiring in 1939. Holmes taught one of the first eugenics courses in the country. During the summer he conducted research at the Marine Biological Laboratory at Woods Hole, Massachusetts. He was internationally known for research on animal **genetics** and behavior and was one of the first researchers to recognize that radiation could cause serious birth defects in animals.

Holmes supported the theory of **social Darwinism** and expressed concern about **race suicide**. The fact that the "highly intelligent" were having fewer children, he felt, was "leading toward extinction of the better endowed stock," and on that account he encouraged **positive eugenics**. He also feared **racial degeneracy** and the decline of democracy due to the "overmultiplication of inferior stocks." Although he conceded that environment was important in determining human characteristics, he argued that "in mental as well as in physical traits it is blood that tells." Holmes wrote several eugenics books reflecting his ideas, among them *The Trend of the Race* (1921), *Studies in Evolution and Eugenics* (1923), and *A Bibliography of Eugenics* (1924). Holmes was active in eugenics and genetics organizations. He was an early member of the **American Eugenics Society** and served as its president (1938–1940). He was a member of the editorial board of its journal, **Eugenics**, during its brief history and a charter member of the **Human Betterment Foundation**. Holmes presented papers at both the **Second** (1921) and **Third** (1932) **International Congress of Eugenics** and was on the governing council of the American Genetics Association formerly the **American Breeders Association**. After his retirement he continued to write articles with eugenics and population themes. He married Cecilia Warfield Skinner (1909) and, following his own positive eugenics advice, the couple had five children. Holmes died at Kaiser Hospital, Oakland, California, two days before his ninety-sixth birthday.

FURTHER READING: Haller, Mark H., *Eugenics* (1984); Paul, Diane B., *Controlling Human Heredity* (1995).

HUMAN BETTERMENT FOUNDATION (1926–1942)

Founded and funded by businessman **E. S. Gosney** at the peak of the early-twentieth-century eugenics movement, the Human Betterment Foundation focused upon **eugenic sterilization** in California. This private organization was formally incorporated in 1928 with twenty-five members. Its purpose was to promote "educational efforts for the protection and betterment of the human family in body, mind, character, and citizenship," and it pressed for state sterilization laws. Members and advisers included noted **birth control, social hygiene**, and eugenics reformers and among the charter members were **Paul Popenoe, David Starr Jordan**, and **Samuel J. Holmes**. Under Popenoe, the foundation's research coordinator, it compiled statistics concerning eugenic sterilization in California over a twenty-five-year period. One study examined the mental, physical, and social effects of sterilizations from July 1909 through 1929, on more than 6,000 "insane and feeble-minded wards of the state." The foundation sponsored publication of numerous tracts and articles in support of eugenic sterilization laws over the later years of the eugenics movement. Its most noted works were a book for the general public, *Sterilization for Human Betterment* (1929), and a collection of scientific papers upon which the popular book was based,

Collected Papers on Eugenic Sterilization in California (1930). It rivaled the **Eugenics Record Office** and the **American Eugenics Society** in terms of influence both in the **United States** and abroad, particularly in **Germany**, where its publications helped shape **Nazi Germany**'s sterilization program in the 1930s. The foundation ceased with the death of Gosney, its founder, in 1942. Its assets were liquidated by his daughter and the proceeds were donated to the California Institution of Technology to establish a research fund.

FURTHER READING: Engs, Ruth Clifford, *The Progressive Era's Health Reform Movement* (2003); Pickens, Donald, *Eugenics and the Progressives* (1968).

HUMAN GENOME PROJECT (1990–2003)

The collective name for several research programs, the Human Genome Project (HGP) "sequenced," or created an ordered set of **DNA** (the genetic code) for the 3 billion pairs of nucleic acids that make up the human **genome** (blueprint for creating an organism). Other aspects of the project included the identification of the approximately 30,000 **genes** (found to be 25,000 in 2004) in human DNA, the storage of collected information in databases, and the dissemination of information to the public. Begun in 1990 and completed in 2003, the project was coordinated by the U.S. Department of Energy and the National Institutes of Health with the National Center for Human Genome Research overseeing the whole effort. Numerous universities and research facilities throughout the United States and other countries including the United Kingdom, France, Germany, Japan, and China also participated. A major force behind this massive study was **James Watson**, director of the **Cold Spring Harbor** Laboratory (CSHL) on Long Island, New York, the home of the eugenics movement in the early twentieth century.

A goal of the HGP was to identify and characterize the genes involved in the most common **genetic diseases**. The project was also aimed at determining those diseases that have a genetic component, such as diabetes, schizophrenia, and Alzheimer's disease, in which a gene or genes creates a predisposition to the illness. Because these and other chronic diseases often cause severe health and social problems, early diagnosis of predisposition could lead to prevention of the condition through changes in lifestyle, medications, **gene therapy**, or other methods. From its beginning, the HGP was fraught with controversy and opposition. Concerns included privacy issues and confidentiality of genetic information gained from **genetic screening and testing**; fairness in use of genetic information by insurance agencies, employers, schools and other organizations; and the psychological impact or discrimination due to an individual's genetic makeup. Because of the potential for changing the genome or DNA segments to cure diseases or change human characteristics, information derived from the HGP is a core aspect of the **new eugenics**.

FURTHER READING: Hawley, R. Scott, and Catherine A. Mori, *The Human Genome* (1999); Kevles, Daniel J., *The Code of Codes* (1992).

HUMAN HEREDITY AND RACE HYGIENE (*Menschliche Erblehre und Rassenhygiene*, 1921)

Written by botanical geneticist Erwin Baur (1875–1933), anthropologist **Eugen Fischer**, and physician **Fritz Lenz**, *Human Heredity and Race Hygiene* became the basic German eugenics text throughout the post–**World War I** and **Nazi German** eras. It went through five editions between 1921 and 1940. Volume one of the third edition, *Menschlichen Erblichkeitslehre*, was translated into English by Eden and Cedar Paul in 1931 as *Human Heredity*, with 734 pages. Its German editions were all published in Munich by J. F. Lehmann, publisher of *Archive of Race Hygiene* and medical, eugenic, and "Nordic supremacy" books. The second volume, *Human Selection and Race Hygiene (Menschliche Auslese und Rassenhygiene)*, was written by Lenz alone.

The first volume of the work was divided into five sections. Section one, "Sketch of the General Theory of Variation and Heredity," written by Baur, discusses genetic theory and focuses on plants and animals. Chapter titles include the "Phenomena of Variation" and "Effects of Inbreeding." Various genetic and inheritance mechanisms, including **Mendelian inheritance** are presented. The second section, "Racial Differences in Mankind," was written by Fischer, who discussed anthropological differences between populations with chapter titles such as "Variable Characters in Human Beings," "Racial Origins and Racial Biology," and "Description of the Races of Man." Sections three through five were written by Lenz and entitled "Morbific Heredity Factors," "Methodology," and "Inheritance of Intellectual Gifts." Chapters in these sections include "The Concepts of Health, Disease and Normality," "Methods for the Study of Human Heredity," and "Inheritance of Particular Talents." Basing his thoughts on works by French diplomat **Arthur de Gobineau**, American nativist **Madison Grant**, and German race theorist **Hans F. K. Günther**, Lenz described a hierarchy of races. Using statistical techniques developed by British statistician **Karl Pearson**, he concluded that temperaments and **intelligence** differ among **races**, with the most intelligent being **Jews** and northern Europeans and the least intelligent, **Africans**. From **family pedigree** and **twin studies**, Lenz proposed that talent and **degeneracy** tended to run in families and rejected environment as being a major factor in shaping human characteristics. A review of the English translation, by American geneticist **Hermann Muller**, considered the section written by Baur to be scientifically accurate but found the ones written by Fischer and Lenz to be less objective.

The second volume was not translated into English. The 1923 edition had 368 pages and was divided into two parts, "Selection in People" and "The Practice of Race Hygiene." Chapter titles in the first section include "Biological Selection" and "Social Selection." They cover the harm to the human race by diseases such as **tuberculosis** and syphilis, as well as **racial poisons** such as **alcohol**, **tobacco**, and **war**. In the second section, chapter titles include "Societal Race Hygiene" and "Personal Race Hygiene." Lenz discussed methods for combating **venereal disease** and **alcoholism**, methods for cutting defective lines of descent such as **eugenic segregation** of the **feebleminded**, preventing marriage among the **unfit**, and encouraging the medical profession toward preventive medicine. He discussed ways a person could live eugenically and proclaimed that the duty of fit youth was to select a healthy mate and to produce many children. The theme of improving the biological efficiency of the nation was the major thread found throughout the German eugenics movement until its decline in the 1940s.

FURTHER READING: Weiss, Sheila Faith, *Race Hygiene and National Efficiency* (1987).

HUNTINGTON, ELLSWORTH (September 16, 1876–October 17, 1947)

A geographer and eugenicist, Huntington researched and wrote on the effect of climate on human development and the progress of civilization. He considered both environmental influences and biological inheritance important (**nature-nurture**) for human development and evolution. Born in Galesburg, Illinois, he was one of six children of a Congregationalist minister from a colonial **Anglo-Saxon** family. The family moved several times when he was a child and settled in Milton, Massachusetts, in 1889. Huntington graduated from Milton High School (1893) and received a B.A. from Beloit College, Wisconsin (1897). He taught and was assistant to the president of Euphrates College in Turkey, where he engaged in geographical fieldwork. In 1901 Huntington returned to the United States and received an M.A. (1902) in physiography from Harvard College. The following year he joined an expedition to the Middle East (1905–1906) and then returned to Harvard (1906–1907), where he wrote *The Pulse of Asia* (1907). In 1907 he went to Yale University as a geography instructor and received a Ph.D. (1909). He went on an expedition funded by the **Carnegie Institution of Washington** (1910–1914), served in the military during **World War I**, and returned to Yale in 1919. He remained at Yale until his retirement in 1945. Over his career he participated in many expeditions around the world and wrote numerous travel books.

Based upon his travels, Huntington suggested that Europeans were "superior" to other **races** and cultures because "heat, cold, storms . . . [shaped] racial morphology, physiology and behavior" through **natural selection**. He proposed that for civilization to advance a temperate climate was necessary because **intelligence** increased and cultures advanced as humans learned to cope with the environment. Huntington argued that all "high" civilizations were found in temperate climates, and as civilization progressed, it moved toward colder climatic regions. These ideas were presented in numerous publications, including *Civilization and Climate* (1915), *World-Power and Evolution* (1919), *Principles of Human Geography* (1920), and *The Character of Races* (1924).

Huntington argued that the rapid increase of the "less able members of the human species" and immigrants from **eastern and southern Europe** were causing **racial degeneracy** leading to the decline of democracy in the United States. He supported **immigration restriction laws** and enjoyed a close association with **nativist** eugenicists **Madison Grant** and **Lothrop Stoddard**. Huntington was a major force in the **American Eugenics Society** for twenty-five years. He served on its advisory council and as its president (1934–1938). In conjunction with the secretary of the society, **Leon Whitney**, he wrote a eugenics catechism, *Tomorrow's Children: the Goal of Eugenics* (1935). Over his career he wrote 28 books, parts of 29 others and more than 240 articles, many which explored the rise and fall of civilizations. He attempted to synthesize his life's work into two volumes. The first, *Mainsprings of Civilization* (1945), was considered controversial, and he died before completing the second volume to defend his position against those who attacked him as an "environmental determinist." Huntington married Rachel Slocum Brewer (1917), with whom he had two sons and one daughter. He died at home of a heart attack.

FURTHER READING: Martin, Geoffrey J., *Ellsworth Huntington* (1973).

I

To wrangle over the question of which is the more important, heredity or environment, is about as idle as . . . which is the more important, the stomach or something to put in the stomach. Man would soon come to grief without either.

Michael F. Guyer, *Being Well-Born* (1916)

IMMIGRATION RESTRICTION LAWS OR MOVEMENT

A movement to restrict immigrants from certain countries emerged in the United States in the 1880s and peaked in the late 1920s. **Nativist**, eugenic, anti**alcohol**, and **public health** interests, in addition to political and economic factors, helped drive the crusade to restrict immigration. Public opinion on immigration was ambivalent. It wavered between a desire for cheap immigrant labor and nativist demands to keep the country free of aliens, who were perceived as the major cause of crime, poverty, disease, and the **degeneration** of old-stock **Anglo American Protestant** values and bloodlines. Increased numbers of **Asians** and **eastern and southern Europeans** came to North America in the post–Civil War era in search of better jobs and opportunities. On the West Coast in the 1880s, out of concerns over crime and lack of assimilation, laws against Chinese immigrants were instituted, and later a gentlemen's agreement in 1907, negotiated between President Theodore Roosevelt and the government of Japan, restricted Japanese immigration. The major focus of the immigration restriction movement in the first decades of the twentieth century, however, was immigrants from eastern and southern Europe, most of whom were **Catholic** or **Jewish**. Until the 1840s most immigrants to the United States tended to be **Protestants** from Britain or northern Europe. They arrived as families, settled on farms, and soon assimilated as Americans. The pattern changed in the pre–Civil War era when poor Irish Catholic immigrants crowded into eastern urban slums, bringing Asiatic cholera in their wake. In the 1890s unskilled young males from southern and eastern Europe gravitated to urban ghettos. Accustomed to a tightly knit, highly structured home country where social status was fixed, these newcomers often found it difficult to adapt to U.S. social mobility and self-reliance. Likewise, the customs of the newcomers were alien to middle-class Protestant values. In the eyes of nativist and other reformers these immigrants were illiterate peasants, making the United States a "dumping ground for the refuse of Europe." Because the new immigrants seemed to resist becoming Americanized, the Anglo-American establishment registered increased animosity toward them.

Fueled by labor and civil unrest, and corrupt politics, middle-class Americans feared a Bolshevik-style political takeover from Russian Jewish immigrants or a papal

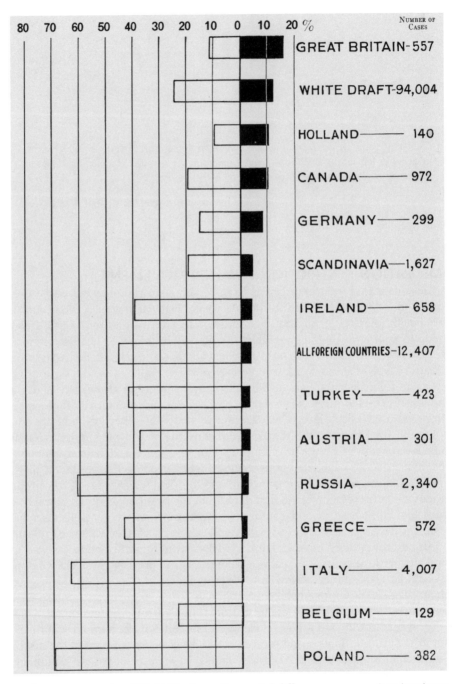

Intelligence tests results of World War I recruits reported differences among various immigrant groups. However, cultural and environmental factors were likely the cause of these differences. (From: *Journal of Heredity* [1922], Main Library, Indiana University, Bloomington, Indiana. Image courtesy of the Digital Library Program.)

takeover from Irish, Italian, and Polish Catholics. In addition, some eugenicists were concerned that these aliens would dilute the genetic makeup of the nation and lead to **racial degeneracy** and **race suicide** of old-line American stock. Between 1914 and 1919 **Harry H. Laughlin**, of the **Eugenics Record Office**, in a series of surveys of state institutions for the **feebleminded, insane**, and **criminals**, as well as leaders of institutions such as **tuberculosis** sanitariums, found that a disproportionally large number of inmates came from southern and eastern Europe, compared with those of northern European stock. **Intelligence tests** of **World War I** recruits showed eastern and southern European immigrants to have lower **IQ** scores than northern Europeans. Results of such studies implied that the new immigrants were harmful to the strength and vitality of the American people. Laughlin's studies and his testimonies before the House Committee on Immigration contributed to passage of the **Johnson-Reed Immigration Restriction Act of 1924**, which terminated the United States' traditional open-door immigration policy. In 1965 the U.S. Congress rescinded the act, allowing millions from Asia and other countries to come to U.S. shores. By the early twenty-first century concerns about immigrants bringing diseases such as HIV or **tuberculosis** to the United States or engaging in terrorist attacks renewed the call for immigration restriction.

FURTHER READING: Burt, Elizabeth V., *The Progressive Era* (2004); Haller, Mark H., *Eugenics* (1984); Kraut, Alan M., *Silent Travelers* (1994); Markel, Howard, *Quarantine!* (1997); McClain, Charles J., *In Search of Equality* (1994).

IN VITRO FERTILIZATION

The process of collecting sperm and eggs, fertilizing the eggs with the sperm, selecting healthy embryos, and implanting them into the uterus is *in vitro* ("in glass") *fertilization* (IVF). This method of **assisted reproduction** was proposed by American geneticist **Hermann Muller** in 1935 as "ectogenesis"—fertilization of an egg outside the body based upon an idea suggested by British biologist J.B.S. Haldane (1892–1964). As a method of **positive eugenics**, Muller suggested that an embryo resulting from "superior" sperm and eggs could be transplanted from one woman into another so that professional women need not be burdened with child-bearing. The first successful IVF was accomplished with a rabbit in 1954. The first human birth by this method using the procedure was **Louise Brown**, in 1978 in Britain. The first successful birth in the United States was in 1981. The term *"test-tube" baby* had been used for several decades to mean **artificial insemination**, but in popular culture its meaning has changed to refer to a child born by IVF. The technique caused controversy as some religious groups, most notably the Roman **Catholic Church**, questioned its ethics and morality. In 1983 the first human embryo from one woman was successfully transferred into another. By the late 1980s the use of donor eggs was a routine procedure that allowed numerous women whose own eggs were not viable due to age, **genetic disease**, or other factors to bear children.

In the IVF process, women are given hormones to stimulate the ovary to produce several eggs. The eggs are extracted and then mixed in a petri dish with sperm from her partner or a donor. In some cases a sperm is injected directly into eggs. Preimplantation diagnosis of the fertilized eggs (zygotes) for genetic diseases or chromosomal defects is often carried out. After several cell divisions two or three healthy embryos are implanted into the uterus of the original woman or into a surrogate woman (in "carrier gestation") to carry the pregnancy to term. IVF is generally used for a woman with blocked, severely damaged, or absent fallopian tubes. However, fewer than 5 percent of infertile couples or single women desiring pregnancy use the procedure. The technique can increase a woman's odds of having multiple births and, in some cases, children with birth defects. IVF is also used to select for gender (preimplantation gender diagnosis) for **family balancing** when a couple wishes specifically to have a male or a female child. The potential of **genetic therapy**, **genetic engineering**, and selecting for specific genetic characteristics in the embryo resulting from IVF is a major aspect of the **new eugenics**.

FURTHER READING: Edwards, R. G., and Patrick C. Steptoe, *A Matter of Life* (1980); Henig, Robin Marantz, *Pandora's Baby* (2004); Singer, Peter, and Deane Wells, *Making Babies* (1985); Vercollone, Carol, Heidi Moss, and Robert Moss, *Helping the Stork* (1997).

INHERITED REALITIES

A concept of the **Jacksonian Era** hereditarian movement, *inherited realities* suggested that mental, moral, intellectual, and physical characteristics were transmitted to offspring. The term began to be used in the United States in by the 1840s. The idea was also found in Europe, in the concepts of **degeneration** and degeneracy theory. Besides physical characteristics, it was proposed that certain tendencies or predispositions, as well as diseases including **tuberculosis, insanity**, heart disease, and gout, were inherited. American health reformers suggested that practices like the use of **alcohol** and **tobacco**; eating hot, spicy foods; and **masturbation** could produce constitutional weaknesses in offspring that, in turn, could be passed on to successive generations—**Lamarckian inheritance of acquired characteristics**. Health reformers **William Alcott** and **Elizabeth Blackwell**, among others, suggested that a person could escape a hereditary predisposition for a disease if he or she obeyed the **laws of health**. These laws included proper diet and exercise and the avoidance of corsets, alcohol, meat, spices, and a dissipated life. Others contended that the "constitution of the offspring" could be improved by marrying a healthy mate free of disease, insanity, "idiocy," and **alcoholism**.

FURTHER READING: Engs, Ruth Clifford, *Clean Living Movements* (2000); Money, John, *The Destroying Angel* (1985); Whorton, James C., *Crusaders for Fitness* (1982).

INQUIRIES INTO HUMAN FACULTY AND ITS DEVELOPMENT (1883)

Written by British scientist **Francis Galton**, *Inquiries into Human Faculty* defines and sets forth the principles of eugenics; *faculty* as used by Galton means *characteristics*. Published in London by Macmillan & Company, this 387-page book laid the groundwork for many social sciences and condensed research topics previously investigated and published by Galton. The purpose of the work was to "elicit the religious significance of the doctrine of evolution" as it applied to the human race. Based upon his cousin **Charles Darwin**'s theory of evolution, Galton held that **natural selection**, also held true for humans. He proposed that "what is termed in Greek, *eugenes*, namely, good in stock, hereditarily [is] endowed with noble qualities. This, and the allied words, *eugeneia*, etc., are equally applicable to men, brutes, and plants."

The publication shows Galton's exceptional range of interests. Short subject titles include "Variety of Human Nature," "Bodily Qualities," "Energy," "Anthropometric Registers," "Statistical Methods," "Criminals and the Insane," "Intellectual Differences," "Nurture and Nature," "History of Twins," "Domestication of Animals," "Objective Efficacy of Prayer," "Selection and Race," "Influence of Man upon Race," "Population," and "Early and Late Marriages." Many of these research topics became integral to the theory and practice of eugenics, including **twin studies**; the **nature-nurture** debate concerning **intelligence**, **insanity**, and other human characteristics; statistical methods; differences among **races**; sociological and psychological measurements; and **intelligence testing**. Because his main premise was that heredity is more important than environment in the formation of human characteristics, he considered eugenics critical for human survival.

FURTHER READING: Gillham, N. W., *A Life of Sir Francis Galton* (2001); Keynes, Milo, (ed.), *Sir Francis Galton* (1993).

INSANITY OR THE INSANE

Mental illness, commonly referred to as *insanity* until the latter half of the twentieth century, was also termed *lunacy, madness,* and *unsoundness of mind*. An individual was referred to as "queer," "peculiar," "strange," or "odd" with less severe forms of mental health problems. From the earliest times it had been noticed that schizophrenia, depression, affective disorders, autism, **alcoholism**, and other mental health problems "ran in families." In 1857 Benedict Augustin Morel (1809–1873), a French physician, suggested that many illnesses, including insanity, were inherited and caused by the process of **degeneration**. He proposed that, based upon **Lamarckian** theory of acquired characteristics, individuals leading dissipated lives could pass undesirable traits to their offspring. Because studies of "degenerate families" in the early twentieth century often showed mental illness in more than one generation, eugenicists reasoned it was inherited. Consequently, they suggested that the insane be segregated during their child-bearing years or undergo **eugenic sterilization** to prevent passing their defect to generations of descendants.

Mental illness is found in roughly 1 to 3 percent of the population and has considerable social costs in terms of treatment, welfare, and loss of production. Research since the mid-twentieth century has suggested that vulnerability to mental illnesses has a **genetic** component. Predisposition to an illness is probably caused by several **genes** acting together or being switched on, or off, by certain environmental factors. Evidence for a genetic component to mental illness comes from **twin studies**. Research has suggested that if one of a pair of identical (monozygot) twins has a mental health problem, there is a 50 to 60 percent chance that the other twin will also develop the condition, compared with a 15 percent chance in nonidentical twins. Other studies have found that children of a parent with mental illness are five to ten times more likely to develop mental health problems compared with the general population. Mental illness also has been linked to creativity. It is more prevalent among notable mathematicians, novelists, painters, and other highly creative individuals than within the population at large. As **genetic screening and testing** of embryos for mental illness becomes perfected, the consequences of eliminating these embryos, who could also become highly creative people, will likely provoke considerable controversy and is an aspect of the **new eugenics**.

FURTHER READING: Friedman, Howard, *Encyclopedia of Mental Health* (1998); Littlewood, Roland, *Pathologies of the West* (2002); Reilly, Philip R., *Abraham Lincoln's DNA* (2000); Reilly, Philip R., *The Surgical Solution* (1991); Stock, Gregory, *Redesigning Humans* (2002).

INTELLIGENCE

The ability of an individual to adapt, learn, and reason is generally defined as *intelligence*. Other terms include *mental ability, mental functioning,* and *cognitive ability*. Indicators of intelligence are generally measured by standardized mental measurement tests. The belief that differences in mental ability are primarily inherited, can be measured, and that some socioeconomic or ethnic/racial groups are innately more intelligent than others became an important justification for **positive** and **negative eugenics** programs in the first half of the twentieth century. In 1904 Charles Spearman (1863–1945), a British statistician, coined the term *g factor* to refer to "general intelligence," which was considered innate cognitive ability measured by all **intelligence tests**. The measure of mental ability, popularized as **IQ**, was an important factor in the eugenics movement because it was believed to be largely inherited. This belief was reflected in programs for **eugenic sterilization** of the **feebleminded** and **immigration restriction laws** in the **United States**. However, over the course of the twentieth and into the twenty-first centuries, the nature of intelligence; whether it can be measured; and whether it is inherited, caused by environmental factors, or a combination of both remained under contentious debate.

Theories concerning the nature of intelligence appear to be cyclical and to coincide with the waning of social and health reform movements. As the **Jacksonian** (1820–1860), **Progressive** (1890–1920), and **Millennial** (1970– ?) health and social reform cycles declined, a genetic base for intelligence, other human characteristics, and social or moral behaviors gained ascendency. In 1869 **Francis Galton** proposed in *Hereditary Genius* that since intellectual capacity appeared to run in families, it must be inherited. By the turn of the twentieth century both heredity and environmental factors were considered important in the development of mental function and abilities. During the upsurge and crest of the eugenics movement (1910–1930), the genetic basis of intelligence was back in favor. From the 1950s through the 1980s environment was considered the major factor in intelligence. By the 1990s the importance of **genetics** reemerged. At the turn of the twenty-first century **twin studies** suggested that both **nature** and **nurture** were important in the development of mental ability. However, the whole area remains controversial.

FURTHER READING: Clark, William R., and Michael Grunstein, *Are We Hardwired?* (2000); Kerr, Anne, and Tom Shakespeare, *Genetic Politics* (2002); Pinker, Steven, *The Blank Slate* (2002); Richardson, Ken, *Making of Intelligence* (2000).

INTELLIGENCE OR IQ TESTS

Tests originally developed to measure the mental age of schoolchildren were coined *in-telligence* or *IQ* (for *intelligence quotient*) *tests* in the early twentieth century. The original test was a ratio between the mental and chronological age of children and measured the average mental level for various age groups. Although several standardized measurements were developed, the popular term for any method that measured mental ability became an *IQ test*. Measures of cognitive ability include verbal, mathematical, and spacial ability; "pencil and paper"; reaction time; reasoning; memory; problem solving and puzzles tests. There is a high degree of association between scores on one test and others.

Intelligence testing began in the late nineteenth century when British naturalist **Francis Galton** unsuccessfully attempted to construct a test to measure cognitive ability. By 1905 French psychologists Alfred Binet (1857–1911) and Théodore Simon (1873–1961) developed a measure that identified quick thinkers and slow learners among schoolchildren. **Henry Goddard** translated their test into English in 1908 and adapted it for children as a screening tool for mental disability. Lewis M. Terman

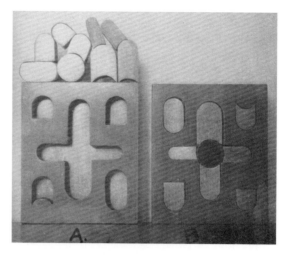

Various tests in the first third of the twentieth century were developed to determine the intelligence of children and adults. They included puzzles, problem solving, spacial relationships, mathematical, and "pencil and paper" tests. (From: *Journal of Heredity* [1914], Main Library, Indiana University, Bloomington, Indiana. Image courtesy of the Digital Library Program.)

(1877–1956) of Stanford University revised the measure, now called the Stanford-Binet, to determine the intelligence of all people, whether mentally disabled or not. Psychologist David Wechsler (1896–1981) also developed a measure for adults that is periodically revised and still in use.

During **World War I** the Army Alpha for English speakers and the Beta test for illiterate or non-English-speaking individuals were developed. These group exams were administered to nearly 2 million recruits for the U.S. Army. By 1917 Goddard had determined that "mentally defective" immigrants could be identified using intelligence tests. He claimed that a large proportion of **eastern and southern European** immigrants were **morons**. These testing results helped further the **immigration restriction movement** in the United States, as mental ability was assumed to be inherited. It also gave credence to **eugenic sterilization** laws to prevent the **unfit** from reproducing, and **positive eugenics** programs to encourage brighter individuals to produce more offspring. However, many social scientists of the era recognized that intelligence tests did not measure the "whole of intelligence." By the mid-1920s IQ tests were widely used in businesses, schools, and governmental agencies for job screening, promotions, educational placement, and college admissions. In the 1970s the use of intelligence tests for job and academic screening fell out of favor, particularly in the United States, out of concern by some that they were culturally biased. The usefulness of IQ tests, and even the meaning of intelligence, continues to be debated.

FURTHER READING: Kevles, Daniel J., *In the Name of Eugenics* (1985); Kline, Wendy, *Building a Better Race* (2001); Lynn, Richard, *Eugenics* (2001); Richardson, Ken, *The Making of Intelligence* (2000).

IQ (Intelligence Quotient)

A measure of mental ability or **intelligence**, the term *IQ* (intelligence quotient) was first coined in 1912 by German psychologist W. L. Stern (1871–1938) and was adopted in 1916 by American psychologist Lewis Terman (1877–1956). When the use of **intelligence tests** became commonplace in the 1920s, IQ became a shorthand phrase to express a person's mental ability. During the most active years of the eugenics movement, intelligence was considered primarily inherited and an individual's IQ test score became an important concept in the enactment of **positive** and **negative eugenics** programs. By the mid-twentieth century, the term *high IQ* referred to someone who was endowed with superior intellectual ability. Organizations formed to foster interaction—and potential reproduction—between highly capable individuals. These included **Mensa** and the Foundation for Germinal Choice. Researchers have noted that the average IQ has risen over the century—the *Flynn effect*—which some suggest might be due to better childhood nutrition.

Variations in IQ scores have been found among different groups since intelligence testing was introduced. During **World War I** Army IQ tests, used to classify recruits for assignments, found differences among socioeconomic classes and racial and ethnic groups. British and German immigrants had the highest scores and Italian and Polish immigrants the lowest. However, later studies suggested that these differences were likely due to culturally biased tests. In general, most intelligence tests in North America, and also within and between other nations and population groups, have shown higher IQ scores associated with more education, wealth, fewer children, and low **crime** rates. Lower scores are associated with little education, **pauperism**, poverty, many offspring, and high crime rates. Studies in the latter half of the twentieth century suggested that people with European ancestry have average IQ scores of 100, compared to 115 for those with eastern European **Jewish** ancestry, 103 for those with East **Asian** ancestry, 85 for those with sub-Saharan **African** ancestry in the United States, and 75 for people living in sub-Saharan Africa. However, other researchers have argued that the differences found among these groups are due to culturally biased tests, racism, lack of educational opportunity, poverty, poor nutrition, and other cultural and environmental factors. The debate concerning the nature of IQ continues in the first decade of the twenty-first century.

FURTHER READING: Herrnstein, Richard J., and Charles Murray, *The Bell Curve* (1994); Jacoby, Russell, and Naomi Glauberman, *The Bell Curve Debate* (1995); Kerr, Anne, and Tom Shakespeare, *Genetic Politics* (2002); Lynn, Richard, *Eugenics* (2001).

J

Great pains are taken to improve our breeds of horses and sheep. . . . is it not absurd for any one [sic] to advance the opinion, that it is too delicate a subject to improve the human race.

Lorenzo Fowler, *Marriage: Its History and Ceremonies* (1850)

JACKSONIAN ERA (1830–1860)

The U.S. Jacksonian Era, sometimes called the "era of the common man" or the Antebellum Reform Era, was characterized by distrust and disillusionment with practicing physicians, clergymen, and politicians. This sentiment, in turn, led to the rise of new health, religious, and political sects that sought to bring the United States to a "golden age," free of **crime**, poverty, drunkenness, and disease. The reforming impulse developed out of the religious fervor of the second **Great Awakening**. Feelings of pietism and belief in the "perfectibility of man" led to efforts to eliminate the evils of society and bring on the Millennium, or the reign of Christ on earth. A **Clean Living Movement** (1820–1860) to eliminate health problems, with moral undertones, evolved out of this religious revivalism. Anti-**alcohol** and **tobacco** agitation, attention to personal hygiene, diet reform, and botanic cures appeared early in this reform era, while hydrotherapy (water cure); **inherited realities**, or hereditarian concerns; women's rights; sanitation; and **public health** issues became more prominent nearer the end of this period. Health reformers including **William Alcott**, Sylvester Graham (1794–1851), **Orson Fowler**, **Lydia Folger Fowler**, and **Elizabeth Blackwell** advocated obeying the **laws of health** and acknowledging inherited realities in an effort to improve the quality of the human race. Religious communities like the **Onieda Community of Perfectionists** and the **Church of Jesus Christ of Latter-day Saints** encouraged practices to produce sturdy offspring. In addition, middle- and upper-class Americans feared that impoverished Irish **Catholic** immigrants who poured into the urban slums beginning in the 1830s and lived amid squalor and disease would thwart the Millennium. This fear led to **nativism**, the formation of the Know-Nothing Party, and efforts to stem the rising influence of the foreign-born on the American way of life.

FURTHER READING: Engs, Ruth Clifford, *Clean Living Movements* (2000); McLoughlin, William G., *Revivals, Awakenings, and Reform* (1978); Whorton, James C., *Crusaders for Fitness* (1982).

JEWS OR JEWISH AND JEWISH AMERICANS

Descendants of the Hebrew people, Jews share a common culture, religion, history, and religious language and trace their roots to the Middle East. Jews are generally classified into three groups: Sephardic, Oriental, and Ashkenaszi. The Ashkenaszi are the largest group, comprising 82 percent of the total Jewish world population. Conflict between Jews and other religious groups in Biblical times, along with Christians blaming Jews for the death of Jesus, have led to intolerance and persecution of Jews over the past 2,000 years. More recent anti-Semitism in Europe had its origins in **Arthur de Gobineau**'s **hierarchy of races** theory and rising "**Nordic** supremacy" ideology during the latter half of the nineteenth century. At that time many European Jews had become successful in business, science, medicine, politics, and other areas. Because of their success and because they often did not mix socially with non-Jews, they were seen by some non-Jews as being power- and money-hungry, manipulative, and conspiring to control world finances, the press, academia, and the arts. This attitude was known as "the Jewish problem." In Germany the Nordic supremacy, **race hygiene** (eugenics), and anti-Semitic movements converged in the 1920s under rising National Socialism. The **Nazi German** regime was obsessed with preventing the intermixing of Jewish and Nordic, or blond, blue-eyed **Aryan** "blood." Sexual activities and marriage between the two **races** were made illegal in 1935. During **World War II** the genocide of 6 million European Jews by the Nazis in their effort to keep the Germanic race pure became known as the Holocaust.

In the United States the Jewish population greatly expanded during the **Progressive Era** (1890–1920). In 1880 U.S. Jews numbered about a quarter of a million. Most were of German origin, well established, and relatively prosperous. The majority practiced Reform Judaism, the more liberal branch of the religion, and had more or less assimilated into American culture. Between 1880 and 1914 approximately 2.5 million impoverished Russian, Polish, and other eastern European Jews emigrated to the United States. They tended to be of the Orthodox, or more conservative, branch of the religion. Most of them flooded into the already overcrowded city tenements, particularly in New York City. They did not readily assimilate and kept to themselves. In 1892 Jewish immigrants were blamed for bringing typhus and a cholera epidemic to New York City. Some of these immigrants were radical Bolsheviks (Communists). Fear of disease and a possible Communist takeover of American democracy and the corruption of traditional rural **Protestant** values helped spawn **nativist** and **immigration restriction movements** that culminated in the **Johnson-Reed Imagination Restriction Act of 1924**, designed to curtail immigrants from eastern Europe.

Jews were active in the early-twentieth-century eugenics movement on both sides of the Atlantic. In **Germany** Jews belonged to the **German Society for Race Hygiene** but increasingly withdrew from the society in the 1920s. Jews were also involved in **genetics** research in the pre-Nazi era. In the United States, liberal Jews supported **birth control**, eugenics, **social hygiene**, and other health-reform efforts of the Progressive Era. Because several **genetic diseases**, such as Tay Sachs disease, in which children are profoundly neurologically damaged and die at an early age, are found almost exclusively among European Jews, Jewish public health and social welfare professionals were interested in genetic information in an effort to prevent these inherited conditions. In the latter part of the twentieth century leaders of the Jewish community supported **genetic screening and testing** programs to identify carriers of genetic diseases. Jews, in contrast to many conservative **Protestants** or Roman **Catholics**, have supported **in vitro fertilization**, **abortion**, birth control, therapeutic **cloning**, and other practices of the **new eugenics**.

FURTHER READING: Carlson, Elof Axle, *The Unfit* (2001); Jacobson, Matthew Frye, *Special Sorrows* (1995); Kraut, Alan M., *Silent Travelers* (1994); Markel, Howard, *Quarantine!* (1997).

JOHNSON, ROSWELL HILL (October 9, 1877–January 17, 1967)

A biologist and geologist, Johnson in later life became a sociologist and marriage counselor. He was an early proponent of eugenics, **birth control**, and sex education. Born in Buffalo, New York, the son of an oil producer and a descendant of old-stock New England families, he attended Brown and Harvard Universities in the late 1890s and graduated with a B.S. from the University of Chicago (1900) and an M.S. from the University of Wisconsin (1903). He taught for several years and worked as a research assistant at the **Carnegie Institution of Washington**'s Station for Experimental Evolution in **Cold Spring Harbor**, New York (1905–1908) under **Charles Davenport**, the leading spirit of the eugenics movement; he also studied geology part time at Columbia University. After working as a consulting oil geologist, he joined the University of Pittsburgh's School of Mines in 1913 and taught geology until 1933, when the department was abolished. He moved to Hawaii, and then to Los Angeles in 1936 to become director of the counseling department of the Institute of Family Relations (headed by eugenics supporter and counselor **Paul Popenoe**) until his retirement in 1959.

Simultaneously with his geological career, Johnson became a pioneer of the eugenics movement and helped popularize the subject in the college setting. He was a member of the **American Breeders Association**'s **Eugenics Section** created in 1906. At the University of Pittsburgh he organized in 1912–1913 one of the first university eugenics courses and advocated eugenics as part of the college curriculum. In 1914 he established a lifelong friendship with Popenoe, editor of the *Journal of Heredity*. Five years later the two men published *Applied Eugenics* (1918), which became a popular classroom text. From 1929 to 1931 Johnson gave a well-liked lecture series for the **American Eugenics Society**, "Some Problems in Eugenics," to help stimulate interest in eugenics courses in colleges and universities around the country. He took active leadership roles in several eugenics organizations and conferences, including the **First National Conference on Race Betterment** (1914) and the **Second International Congress of Eugenics** (1921). He was a founding member of the **American Eugenics Society** (AES), its second president (1926–1927), and a director for a number of years. Johnson was on the editorial board of *Eugenics* (1929–1931), the official journal of the AES. He was also involved with the **Eugenics Research Association**. Johnson supported methods of both **positive** and **negative eugenics** and argued that "our most pressing problem is to increase the birth rate from the superior and to decrease that from the inferior." He promoted **social hygiene** education to prevent **venereal diseases**. In his later years he published numerous works on marriage counseling and was considered an expert in the area. Over his lifetime he had three wives and fathered five children. His marriages were to Mary Simmons (1900), Mary Brenk (1937), and Lois Blakey. He died in Los Angeles, at his home.

FURTHER READING: Allen, Garland E. "The Eugenics Record Office at Cold Spring Harbor, 1910–1940" (1986); Paul, Diane B., *Controlling Human Heredity* (1995).

JOHNSON-REED IMMIGRATION RESTRICTION ACT OF 1924

The culmination of a series of U.S. state and federal laws that had begun in the 1880s, the Immigration Restriction Act of 1924 established a quota system based upon country of origin and reversed the United States' traditional open-door immigration policy. In the immediate pre– and post–**World War I** years, a rapid influx of **Catholic** and **Jewish** immigrants poured into the country from **eastern and southern Europe**, triggering a surge of **nativism** among **Anglo Americans**. Nativist, prohibition, **public health**, and eugenic interests called for restriction of these "undesirables." To address the "immigration problem," Albert Johnson, chair of the House Committee on Immigration and Naturalization, was charged with introducing a new immigration restriction bill. Leaders of the nativist branch of the eugenics movement including **Madison Grant** and **Harry H. Laughlin** gave input into this process. However, there are contested views as to the significance of their roles in the passage of the act. Laughlin testified as an expert witness in 1920. He argued, based upon his 1914 survey of **feebleminded** residents of state institutions, that immigrants from southern and eastern Europe were hindering rather than helping U.S. society as higher proportions of immigrants, compared with native-born Americans were institutionalized. In 1921 Congress passed and President Warren Harding signed a bill that created a quota system. It limited the annual immigration of people from each European country to 3 percent of the total foreign-born persons from that respective country in the 1910 U.S. census. This law was passed as a temporary measure. After public hearings, more surveys, and compromises, Senator David A. Reed's "national origins plan" was overwhelmingly passed by both houses. Signed by President Calvin Coolidge on May 26, 1924, the measure shifted the base year for the immigration quota from 1910 to 1890, favoring northern European immigrants. Due to various delays, the law did not take effect until July 1, 1929; it remained in effect until 1965.

FURTHER READING: Burt, Elizabeth V., *The Progressive Era* (2004); Higham, John, *Strangers in the Land* (1955); Jacobson, Matthew Frye, *Special Sorrows* (1995); Kraut, Alan M., *Silent Travelers* (1994).

JORDAN, DAVID STARR (January 19, 1851–September 19, 1931)

A renowned biologist and educational leader, Jordan was a powerful and pivotal leader of many health reform movements of the **Progressive Era** (1890–1920). In particular, he was a leader of the eugenics movement and integrated eugenic concerns into other health and social causes. Jordan, the fourth of five children, was born in Gainesville, New York, the son of two teachers from proud old-stock **Anglo American** families. As a child he was educated at home and in a local ungraded school. At fourteen he enrolled at Gainesville Seminary, and at seventeen he taught in a village school near home for one term before entering Cornell University. His knowledge of biology was so advanced that he was appointed an instructor in his junior year and graduated with an M.S. (1872) instead of a B.S. He spent summers at the Penikese, New York, biological station. For most of his adult life he generally spent each summer collecting fish and other specimens from various parts of the world. Following graduation he held a series of short teaching positions and received an M.D. from Indiana Medical College (1875). In 1879 he became head of biological science at Indiana University in Bloomington and taught until 1885, after which he served as university president (1885–1891). Subsequently he became the first president of the recently established Leland Stanford Junior University, in Palo Alto, California (1891–1913) and built this institution into the world-renowned Stanford University. Over his lifetime Jordan earned an international reputation in ichthyology (the study of fishes) and won many honorary awards. However, at the turn of the twenty-first century some evidence suggested that he may have covered up the possible murder of the primary benefactor to Stanford University.

A pioneer of the eugenics movement, Jordan held leadership positions in the major eugenics organizations. In 1906 he chaired the eugenics committee of the **American Breeders Association**, renamed the **American Breeders Association Eugenics Section** in 1910. He remained its chair, in addition to chairing the board of the **Eugenics Record Office**. He was president of the board of the **Eugenics Registry**, which evolved from the **Second**

David Starr Jordan, who was president of Indiana University and Lealand Stanford Junior University (later Stanford University), was a pioneer leader of the early eugenics movement. (Courtesy Indiana University Archives, Bloomington, Indiana.)

National Conference on Race Betterment in 1915, and was an early member of the **American Eugenics Society**. Jordan supported **eugenic sterilization**, **segregation**, and **marriage-restriction laws**. He feared **race suicide** in the Western nations as a result of **war** due to the fittest men being drafted while the **unfit** remained at home to reproduce. Jordan wrote many eugenics articles and books including *The Blood of the Nation* (1902) and *The Heredity of Richard Roe* (1911). His **nativistic**, pro-American leanings intertwined with eugenics and led to his opposition to unrestricted immigration. He promoted **social hygiene** education to prevent **venereal diseases** and considered their transmission a "hideous and dastardly crime." In his latter years Jordan became a crusader for international peace. He was a prolific writer, publishing over a thousand articles and books on a variety of subjects, including a two-volume autobiography, *The Days of a Man* (1922). He married Susan Bowen in 1875. She died ten years later and in 1887 he married Jessie Knight. The two marriages yielded six children. Jordan died of a stroke related to heart disease and diabetes at his home at Stanford.

FURTHER READING: Carlson, Elof Axle, *The Unfit* (2001); Cutler, Robert W. P., *The Mysterious Death of Jane Stanford* (2003).

JOURNAL OF HEREDITY (*American Breeders Magazine* 1910–1914; 1914–present)

First published in 1910 as the *American Breeders Magazine, Journal of Heredity* was one of the first U.S. journals to include eugenics as a scientific field of study. It served as the official organ of the **American Breeders Association** and was not a major popularizer of eugenics as it was aimed at professionals and published scientific articles concerning genetic research. An initial purpose of the periodical was to bring new information concerning the developing field of **genetics** to plant and animal breeders. By the second volume the subtitle of the journal was *A Journal of Genetics and Eugenics*. Articles in its first four years included a range of topics, for example, "Increasing Protein and Fat in Corn," "Inheritance of 'Acquired Epilepsy,'" "A Theory of Mendelian Phenomena," and "Twins, Heredity, Eugenics." Because many members were not livestock breeders and to reflect the growth of eugenics and genetics, the association voted in November 1913 to change its name to the American Genetic Association, and that of its official publication to *Journal of Heredity*. Eugenics advocate **Paul Popenoe**, who later became a director of the **Human Betterment Foundation**, was its first editor. After Popenoe left, subsequent editors were genetics researchers. The new journal (vol. 5) was larger in size, and its subtitle was changed to *A Monthly Publication Devoted to Plant Breeding, Animal Breeding, and Eugenics*. It contained many articles concerning plant and animal improvement, such as "Inbreeding in Dogs" and "Alfalfa Hybridization," as well as topics in human genetics and eugenics, such as "Skin Color of Mulattoes" and "Eugenics and Genius."

During its first two decades of publication, the journal published eugenics articles and kept readers in touch with the eugenics movement in **Britain** and **Germany**. Its articles, likewise, were often abstracted by the British journal *Eugenics Review* and the German journal *Archive for Racial and Social Biology*. In 1919 the journal devoted an entire issue to **twin studies** in an effort to determine the role that heredity and environment (**nature-nurture**) played in human development. By 1923 drawings of chromosome patterns in fruit flies and, later, other organisms appeared in the periodical. Research was also published concerning the inheritance of certain human behaviors such as **pauperism**, wanderlust, and **criminality**. However, these complex behaviors were found not to be the result of simple **Mendelian inheritance** principles, and the findings began to weaken the underpinnings of the eugenics movement. By the 1930s the Journal shifted away from eugenics to genetic research. As the field of genetics changed over the course of the twentieth century and new research avenues were explored, the journal reflected these changes. By the early twenty-first century the *Journal of Heredity* included articles on rapidly advancing fields such as genomics and the **genome** (blueprint of an organism), **gene** mapping, animal models of human disease, **genetic engineering** in plants and animals, conservation genetics of endangered species, molecular evolution, and genetic epidemiology. By the twenty-first century the journal had established an electronic version on the Internet.

JUKES, THE: A STUDY IN CRIME, PAUPERISM, DISEASE, AND HEREDITY (1877)

The first detailed study to explore the possible inheritance of antisocial behaviors among a kinship group, *The Jukes*, by social reformer **Richard Dugdale**, served as a model for **family history and pedigree studies** of the early-twentieth-century eugenics movement. This report and its follow-up almost forty years later, *The Jukes in 1915* (1916) helped to popularize eugenics and became symbolic of the assumption that "bad breeding" led to **racial degeneracy** and a downward trend in human attainment and ability. Elisha Harris (1824–1884), a public health physician in New York State, discovered six generations of one family lineage that had produced criminals and paupers. Harris delivered his report to the Prison Association of New York State in 1874. Dugdale, an associate of Harris, was asked to inspect surrounding jails in pursuing the family's history. He inspected county and state records of this lineage of so-called criminals, whom he called the Jukes, taking the name from the term *juke*, for someone who did not like to be tied down. Members of the family were descendants of original settlers of the area. These uneducated subsistence farmers and hunters, who had become outcasts when the region grew popular among the middle class, became identified with **crime, prostitution, venereal disease**, and **alcoholism**.

Dugdale reviewed the history of 709 individuals with the Juke lineage. "Of 535 children born 335 were legitimate, 106 illegitimate and 84 of unknown" [*sic*] parentage. He suggested that fornication (sex outside of marriage) was the "backbone" of the social problems found within the family. Dugdale reported that 280 family members were **paupers** and 149 criminal offenders. In his conclusions he balanced heredity and environment (**nature-nurture**) as factors in explaining the family's high rates of social dysfunction. While children might inherit tendencies for a particular social problem, he suggested, their childhood environment could reinforce the trait. To prevent this, Dugdale argued for a change in environment including separating children from parents and placing them in good families or institutions in which they could receive industrial training and form good work habits.

The Jukes became so popular that it went through seven editions over the next twenty-five years. When it was first published it was considered an outstanding example of the scientific method applied to a sociological investigation. However, from the beginning it was subject to two different interpretations. One point of view suggested that environmental factors such as poor family backgrounds, lack of education, poverty, and disease could lead to antisocial behaviors. The second claimed that *The Jukes* was a eugenics tract in which poverty and crime were rampant in the family due to heredity. The hereditarian interpretation of the report became prevalent, and the study emerged as a template for other family-history investigations at the turn of the twentieth century, including *The Tribe of Ishmael*, *The Kallikak Family*, *The Hill Folk*, and *The Nam Family*.

FURTHER READING: Carlson, Elof Axle, *The Unfit* (2001); Dugdale, Richard Louis, *"The Jukes"* (1970); Haller, Mark H., *Eugenics* (1984); Kevles, Daniel J., *In the Name of Eugenics* (1985); Rafter, Nicole Hahn, (ed.), *White Trash* (1988).

JUKES IN 1915, THE (1916)

A followup investigation of **Richard Dugdale's** original study *The Jukes* was researched by **Arthur Estabrook**, a eugenics **field worker** at the **Eugenics Record Office** (ERO). Estabrook's report, *The Jukes in 1915*, helped to popularize the burgeoning eugenics movement. In 1911 Dugdale's original research notes were found in the basement of the offices of the Prison Association of New York State and were given to the ERO. In 1912 Estabrook began investigating the living descendants of the Jukes fam-

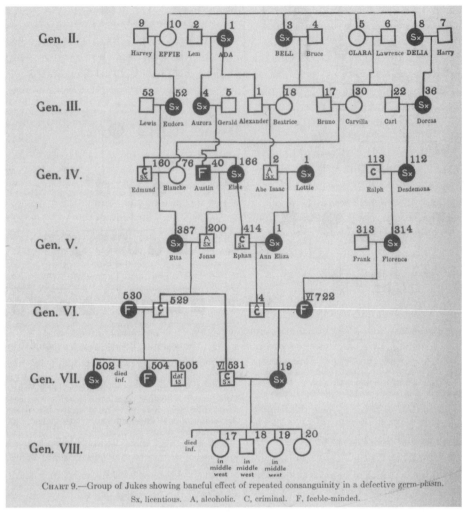

CHART 9.—Group of Jukes showing baneful effect of repeated consanguinity in a defective germ-plasm. Sx, licentious. A, alcoholic. C, criminal. F, feeble-minded.

The "Jukes" family pedigree with designated alcoholics, feebleminded, and criminals over several generations, along with other genealogy studies, provided a "scientific basis" for eugenic sterilization, segregation, and immigration restriction laws. (From: *The Jukes in 1915* [1916], Main Library, Indiana University, Bloomington, Indiana. Image courtesy of the Digital Library Program.)

ily who resided in fourteen states. He discovered that over 130 years the family had increased to a group of 2,094 people, of whom 1,258 were alive in 1915. He found among the kinship 366 paupers, 171 criminals, 277 "harlots," 282 "intemperates," 107 "mentally deficient," 660 with a sexually transmitted disease, and 1,650 earning low wages. With 378 children having died before age five, the mortality rate for that group was higher than the average population's. Estabrook argued that the financial costs to society from such a kinship were high in terms of support through social services. Based upon this study, he concluded that **feeblemindedness**, **criminality**, laziness, **alcoholism**, and **pauperism** were inherited but that a change of environment in some cases was beneficial inasmuch as "heredity, whether good or bad, has its complemental factor in environment . . . [however] the constitution must be adequate before we can attain the perfect individual, socially and eugenically." The report argued for **eugenics sterilization** or **segregation** (permanent custodial care) of feebleminded men and women of child-bearing age, in addition to sterilization of those with defects that society deemed should be eliminated. He noted that, of the approximately 600 living feebleminded and epileptic Jukes, only three were in custodial care. The **Carnegie Institution of Washington** published this study in 1916. The 85-page report became a standard reference for eugenics research and education supporting the hereditarian concept of social behavior. Over time the Jukes became the symbol of degenerate families and became symbolic of the importance of eugenic sterilization and institutionalization, in contrast to social welfare changes as a method to improve society.

FURTHER READING: Carlson, Elof Axle, *The Unfit* (2001); Estabrook, Arthur H., *The Jukes in 1915* (1916); Paul, Diane B., *Controlling Human Heredity* (1995); Rafter, Nicole Hahn, (ed.), *White Trash* (1988).

K

If man is not intelligent enough to take his own evolution in hand and direct it through wise personal selection according to the teachings of eugenics, then nature will do it for him in the same old crude, ruthless way; war, famine and pestilence will become the final arbiters.

Micheal F. Guyer, *Eugenics* I (October 1928)

KAISER WILHELM INSTITUTES, GERMANY (1927–1945; Max Planck Institutes, 1947–)

The leading organization for eugenics in Germany was the Kaiser Wilhelm Institute (KWI) for Anthropology, Human Heredity, and Eugenics (*Kaiser Wilhelm-Institut für Anthropologie, Menschliche Erblehre und Eugenik*) in Berlin. It was established as a research center to study the "nature of man" and integrate anthropological, social, and biological research with eugenics. Other institutes involved with eugenic research included the Kaiser Wilhelm Institute for Psychiatry in Munich, which studied the inheritance of mental illness, and the Institute for Heredity and Race Hygiene at the University of Frankfurt, established under the **Nazi German** regime. The KWI in Berlin corresponded to the **Eugenics Record Office** in the United States and the **Galton Laboratory** in Britain.

In 1911 the Kaiser Wilhelm Society was established by governmental and industrial officials interested in creating scientific research institutes in **Germany** supported by both private and governmental funding. In the post–**World War I** years several forces came together to create a national eugenics research institution in Germany. Leaders of the Weimar Republic government were concerned that the country had lost its fittest youth in the war. They were concerned that a decline in the birthrate and increased health and social problems were leading to a lack of fitness and efficiency in the German population. In 1925 Hermann Muckermann (1877–1962), a Jesuit biologist, social welfare reformer, and **race hygiene** supporter, lobbied to raise funds for a national institute to address these problems. A research institute modeled after the Kaiser Wilhelm Institute for Physics was approved in 1926 by the Ministry of Interior and the Kaiser Wilhelm Society.

In September 1927 the Kaiser Wilhelm Institute for Anthropology, Human Heredity, and Eugenics was opened in Berlin-Dahlem in conjunction with the International Congress of Heredity. The institute's headquarters was partially funded by the Rockefeller Foundation. At the opening celebration American eugenics leader **Charles Davenport** was the featured speaker. Anthropologist **Eugen Fischer** was named director and also head of the Anthropology Division. His research focused on the genetics of human characteristics. Muckermann headed the Department of Eugenics and studied "normal" and fit families and rural populations with low infant mortality rates. Medical geneticist Otmar Freiherr von Verschuer (1896–1969) led the Department of Human Heredity and focused on **twin studies** in order to determine the inheritance of mental illness, genetic diseases, **intelligence**, and cancer. Under Fischer, Verschuer, and Muckermann the institute also emphasized educating and training physicians in eugenics.

After the Nazis seized power in 1933 the Kaiser Wilhelm institutes were mandated to cooperate with the government in order to obtain funding. In the process they somewhat lost their independence as research centers. After the sterilization law of 1933 the institute provided expert witnessess to the sterilization courts, undertook **family history** (genealogical) profiles of the mentally ill, and identified characteristics to determine people of **Aryan** and other **races**. Twin studies continued in an effort to determine who should be sterilized in an effort to reduce the financial cost of caring for the **insane** and **feebleminded**. One of the initial founders of the KWI in Berlin, Muckermann, was considered too liberal by the Nazis and was dismissed from his position, which was filled by physician **Fritz Lenz**. When Verschuer was appointed in 1935 as director of the Third Reich Institute for Heredity and Race Hygiene at the University of Frankfurt, Lenz replaced him as head of the Department of Human Heredity in Berlin, which was combined with the Department of Eugenics. By the late 1930s physical anthropology and human genetics began to split into two separate fields, and the Berlin institute provided a link among race hygiene, anthropology, and human **genetics**.

Upon Fischer's retirement in 1942 Verschuer upon Fischer's recommendation was appointed director of the Berlin Institute, a position he retained until the end of the war in 1945. On Fischer's seventieth birthday, in 1944, the institute was renamed the Eugen Fischer Institute. From 1943 to 1945 Josef Mengele (1911–1979?), Verschuer's former medical student and assistant in Frankfurt, was posted to Auschwitz as camp physician, and sent specimens to Verschuer for his twin research studies. Many records and materials from the institute were deliberately destroyed at the end of the war leading to debate concerning what information the scientists had concerning the origins of these specimens or their awareness of the mass killings of **Jews** and others in the concentration camps.

In the postwar years, because the Kaiser Wilhelm Society (KWS) could no longer support the Berlin institute, it was taken over by the German Academy of Sciences. With British support the KWS was reorganized as the Max Planck Society (*Max Planck Gesellschaft*) in 1947. In 1948 the old Berlin KWI for Anthropology was renamed the Max Planck Institute for Applied Anthropology. Following the de-Nazification process many race hygienists were reinstated as geneticists to academic positions. Verschuer attempted to resume his power as director of the institute, but political pressure from other scientists prevented his reinstatement. He founded the Institute of Human Genetics in Münster to study the genetics of human diseases and taught until his death in 1969. Lenz obtained a position in the Department of Human Genetics at the University of Göttingen in 1946. Muckermann was given an appointment in anthropology at the institute and in 1949 was made its director. In 1954 the German Institute for Psychiatric Research in Munich was incorporated into the Max Planck Society. By the end of the twentieth century the various institutes of the Max Planck Society were internationally acknowledged as prestigious scientific research centers in many fields of science including **genomics**.

FURTHER READING: Kühl, Stefan, *The Nazi Connection* (1994); Weindling, Paul, *Health, Race and German Politics between National Unification and Nazism 1870–1945* (1989); Weingart, Peter, "German Eugenics between Science and Politics" (1989); Weiss, Sheila Faith, *Race Hygiene and National Efficiency* (1987).

KALLIKAK FAMILY, THE: A STUDY IN THE HEREDITY OF FEEBLE-MINDEDNESS (1912)

A classic **family history and pedigree study** of the second decade of the twentieth century, the *Kallikak Family*, by psychologist **Henry H. Goddard**, helped to support the **eugenics sterilization** and **immigration restriction movements** in the **United States**. Published by the Macmillan Company in New York City, the 121-page book was widely distributed. It went through twelve editions and was translated into German. For the study Goddard used his new **intelligence or IQ test**, the **family and history pedigree** method developed by geneticist **Charles Davenport**, and information concerning the rediscovered theory of **Mendelian inheritance**. Goddard claimed that he was able to trace back six generations of a family he named the *Kallikaks*, derived from the Greek words *kalos* (good) and *kakos* (bad). The study purported to show two branches of the same family. One branch was descended from "Martin Kallikak," "of a good family," and a "nameless feebleminded girl." In this branch 480 descendants were traced; only 46 were considered normal and the rest were found to exhibit **feeblemindedness** or to be **alcoholics, criminals, paupers**, and **prostitutes**, and were considered "not good members of society." The other branch was descended from Martin and an upstanding "Quakeress" he married. Of this branch 496 descendants were traced. Of this "legitimate" family, many were "upstanding citizens and professionals."

Goddard described in detail one member of the **dysgenic** side of the family, "Deborah Kallikak," an illegitimate girl who lived for a number of years at the Training School for Backward and Feeble-minded Boys and Girls in Vineland, New Jersey, where Goddard was director. At age twenty-two Deborah was described as an attractive young woman who was able to sight-read music, play the coronet, garden beautifully, sew clothes, and carve wood, but was noisy and a poor speller and reader. Her teachers were reluctant to admit that she was feebleminded. Goddard

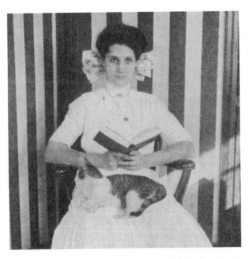

Photo of "Deborah Kallikak." The "Kallikak" family history purported to show the hereditary nature of mental disabilities and other social problems. It studied two lines of a family, one with "defective" and the other with "upstanding" members. This publication among others gave a "scientific basis" for eugenic sterilization, segregation, and immigration restriction laws. (From: *Kallikak Family* [1912], Main Library, Indiana University, Bloomington, Indiana. Image courtesy of the Digital Library Program).

suggested that this was the "menace of the high-grade feebleminded," or **moron**, inasmuch as they were lacking "morals" so were likely to produce many "illegitimate" children who would also be **degenerate** due to their "bad heredity." Goddard concluded that the "Kallikak family presents a natural experiment in heredity" and that feeblemindedness was likely inherited through simple Mendelian inheritance. He recommended that the **unfit** be prevented from reproducing. The study became widely known in many countries; however, by the mid-twentieth century this and other family history studies such as *The Dacks* and *The Nams* were considered flawed because most human characteristics, other then some **genetic diseases**, were found not to be based upon the rules of simple Mendelian inheritance.

FURTHER READING: Carlson, Elof Axle, *The Unfit* (2001); Goddard, Henry H., *The Kallikak Family* (1972); Kline, Wendy, *Building a Better Race* (2001); Paul, Diane B., *Controlling Human Heredity* (1995).

KELLOGG, JOHN HARVEY (February 26, 1852–December 14, 1943)

A physician and health reformer, Kellogg was one of the most influential leaders of the **Progressive Era**'s **Clean Living** and eugenics movements in the **United States**. His involvement and leadership in many reform activities infused eugenics as an underlying concept into many health crusades, including prohibition, antitobacco, **physical education**, personal hygiene, diet and nutrition, **tuberculosis**, **social hygiene**, **public health**, and sanitation. Born in Tyrone Township, Missouri, the son of a devout Seventh-day Adventist broom maker, Kellogg was a sickly child and educated at home by his schoolteacher mother. At age twelve he apprenticed at the Adventist printing shop, where he adopted vegetarianism. Kellogg's health improved and he rose to editorial assistant at the publishing house, taught school, and attended several colleges. He spent a year at the University of Michigan Medical School and transferred to Bellevue Hospital Medical College in New York City, where he received his M.D. (1875). In 1876 he became medical superintendent of the Western Health Reform Institute in Battle Creek, Michigan, which the Adventists had founded ten years earlier. Kellogg renamed it the Battle Creek Sanitarium, combining aspects of a European spa, hydrotherapy institution, and hospital. He remained as its director for his lifetime.

To further the eugenics cause Kellogg in 1906 founded the **Race Betterment Foundation**, an organization that over its lifetime sponsored conferences on race betterment. At the **First National Conference on Race Betterment** in 1914, Kellogg encouraged eugenic reform by suggesting the establishment of a eugenics registry "to establish a race of human thoroughbreds." The **Eugenics Registry** was established in 1915, under his guidance, in partnership with the **Eugenics Record Office** at **Cold Spring Harbor**, New York; Kellogg was secretary of the board. He was also a financial sponsor of the **Third International Congress of Eugenics** in New York City in 1932. A prolific author, Kellogg incorporated his health and eugenics ideas into numerous books and tracts. However, long after biologists had discarded the doctrine of **Lamarckian inheritance** of acquired characteristics, Kellogg continued to suggest that use of **alcohol**, **tobacco**, tea, and coffee, improper methods of dress, and unhealthy foods led to **feeblemindedness**, **insanity**, **crime**, and **pauperism** and hoped that evidence some day would demonstrate that the adoption of correct health habits would improve future generations. Kellogg married Ella Ervilla Eaton in 1879, but their marriage reputedly was never consummated. With no children of their own, they raised

Physician John Harvey Kellogg, in white in the center of the photo, was influential in many health reform crusades of his day including eugenics. (From: *Proceedings of the First National Conference of Race Betterment* [1914], Main Library, Indiana University, Bloomington, Indiana. Image courtesy of the Digital Library Program.)

over forty foster children. Kellogg actively promoted eugenic and other health causes until the end of his life. He died in Battle Creek from pneumonia after a long, reform-filled life.

FURTHER READING: Boyle, T. Coraghessan, *The Road to Wellville* (1993); Carson, Gerald, *Cornflake Crusade* (1957); Kellogg, J. H., *Plain Facts for Old and Young* (1974); Money, John, *The Destroying Angel* (1985); Schwarz, Richard William, *John Harvey Kellogg, M.D.* (1981).

KEY, WILHELMINE MARIE ENTEMAN (February 22, 1872–January 31, 1955)

One of the few women eugenics researchers in the **United States**, Key conducted **family history studies** during the early-twentieth-century eugenics movement. Like most other eugenicists of the era, she argued that human traits and characteristics were inherited, and supported **negative eugenic** practices. Born in Wisconsin, she graduated from the University of Wisconsin (1894) in zoology and then taught for four years in Green Bay. She attended the University of Chicago, where she studied with **Charles Davenport**, who later became the pivotal leader of the eugenics movement in the United States. In 1901 she received a Ph.D. in biology, one of the first women to gain a doctorate at the university, and stayed an additional year as a biology assistant (1903–1904). She then taught biology at New Mexico Normal College and Belmont College (1907–1909) before becoming professor of biology at Lombard College, Galesburg, Illinois (1910–1912). In 1912 she became a eugenics **field worker** for Davenport at the **Eugenics Record Office** (ERO) at **Cold Spring Harbor**, New York. Key became the archivist for the ERO (1918–1920) and then head of biology and eugenics at the **Race Betterment Foundation** at Battle Creek, Michigan (1920–1925).

In the fall of 1914 Key carried out a survey of **feeblemindedness** in northeastern Pennsylvania for the Public Charities Association of Pennsylvania. She then conducted a similar study of rural slum dwellers in northwestern Pennsylvania (1915–1917), published as *Heredity and Social Fitness* (1920). She hypothesized that there was a synergistic relationship between poor heredity and poor environment, but considered heredity more critical. As a representative of the Race Betterment Foundation she presented a paper, "Heritable Factors in Human Fitness and Their Social Control," at the **Second International Congress of Eugenics** (1921) in New York City, in which she argued that "the foundations for national power and fitness . . . are biological." From 1919 to 1921 she wrote a series of five articles in *The Journal of Heredity* on heredity's link to both outstanding and **degenerate** family lines. She married Francis Key (c. 1904), who died early in their marriage; they had no children. Upon her death she left an endowment to the American Genetics Association to support lectures for the "implementation of **genetics** for human welfare and improvement."

FURTHER READING: Bix, Amy Sue, "Experiences and Voices of Eugenics Fieldworkers" (1997).

KNOPF, S(IGARD) ADOLPHUS (November 27, 1857–July 15, 1940)

A physician and **public health** leader, Knopf championed eugenics during the early-twentieth-century eugenics movement in the **United States**. He recommended environmental changes as methods for race improvement, along with **negative eugenics**. Unlike many eugenics supporters of the era he supported **birth control**. Knopf was born at Halle an der Saale, Germany, and received his early education in local schools. He moved to the United States in 1880, taught languages in Los Angeles, was a student at the University of Southern California (1884–1886), and received his M.D. from Bellevue Hospital Medical School, New York City. Returning to Los Angeles, he practiced medicine for two years, and then in 1890 went to Paris, where he graduated from the University of Paris with an A.B. and a B.S. (1891). Knopf worked in Paris hospitals for four years while attending a medical school in France, where he received his second medical degree (1895). After graduation he worked in a tuberculosis hospital. Upon his return to the United States in 1896, he established a practice in New York City as a specialist in the disease and was a pioneer in the U.S. **Progressive Era**'s tuberculosis movement. Over the next twenty years he instructed and consulted in many hospitals and sanitariums in the New York City and nearby states.

Knopf became involved in the **social hygiene**, eugenics, antisaloon, and birth control movements. By the second decade of the twentieth century he viewed birth control and eugenics as interlinking concepts and saw **tuberculosis** and **alcohol** as **racial poisons** that contributed to **racial degeneracy**. He argued that large families were associated with poverty, tuberculosis, **alcoholism**, **venereal disease**, and **feeblemindedness**. To counter these problems and to produce a "better population," Knopf favored the establishment of public birth-control clinics and the rehabilitation of **prostitutes**. He detailed his philosophy in *Birth Control in Its Medical, Social, Economic, and Moral Aspects* (1917). A supporter of eugenics, he was a member of the organizing committee for the **First National Conference on Race Betterment** (1914). At the conference Knopf suggested that tubercular parents who bore children were criminals and recommended education to encourage them to refrain from parenthood until they were cured. He championed **eugenic sterilization** laws and mandatory sterilization for patients who "procreate willfully." At the **Second International Congress of Eugenics** (1921) he detailed how eugenics was related to the "tuberculosis problem." A prolific writer throughout his life, Knopf wrote over 400 books and pamphlets. Many awards and honors were conferred upon him. He married Perle Nora Dyar (1889), who died in 1931. In 1935 he married Julia Marie Frederick. No children resulted from either marriage. Knopf died in a New York City hospital form complications due to surgery.

FURTHER READING: Bates, Barbara, *Bargaining for Life* (1992); Dubos, René, and Jean Dubos, *The White Plague* (1987); Teller, Michael E., *The Tuberculosis Movement* (1988).

L

Our knowledge is now so accurate that it is possible to predict almost exactly the kind of children that will be born to parents whose heredity and mental habits are known.

Norman Barnesby, "Eugenics and the Child," *The Forum* 49 (March) 1913

LAMARCK, JEAN-BAPTISTE (August or September 1, 1744–December 18, 1829)
A French botanist and originator of the theory of inheritance of acquired characteristics, Lamarck lent a theoretical basis to hereditarian concerns of the mid-nineteenth century and the initial surge of the early-twentieth-century eugenics movement. Born at Bazentin-le-Petit, Picardy, in northern France, the son of a miliary officer from a noble family, Lamarck studied for the priesthood but upon his father's death in 1760 joined the French army. Due to illness Lamarck was forced to leave the military in 1768 and went to Paris to study medicine while working at a bank to support himself. He also developed an interest in meteorology, chemistry, botany, and shell collecting. In 1778 Lamarck wrote *Flore française* (*French Plants*), a catalog of French plants that established his scientific reputation. This work brought him to the attention of French naturalist Georges de Buffon, who helped him obtain a job at the *Jardin des Plantes* (Botanical Gardens) as a botanist for the king. As part of his duties he traveled from 1780 to 1782 through Europe and collected botanical specimens. Lamarck continued to work at the garden until the French Revolution, when it was closed. When the Botanical Gardens were reopened in 1793 after the revolution as the National Museum of Natural History Lamarck was appointed a professor of zoology and placed in charge of organizing the museum's collection of invertebrates. From 1784 to 1792 he published articles on botany, his primary interest.

Drawing from chemistry, meteorology, invertebratology, and geology, Lamarck crafted a theory of inheritance and evolution based upon environmental influences, and he was the first scientist to acknowledge the adaptability of organisms. Lamarck first mentioned his theory around 1800, detailing it in *Philosophie zoologique* (*Zoological Philosophy*, 1809). To explain how complex life-forms evolved from simpler ones, Lamarck formulated four laws. His last and most famous law argues that all characteristics acquired by an organism from its environment can be transmitted to offspring. In his final work on the subject, *Histoire naturelle des animaux sans vertèbres* (*Natural History of Invertebrate Animals*, 1815), he further developed his theory. By the mid-nineteenth century **Lamarckian inheritance** of acquired characteristics was widely considered an accepted mechanism of inheritance. By the mid-1920s it was disproved by the new field of **genetics** based upon the work of **Gregor Mendel**. Although

Lamarck's botany books were admired during his lifetime, other than among the French his theory of inheritance was largely ignored. Near the end of his life he lost his eyesight but continued writing by dictating material to his daughter. He died in poverty in Paris.

FURTHER READING: Corsi, Pietro, *The Age of Lamarck* (1988); Jordanova, L. J., *Lamarck* (1984).

LAMARCKIAN INHERITANCE OF ACQUIRED CHARACTERISTICS

The transmission of environmentally influenced physical, mental, moral, and social traits to descendants is called the *Lamarckian theory of acquired characteristics*, or *Lamarckian inheritance*. The doctrine was accepted throughout the nineteenth century but had lost much support by the second decade of the twentieth century. Although the concept goes back to antiquity **Jean-Baptiste Lamarck** published the first scientific support for the theory in 1809. Lamarck noticed that plants had different appearances when seeds were planted in different environments. He also noted that wild seeds that were cultivated produced different traits. These observations led him to propose that environment can influence hereditary characteristics. For example, if a giraffe was forced to stretch its neck to get food, over time it would develop a longer neck. Applied to humans, acquired characteristics could range from athletic and musical ability to **alcoholism** and **pauperism**. Relying upon Lamarck's theory, social reformers argued that an impoverished or morally corrupt environment was responsible for bad heredity and tended to support reforms such as education, **public health**, and better housing in an effort to eliminate unwanted characteristics. By the late 1930s, after the rediscovery of monk **Gregor Mendel**'s mechanism of inheritance (**Mendelism**) in 1901, most researchers, other than a few Russian biologists, had dismissed the theory. However, this concept, along with **natural selection** and **degeneracy** theories, was a central tenet of eugenics on both sides of the Atlantic at the turn of the twentieth century.

FURTHER READING: Carlson, Elof Axle, *TheUnfit* (2001); Corsi, Pietro, *The Age of Lamarck* (1988); Jordanova, L. J., *Lamarck* (1984); Paul, Diane B., *Controlling Human Heredity* (1995).

LAUGHLIN, HARRY HAMILTON (March 11, 1880–January 26, 1943)

A central figure of the eugenics movement in the **United States**, Laughlin was a champion of immigration restriction and sterilization laws and a leader of the **nativistic** faction of the movement. One of five brothers, Laughlin was born in Oskaloosa, Iowa, to a family that traced its ancestry to colonial times. His father was a professor and a Disciples of Christ minister. After attending local schools Laughlin graduated from North Missouri State Normal School (1900), taught biology, took biology courses at Iowa State University, and served as a high school principal in the Kirksville, Missouri, area (1900–1905). Laughlin became superintendent of schools and taught agriculture courses at the Kirksville Normal School. He was interested in breeding experiments, and after contacting geneticist and eugenics leader **Charles Davenport** at **Cold Springs Harbor**, New York, he was invited to become superintendent of the new **Eugenics Record Office** (ERO), to teach field collection techniques and analyze the data. Laughlin remained at the ERO for twenty-nine years. During **World War I** he completed his doctoral work in biology at Princeton University (1917).

Between 1910 and 1939 Laughlin was involved with most aspects of the eugenics movement, including all the major eugenics organizations and conferences. He was secretary of the **American Breeders Association** Committee on Sterilization, a member of the **Galton Society**, and an organizer of the **Second** and **Third International Congresses of Eugenics**. He was a founding member of both the **Eugenics Research Association** and the **American Eugenics Society**, for which he served a term as president (1927–1929). For these two organizations, respectively, he was the coeditor of *Eugenical News* and on the editorial board of *Eugenics*. He gave presentations at the **First** and **Third National Conferences on Race Betterment**.

In the pre–World War I years Laughlin collected extensive data on sterilization and began to promote **eugenic sterilization laws** for the **unfit**. His research resulted in *Eugenical Sterilization in the United States* (1922), which included a model sterilization law.

Harry H. Laughlin, superintendent of the Eugenics Record Office, championed eugenic sterilization and immigration restriction to prevent racial degeneracy and the downfall of civilization. (Image courtesy of Harry H. Laughlin Papers, Pickler Memorial Library, Truman State University, and the CSHL Eugenics Web site.)

With this report Laughlin became the nation's foremost authority on sterilization. He testified at the **Buck v. Bell** proceedings, in which the U.S. Supreme Court upheld a state's right to sterilize **feebleminded** and other "defectives." Besides advocacy for sterilization Laughlin championed **immigration restriction**. In 1916 he surveyed 636 state institutions and their populations. Studies published from his research suggested that a high proportion of inmates were recent immigrants. These publications resulted in Laughlin's being asked to be an "expert eugenical witness" at hearings before the House Committee on Immigration and Naturalization, where he maintained that a disproportionate share of "socially inadequate citizens" were of **eastern and southern European** extraction. His reports played a crucial role in passage of the **Johnson-Reed Immigration Restriction Act of 1924**. In the 1930s the **Carnegie Institution of Washington**, which supported the ERO, became embarrassed by Laughlin's support of **Nazi Germany**'s sterilization program. It investigated his research and in 1935 found the ERO's collection of records unsatisfactory for the study of human **genetics**. In 1936 Laughlin was awarded an honorary doctorate by the University of Heidelberg, Germany, causing further embarrassment to the institution. After involved negotiations Laughlin was asked to resign in 1937, and he retired to Kirksville, Missouri. He married Pansy Bowen (1902), but they remained childless. Laughlin had epilepsy, one of the very traits he and other eugenicists wanted to eliminate from the population. He died in Kirksville six years after his forced retirement.

FURTHER READING: Haller, Mark H. *Eugenics* (1984); Kevles, Daniel J., *In the Name of Eugenics* (1985); Reilly, Philip R., *The Surgical Solution* (1991).

LAWS OF HEALTH

A healthy lifestyle that included exercise, proper diet, and avoidance of toxic substances and sexual immorality in order to produce fit and healthy offspring and to improve the human race embodied the laws of health. Its tenets were an underlying principle of the mid-nineteenth-century hereditarian and **Clean Living Movements** of the **Jacksonian Era** in the United States. During this era **Protestant** health reformers, basing their ideas on the theory of **Lamarckian inheritance of acquired characteristics** (which in the United States was sometimes termed **inherited realities**) suggested that humankind had degenerated since "the time of Adam and Eve." This degeneration was illustrated by rampant **alcoholism, pauperism, tuberculosis**, and general debility found in the nation. Reformers were concerned that a Christian millennium of perfect and healthy humans might not occur if the laws of health continued to be ignored. By the late 1840s obeying these laws included selecting the right marriage partner; physical exercise; proper diet; and the avoidance of **masturbation**, "immoral" sexual behavior, **alcohol, tobacco**, and other stimulants. These laws became incorporated into newly formed religious sects, including the **Church of Jesus Christ of Latter-day Saints** and the Seventh-day Adventists. Reformers of the era, like **William Alcott, Orson** and **Lydia Folger Fowler**, and **Elizabeth Blackwell**, discussed the importance of obeying the laws of health to prevent "**degeneracy** of the race." Health ideals again surged at the turn of the twentieth century in the **Progressive Era**'s health and social reform movements in the United States and in the health and vitality movements in **Germany** and **Britain**.

FURTHER READING: Engs, Ruth Clifford, *Clean Living Movements* (2000); Money, John, *The Destroying Angel* (1985); Nissenbaum, Stephen, *Sex, Diet, and Debility in Jacksonian America* (1988).

LEBENSBORN PROGRAM (December 12, 1935–c. 1945)

Nazi Germany's **positive eugenics** program to breed a **Nordic** "**master race**" was coined *Lebensborn*. The term, from *Leben* ("life") and *born* ("spring" or "fount"), is translated as the "wellspring" or "fount" of life. Nazi SS and Wehrmarcht officers and both married and single German women who had passed "racial purity" exams were encouraged to produce children in order to increase the population of the Nordic, or, as termed by the Nazis, **Aryan race**. Light hair and eyes were preferred, and a "pure" Germanic lineage had to be traced back at least three generations. Teenage girls were encouraged to produce a "child for the Führer," called "children of the SS" by ordinary Germans. The Lebensborn maternity facilities offered a place where expectant mothers could have children in privacy and safety; records were carefully guarded. After 1940 the program also included kidnapping children with Nordic characteristics from eastern European countries to be raised by German families and encouraging the mating of "racially pure" German soldiers with selected women in Scandinavian and other northern European countries. Children born in the Lebensborn maternity facilities were raised by their mothers or given to German families for adoption. The program was headed by Heinrich Himmler (1900–1945), head of the SS and the Gestapo.

FURTHER READING: Clay, Catrine, and Michael Leapman, *Master Race* (1995); Henry, Clarissa, and Marc Hillel, *Children of the SS* (1976).

LENZ, FRITZ (March 9, 1887–July 6, 1976)

The first professor of race hygiene in Germany, Lenz was a prolific writer and leader of the **German eugenics** movement in the post–**World War I** and **Nazi German** eras. Like other German eugenicists of the time, Lenz was concerned with improving the health and national efficiency of the German people. However, unlike most eugenicists, he supported Nordicism (superiority of the **Nordic race**) and accepted **Arthur de Gobineau**'s theory of the **hierarchy of races** in intelligence and cultural productivity. The son of a landowner, Lenz was born in Pflugrade im Pomerania. While studying medicine at the University of Freiburg, in Freiburg im Breisgau (1906–1912) he came under the tutelage of eugenics pioneer **Alfred Ploetz**. He also studied with anthropologist **Eugen Fischer** and geneticist August Weismann (1834–1914), whose research supported **Mendelian inheritance**. After finishing his degree Lenz moved to Munich, where he worked at the Institute for Hygiene. In 1919 he began to teach a eugenics course entitled Social and Racial Hygiene at Munich University. In 1923 he was appointed associate professor of racial hygiene, the first such position in Germany. He remained in Munich until 1933, when he was appointed a full professor of eugenics at the University of Berlin and head of the Department of Eugenics at the **Kaiser Wilhelm Institute**. In 1946 Lenz left Berlin and became chair of the Department of Human Genetics at the George August University in Göttingen, where he remained until his retirement in 1955.

Lenz was involved in German eugenics organizations and publications. He was a founding member of the Freiburg chapter of the **German Society for Race Hygiene**, served as its secretary, and became a leader in the Munich chapter of the society in the 1920s. In 1913 he was appointed to the editorial board and became coeditor of the noted eugenics journal *Archive for Racial and Social Biology*. In 1933 he stepped down as coeditor but remained on the board until the demise of the journal in 1944. In 1921 Lenz, along with Fischer and botanical geneticist Erwin Baur (1875–1933), published a popular **race hygiene** text, *Human Heredity and Race Hygiene* (*Menschliche Erblehre und Rassenhygiene*). Lenz was sole author of the second volume, which concerned eugenic methods. The work brought him to the forefront of the German eugenics movement.

Lenz was a political conservative who began to support National Socialism in the 1920s. When the Nazis came to power in 1933 he served as a member of the Ministry of Interior Expert Committee for Population and Race Policy. He helped formulate public policy that was designed to support large, healthy German families. Lenz promoted both **positive eugenics** (increasing the birth rate among the fit) and **negative eugenics** programs (such as **eugenic sterilization** for the **unfit**). He was a member of the Nazi "expert committee" to draft a **eugenic sterilization** law and served on the "Eugenic Sterilization Court." He applied for Nazi party membership in 1937 and was a member of the committee that drafted euthanasia law. It is not known, however, whether or not he knew of the secret October 1939 Nazi decree that allowed euthanasia of mentally ill and disabled individuals. Lenz wrote over 600 publications, most of which dealt with eugenic reform. He married Emmy Weitz (1915), with whom he had three sons, and a year after her death he married Kara Borries (1929), with whom he had a son and a daughter. Lenz died in Göttingen after an active retirement.

FURTHER READING: Weindling, Paul, *Health, Race and German Politics between National Unification and Nazism, 1870–1945* (1989); Weiss, Sheila Faith, "Race and Class in Fritz Lenz's Eugenics" (1992).

LITTLE, C(LARENCE) C(OOK) (October 6, 1888–December 22, 1971)

A university president, geneticist, and cancer researcher, Little was a leader in the **social hygiene**, birth control, and eugenics movements. He was born in Brookline, Massachusetts, to an old-stock **Anglo-American** Episcopalian family. His father was an architect and businessman. Little attended private schools, graduated from Harvard University with both an A.B. (1910) and a Ph.D. (1914), and took administrative positions at the university. He also bred mice and over his career developed special strains for cancer and other research. From 1921 to 1922 he served as assistant director of the Station for Experimental Evolution of the **Carnegie Institution of Washington** under the direction of **Charles Davenport**. Little was president of the University of Maine (1922–1925) and the University of Michigan (1925–1929). He was forced to resign the Michigan presidency because he supported the teaching of **birth control** and opposed **alcohol** and automobiles on campus. He also believed women should have a curriculum that would prepare them for motherhood. In 1929 he was appointed head of the American Society for the Control of Cancer (renamed the American Cancer Society in 1944), a post he retained until retirement from the society in 1945. In 1929 he also founded the Roscoe B. Jackson Experimental Laboratory in Bar Harbor, Maine; he was its director until his retirement in 1956.

Over his lifetime Little belonged to numerous eugenics and other health organizations. Because he was committed to birth control as a eugenics measure, he served as scientific director to **Margaret Sanger's American Birth Control League** for twenty years (1925–1945). He was general secretary of the **Second International Congress of Eugenics** and a member of the **Galton Society** and the **Eugenics Research Association**. He was president of the **Third National Conference on Race Betterment** and edited its proceedings. He was briefly president of the **American Eugenics Society** (1928–1929) and, later, vice president of the **American Social Hygiene Association**. Little wrote several books and articles. He married Katharine Andrew (1911), with whom he had three children. In 1929 he divorced her, and the following year he married his laboratory assistant, Beatrice Johnson, with whom he had two children. Little died near Bar Harbor, Maine, of a heart attack.

FURTHER READING: Engs, Ruth Clifford, *The Progressive Era's Health Reform Movement* (2003).

*There is nothing utopian in hoping for the time to come when men and women will con-
sult a wise sanitarian before entering into the marriage relation.*

Martin Luther Holbrook, *Marriage and Parentage* (1882)

MASTER RACE

The concept of an aggregate of individuals of a certain "**race**," ethnic, or cultural
group that is endowed with supposed physical, moral, mental and intellectual "supe-
riorities" has been termed a *master race*. The concept goes back to antiquity. Plato, in
the *Republic*, argued that the "guardians" of the state, the most superior individuals,
were the best qualified for leadership and recommended selective breeding among
them to produce a superior race. In the nineteenth century the concept of a master
race was suggested by **Arthur de Gobineau**, who argued that cultures **degenerate** when
races mix. In the late nineteenth and early twentieth centuries the concept was linked
to the belief of the **superman**, or superior human. Nineteenth-century racial theory
argued that those with Teutonic or **Nordic** inheritance were innately superior to all
other ethnic/cultural groups on the grounds that they constituted the most "advanced"
civilizations and thus were born leaders. **Social Darwinism** theory suggested that because these individuals were the fittest they survived and thus advanced society. This
idea formed the underpinning of the **nativistic** faction of the U.S. eugenics movement.
Nativist eugenics supporters **Madison Grant** and **Lothrop Stoddard** argued that the
Anglo-Saxon and Nordic races were dying out (**race suicide**) as a result of lack of reproduction, interbreeding with "inferior races," and the death of the fittest in **World
War I**. Eugenic marriages and **immigration restriction** programs were recommended
to improve the race in the United States.

Influenced by both European and American thought concerning a Nordic master
race, **Nazi Germany** turned the concept into dogma (*Herrenrasse*, *Herrenvolk*). Under
Nazism the racial ideal was the blond, blue-eyed Nordic, or **Aryan**. The *Untermenschen*,
or racially inferior, included **Jews** and Slavic groups, such as Czechs, Poles, and Russians. In order to perfect society, achieve "racial purity," and propagate a master race,
marriages were encouraged among the most intelligent and beautiful. After the **World
War II** Nazi atrocities that became known as the Holocaust the ideology of a master
race receded. However, as a reaction to the Civil Rights Movement of the 1960s in the
United States and the rising concept of multiculturalism, diversity, and increased mass
immigration from developing countries over the rest of the century, white supremacist groups such as the Ku Klux Klan reemerged with a master race ideology. By the

turn of the twenty–first century, as a result of increased knowledge of the human **genome** and **genetic engineering**, it became theoretically possible to selectively breed for race, gender, and other human characteristics. These continuing advances take the concept of creating a group of superhumans out of the realm of science fiction and place it within scientific possibility, where it is a concern of the **new eugenics**.

FURTHER READING: Cecil, Robert, *The Myth of the Master Race* (1972); Clay, Catrine, and Michael Leapman, *Master Race* (1995); Gobineau, Arthur, Comte de, *The Moral and Intellectual Diversity of Races* (1984); Henry, Clarissa, and Marc Hillel, *Children of the SS* (1976).

MASTURBATION

The practice of sexual self-stimulation, masturbation, was termed *self-abuse* or *self-pollution* in the nineteenth and very early twentieth centuries. *Onanism* was the clinical medical term for this behavior. Because inmates in asylums and institutions masturbated, it was thought to be a major cause of **insanity, alcoholism, feeblemindedness, tuberculosis**, disease, **crime, pauperism**, and other **dysgenic** health and social problems that lead to **racial degeneracy**. The supposed damaging nature of masturbation was first discussed in an anonymous 1710 French pamphlet, *Onania*. By the 1830s it was considered a vice that lead to disease. Medical students were taught that it was a primary cause of **degeneracy**. Health reformers of the **Jacksonian Era** (1830–1860) including **William Alcott** and Sylvester Graham wrote against the practice. Concerns were raised that debility produced by masturbation could be passed to offspring via the mechanism of **Lamarckian inheritance of acquired characteristics**. Many treatments such as vegetarian diets and cold baths were tried to prevent the habit. In the late 1890s vasectomy (male sterilization) was first tried as a cure for masturbation by Indiana reformatory physician **Harry Sharp**, and the proclaimed "success" of this surgery led to "voluntary" sterilization of inmates. Some eugenic reformers championed **eugenic sterilization** to prevent other **unfit** individuals from reproducing. Although no evidence has linked insanity or physical degeneration with masturbation, many religions consider it a sin or vice.

FURTHER READING: Carlson, Elof Axle, *The Unfit* (2001); Reilly, Philip R., *The Surgical Solution* (1991).

MENDEL, GREGOR JOHANN (July 22, 1822–January 6, 1884)

Revered as the father of **genetics**, Mendel, a monk in what is now the Czech Republic, first demonstrated the laws of heredity. These rules for inheritance, later termed *Mendelism* or *Mendel's laws of inheritance*, became the basis of genetics. Mendel was born in Heizendorf, Moravia, one of several children, to a peasant farmer of southern German ancestry. As a child he showed interest in the natural sciences and studied at a public school where his brilliance was recognized. He studied for two years at the Philosophic Institute in Olmütz, but due to his limited financial resources he entered St. Thomas Monastery, an Augustinian Order, in Brno, Austria, in 1843. Mendel was given the name Gregor and ordained as a priest in 1847. He subsequently was assigned pastoral duties, but was found more suitable for teaching. In 1849 he became an assistant to a physics teacher in a neighboring community but failed the qualifying examination for the teacher certification. Mendel was then sent to the University of Vienna in 1851 for further education in mathematics and biology. He returned in 1854 to Brno to teach in the local technical high school. In 1856 he again failed the teacher's certification examination and remained an assistant teacher. During his teaching years he studied a variety of natural phenomena including sunspots and bees. He made microscopic slides and recorded daily weather observations. He also was familiar with the writings of **Charles Darwin**. In 1868 Mendel was elected abbot of the monastery. Due to its large holdings and wealth he became involved with its many administrative activities and had little time for scientific endeavors.

After observing that plants did not appear to change characteristics in different environmental conditions Mendel began experimenting with several traits found in the pea pod. Between 1856 and 1863 he cultivated and tested some 28,000 pea plants. Since the peas were planted next to each other, he concluded that their characteristics were not influenced by the environment. He observed what he termed "dominance and segregation

Austrian monk Gregor Mendel discovered the basic principles of inheritance that became the basis for genetics while breeding peas in a monastery garden during the mid-nineteenth century. (From: *American Breeders Magazine* [1910], Main Library, Indiana University, Bloomington, Indiana. Image courtesy of the Digital Library Program.)

of characters" in the forms of plants and discovered that traits were inherited in certain numerical ratios, which led to a theory of three basic laws of inheritance. Mendel presented his results to the local science society and published "Experiments with Plant Hybrids" in the society's *Proceedings* in 1866. However, when he could not confirm his observations in another plant he gave up this research. His paper was ignored and not widely distributed because most scientists of the day accepted **Lamarckian** theory of the inheritance acquired characteristics as the key to heredity. In 1900 Mendel's theory was rediscovered. It became the basis of the early-twentieth-century eugenics movement and the field of genetics.

FURTHER READING: Henig, Robin Marantz, *The Monk in the Garden* (2000); Paul, Diane B., *Controlling Human Heredity* (1995).

MENDELISM OR MENDELIAN INHERITANCE

The laws of inheritance discovered by the Austrian monk **Gregor Johann Mendel** in the 1860s, or *Mendelism*, demonstrate that the transmission of specific traits are passed from one generation to another in a predictable mathematical ratio. These rules of inheritance have also been termed *Mendel's laws of inheritance* and *Mendelian theory of heredity*. Mendel found that each parent strain of a variety of pea plant contributed a hereditary unit (now termed a **gene**). The resulting hybrid only expressed one of the traits, called the *dominant* form. The one not expressed became known as the *recessive* form. When hybrids were self-fertilized the recessive traits were then expressed in a specific ratio of 1 to 4. These findings led Mendel to suggest three basic laws of heredity: (1) hereditary factors do not combine, but are passed intact to offspring; (2) each member of the parental generation transmits only half of its hereditary factors; and (3) different offspring from the same parents receive different sets of hereditary factors. These laws became the basis of classic **genetics**.

Mendel's theory was ignored for thirty-four years because scientific thought was focused on **Darwinian** evolution and the adaptation of species to changes in the environment. In addition, the theory of **Lamarckian inheritance of acquired characteristics** through environmental influences was the accepted theory of inheritance. Mendel's theory, however, accounted for the transmission of traits and stability within a species. In 1900 four Europeans—Hugo De Vries of Holland, William Bateson of England, Karl Correns of Germany, and Erich Tschermak von Seysenegg of Austria—independently rediscovered Mendel's findings. The popularization of Mendelian theory, along with **social Darwinism**, led to the belief that a wide variety of complex human health, moral, and social conditions could be inherited and were not, as implied by the Lamarckian theory, caused only by poor environmental circumstances. The **nature-nurture** debate concerning human development shifted from environmental to inherited causes, a change that influenced the rise of the eugenics movement during the first two decades of the twentieth century.

FURTHER READING: Henig, Robin Marantz, *The Monk in the Garden* (2000); Paul, Diane B., *Controlling Human Heredity* (1995).

MENSA (1946–)

An organization for people with "high **IQs**," Mensa requires for membership a score in the top 2 percent of a standardized **intelligence test**. The organization acts as a force for **positive eugenics** inasmuch as intellectually gifted individuals sometimes find compatible marriage partners through involvement with its activities. The Latin term *mensa* refers to "mind," "table," or "month," suggesting a monthly meeting for discussion around a table. Mensa was founded in Britain by Lancelot Lionel Ware, an attorney and researcher, and Roland Berrill, an Australian lawyer. One of Mensa's first presidents was psychologist and eugenics supporter Cyril Burt (1883–1971), who conducted **twin studies** on the hereditary nature of **intelligence**. Mensa was exported to the United States in 1960. It expanded under the guidance of Russian immigrant Vicktor Serebriakoff. The aims of the organization are to identify and foster human intelligence for the benefit of humanity; to encourage research into the nature, characteristics, and uses of intelligence; and to provide a stimulating intellectual and social environment for members.

FURTHER READING: Mensa, *A History of Mensa* (1990).

MENTAL DEFECTIVES IN INDIANA (1916–1922)

A series of bulletins reporting information gathered from a eugenics field study in Indiana was published as *Mental Defectives in Indiana*. The survey was undertaken to identify the "uncared-for insane, epileptic, [and] feeble minded" in the state of Indiana as a basis for treatment, care, and prevention programs. The study was carried out by eugenics **field workers** under the supervision of **Arthur H. Estabrook** of the

Like MENSA members, who are required to prove their superior intelligence through intelligence tests, these young adults at the First National Conference of Race Betterment in 1912 entered contests, were tested, and then judged superior in intelligence, health, and physical ability. (From: *Proceedings of the First National Conference of Race Betterment* [1914], Main Library, Indiana University, Bloomington, Indiana. Image courtesy of the Digital Library Program.)

Eugenics Record Office. On April 25, 1915, the Board of State Charities of Indiana adopted a resolution arguing that "mentally defective" individuals caused staggering social and financial burdens to the state and were the major cause of "pauperism, degeneracy, and crime." To investigate the problem, the board recommended that a committee be formed. The governor appointed the Indiana Committee on Mental Defectives, which first met December 17, 1915, to undertake a study. During the first year two counties were surveyed, and the committee published its first report on November 10, 1916. The 1917 state legislature appropriated funds to continue the work, and as a result eight more counties were surveyed during 1917, 1918, and 1919. The committee's second report was published March 6, 1919. A special session of the legislature appropriated funding for a survey of one more county and the collection of data on the mental ability of criminals, orphans, and schoolchildren. This third report was published July 31, 1922. The bulletin included cumulative data from all eleven counties. It was reprinted for the public the following year.

The bulletins reported that individuals were more likely to have mental illness and disability if they lived in poor urban or isolated rural environments and were less likely to have problems if they lived under more prosperous conditions. It was estimated that 1.2 percent of the population of Indiana was mentally defective. About 3 percent of the school population was found to be "definitely **feebleminded**" and another 3 percent "borderline in intelligence." The committee suggested that both mentally inferior and superior children be identified and placed in special classes, and that teachers be trained to work with the mentally disabled. It recommended the construction of additional treatment institutions and "farm colonies" and the establishment of traveling psychiatric clinics and social work services for young criminals. This survey, like other state studies, helped secure the concept that the mentally ill and disabled were a menace to society and should be controlled. These measures were instigated to prevent them from "propagating their own kind," thus leading to more **crime**, **prostitution**, **pauperism**, **venereal disease**, and an increase in public welfare costs.

FURTHER READING: Carlson, Elof Axle, *The Unfit* (2001); Ingalls, Robert P., *Mental Retardation* (1986).

MENTAL ILLNESS
See Insanity *or* the Insane

MENTALLY RETARDED
See Feebleminded *or* Feeblemindedness

MILLENNIUM ERA (1970s–)

During the last decades of the twentieth century a surge of social and health reform movements emerged in the United States. This era has been termed the *Millennium Era* or *Age*, the Fourth **Great Awakening**, or a **Clean Living Movement** for issues encompassing health reform with moral undertones. The Millennium reform era resulted from reactions to social, cultural, economic, and technological changes during the post–**World War II** years. It had roots in the 1960s with a religious resurgence, plus civil rights, women's rights, and antismoking movements. During the early 1970s concern with fitness and improved diet emerged, followed by agitation against drunken driving in the late 1970s. Alternative religions and medicine also became more popular during this decade. During the 1980s antidrug, anti**alcohol** and anti**tobacco** campaigns intensified. By the late 1980s the **new eugenics** had emerged, when heredity, not just the environment, was again seen as important to the development of human characteristics (the **nature-nurture** debate).

"Reformers" in the Millennium era, unlike the individuals of the **Jacksonian** and **Progressive Eras**, were often well-financed, large grass-roots—or even governmental—organizations founded, or influenced, by middle-class activists. Some reformers began to blame "the system" rather than encouraging individuals to take personal responsibility for their behaviors. This attitude led to reform campaigns that included litigation against tobacco interests, alcoholic-beverage retailers, and pharmaceuticals manufacturers, food processors, and other corporations. Legislation was passed to change public policy regarding drinking, smoking, driving automobiles, and other behaviors. The era was also characterized by countercrusade activities. Women's liberation was countered by a "pro-family" movement; the use of marijuana and other drugs was followed by a "war on drugs"; nonmarital sexual activity was challenged by a new "purity" movement; the legal right to obtain an abortion was met with a "pro-life" movement against abortion; and the use of newly developed **genetic engineering** and **assisted reproduction** technology was countered with laws to prevent the use of some procedures emerging from this technology.

FURTHER READING: Engs, Ruth Clifford, *Clean Living Movements* (2000); McLoughlin, William G., *Revivals, Awakenings, and Reform* (1978); Strauss, William, and Neil Howe, *The Fourth Turning* (1997).

MONGOLOID

See Asians *or* Asian Americans

MONGREL VIRGINIANS: THE WIN TRIBE (1926)

Written by eugenics researcher **Arthur H. Estabrook** of the Station for Experimental Evolution, **Cold Spring Harbor**, New York, and Ivan E. McDougle (1892–1955) of Goucher College, *The Mongrel Virginians* was one of few **family pedigree studies** that examined a southern "mixed race" of "Indians, Negroes, and whites." The 205-page book was published in Baltimore by the Williams and Wilkins Company. It was divided into three parts: The "Win Tribe"; "Other Mixed Areas," which discussed similar mixed-race kinships in other regions; and "General Summary." The first part contains sixteen chapters including "The White Brown Family," "The Indian Browns," "Population," "Fecundity," "Consanguinity," "Legitimacy," "Sex Mores," "Alcoholism," "Venereal Disease," and "Tuberculosis." The study, undertaken from 1923 to 1925, investigated 570 individuals who lived in the foothills of the Blue Ridge Mountains of Virginia and were considered neither black nor white. The "Win Tribe" claimed to be of "Indian and white" descent only, although court records suggested that the original Indian and white mixture had interbred with freed slaves. The main source of income of the Wins was tobacco farming and sharecropping. Information about the kinship was gathered from census records and interviews of family members, neighbors, teachers, and other professionals. The authors reported that the Wins were "below average" mentally and socially, lacked academic ability, were not industrious, and that 80 percent had been born "illegitimately." Little **alcoholism** or **venereal disease** was found. They reported that groups of men would walk without talking in single file along the road, "Indian style," and that "as is well known, the Negro is full of music . . . but no Win has every shown any semblance of ability in this line." The authors concluded that the "Indian temperament" was dominant in this kinship. The work supported **race** stereotypes and the **nativist** branch of the eugenics movement in the **United States**.

FURTHER READING: Carlson, Elof Axle, *The Unfit* (2001); Rafter, Nicole Hahn, (ed.), *White Trash* (1988).

MORMONS

See Church of Jesus Christ of Latter-day Saints

MORONS

Coined around 1910 by **Henry Goddard**, a psychologist, this diagnostic category identified a person with a minor degree of mental disability. *Morons* were defined as higher functioning **feebleminded**. Since scientific opinion held that intelligence was primarily inherited, Goddard and other eugenicists during the early twentieth century considered this group a "menace to society," and because they were linked to "moral deficiency" and sexual permissiveness and "reproduced their own kind," morons were thought to lead to the decay of civilization. Goddard adopted "moron" from the Greek word for "foolish," *moronia*. Individuals diagnosed as morons tended to look "normal," were slow learners, and were found in about 2 percent of schoolchildren. They had a mental age between eight and twelve on the Binet Measuring Scale for **Intelligence**. Binet's original scale had included only two gradations of mental deficiency: the "imbecile," with a mental age of three to seven years, and the "idiot," with a mental age of two or younger. Since no term existed for a higher level of mental disability the term *moron* was created. Unlike more seriously impaired individuals, who were generally institutionalized, higher-functioning people often survived on their own. This concerned social welfare and health professionals as the newly developed **intelligence tests** revealed that a high percentage of **prostitutes**, "unwed mothers," "loose women," **criminals**, and **paupers** could be classified as morons. To prevent **racial degeneracy** and the genetic debilitation of society, many eugenics supporters and social welfare professionals suggested **eugenic sterilization** or **segregation** in state-run institutions. In the latter half of the twentieth century, the term *moron*, along with *idiot* and *imbecile*, fell out of use as scientific categories of mental disability. The diagnostic categories were replaced by terms such as *mildly, moderately, severely,* or *profoundly mentally retarded, impaired,* or *developmentally disabled*.

FURTHER READING: Ingalls, Robert P., *Mental Retardation* (1986); Kline, Wendy, *Building a Better Race* (2001).

MULLER, HERMANN JOSEPH (December 21, 1890–April 5, 1967)

The first geneticist to induce mutations (change in genetic material) in an organism by X-ray, Muller was concerned about the **dysgenic** effect of radiation on the human race and promoted **positive eugenics**. He was born in New York City, one of four living children of a mixed **Catholic**, **Jewish**, and **Protestant** family. His father, a part owner of a bronze artwork shop, died when Hermann was ten. From childhood Muller was interested in science and evolution and as a student helped to organize one of the first school science clubs. After graduating from Morris High School, he attended Columbia University on scholarship and graduated with a B.A. (1910). He then received a M.A. in physiology (1911) from Cornell Medical School and returned to Columbia in 1912 to join geneticist Thomas Hunt Morgan (1866–1945) in studying genetic inheritance in *Drosophilia* (fruit flies). Muller was a faculty member in biology at the Rice Institute (1915–1918), received his Ph.D. from Columbia (1916), returned to Rice as a zoology instructor (1918–1920), and became a faculty member at the University of Texas (1920–1932). He received a Guggenheim fellowship for research in Germany (1932–1933), but failed to complete the research after becoming concerned about the **Nazi German** regime. Having embraced socialism, he was invited to become senior geneticist at the Academy of Sciences, Moscow (1933–1937). He subsequently left the academy due to a disagreement with Trofim Denisovich Lysenko (1898–1976), who promoted a neo-**Lamarckian** theory of acquired characteristics, which had become the official Communist view of inheritance. Muller was a guest lecturer at the University of Edinburgh (1937–1940) and a research associate and visiting professor at Amherst College (1940–1945). He finished his career as a faculty member at Indiana University (1945–1964).

While in graduate school, Muller began to experiment with ways of causing mutations in the fruit fly. In the 1920s he found that X-rays could increase the rate of mutations to 150 times their spontaneous rate. Based upon his findings, he proclaimed in 1926 that the **gene** was the basic hereditary unit of life, a conclusion that met with controversy. His findings, detailed at the International Congress of Genetics in Berlin in 1927, resulted in the Nobel Prize for Physiology or Medicine in 1946. Due to fears that radiation would increase lethal mutations in humans, he campaigned in the 1950s against the use of nuclear bomb tests and advocated protection from medical and industrial exposure to radiation.

Muller promoted **positive eugenics**, although he refused to join any eugenics organization. He presented a paper at the **Third International Congress of Eugenics** (1932) but criticized eugenicists for unsupported hereditary claims. He argued that environment was important for the development of human characteristics and encouraged the "fit" to reproduce. He expanded his eugenics ideals in *Out of the Night* (1935), a work with a socialist slant. Muller proposed that human evolution could be directed through several methods, including

Hermann Muller, a geneticist, recommended positive eugenics measures including artificial insemination with sperm from intellectually superior men as a method for improving the human race through selective breeding. (From: *Indiana Alumni Magazine.* Image courtesy of Indiana University Archives, Bloomington, Indiana.)

"children of choice," or voluntary **artificial insemination** of the best sperm for couples in which the male was sterile; **birth control**; and **abortion** on demand. He discussed the concept of "parthenogenecis" (**cloning**) and advocated equal rights for women and child care centers. In the 1954 he suggested that because individuals who were now kept alive by modern medicine would have died through **natural selection** (survival of the fittest), the genetic quality of the human race was being lowered. Therefore he advocated "artificial selection" to improve the human race. Muller coined the term **germinal choice** for this concept in 1960. Women with infertile husbands could select sperm, held in **sperm banks**, of highly intelligent and talented men. His ideas often brought severe criticism and condemnation from groups such as the Roman Catholic Church. Over the course of his career Muller was involved with many genetics organizations, awarded many prizes, served as a member of several editorial boards including the *Journal of Heredity*, and published numerous papers. He married Jessie Marie Jacobs (1923); they had one child and divorced in 1934. He married Dorothea Kantorowica (1939), and they also had one child. Muller died of congestive heart failure in Indianapolis, Indiana, before he could complete a book on germinal choice.

FURTHER READING: Carlson, Elof A., *Genes, Radiation, and Society* (1981); Muller, Hermann, *Out of the Night* (1984).

n

We believe the time will come when involuntary and random propagation will cease, and when scientific combination will be applied to human generation as freely and success-fully as it is to that of other animals.

John Humphrey Noyes, *Male Continence* (1872)

NAM FAMILY, THE: A STUDY IN CACOGENICS (1912)

The second of a series of research reports by the **Eugenics Record Office** (ERO), *The Nam Family* investigated the "hereditary nature" of undesirable social behaviors in a kinship. Both this report and *The Hill Folk*, the ERO's first study, were published in August 1912. Authors of *The Nam Family* were **Arthur H. Estabrook**, a eugenics **field worker**, and **Charles Davenport**, director of the ERO. The 85-page study, ERO Memoir No. 2, was funded by John D. Rockefeller Jr.; it was reprinted in 1968. This report, along with other family histories including *The Jukes in 1915* and *The Kallikaks*, helped popularize and give a "scientific basis" to **negative eugenics**, the major thrust of the early-twentieth-century eugenics movement in the **United States**.

The Nam family was reported to have been the descendants of a union between a "roving Dutchman" who had wandered into western Massachusetts from the Hudson Valley and an "Indian princess." The authors reported that local historical accounts suggested that around 1760 the family was "wealthy in land, having inherited it from their Indian ancestor." However, they were considered "vagabonds," surviving by sub-sistence farming, fishing, and hunting. Five children of one branch of the family left Massachusetts around 1800 and migrated into New York State. One bought a 160-acre farm, and many surviving descendants still lived near this tract of land, called Nam Hollow. Although some of the descendants became prominent in nearby com-munities, the majority were impoverished generation after generation.

Estabrook collected data from seven generations of Nams. Of the 784 family mem-bers traced, the authors reported that 88 percent of females and 90 percent of males were "given to drinking in excess." The authors noted, "In pairings with two alcoholic parents all children were in turn alcoholic. In pairings where both parents are tem-perate 82 percent were temperate." Although only one case of syphilis had been iden-tified among this kinship, 232 women and 199 men were identified as licentious (engaged in sexual relations outside of marriage). About 20.2 percent of children (180), were born out of wedlock. Eleven Nams had been in the county poorhouse and 40 had been sent to prison. However, the expenditure of public funds for the Nams was low in contrast to other **degenerate** families because they "lived outside of public control" in an isolated area. The authors also compared the Nams to the Jukes. A similar percentage of both groups were born out of wedlock. They found "harlotry" more common among the Nams and **prostitution** more common among the Jukes. Syphilis was less frequent among the Nams than the Jukes, and a lower percentage of Nams had been in the poorhouse.

The authors concluded that heredity was a critical factor in determining the lack of ambition, **alcoholism**, and licentiousness among the Nam kinship, but thought that environment might be important in some cases. Estabrook and Davenport suggested placing brighter children in "better families," but for the most part recommended **eugenic segregation** in a state institution through the reproductive years. They also recommended mandatory **eugenic sterilization** but realized that this was a controversial choice. The British *Eugenics Review* criticized the authors for not writing "with more sympathy for their unfortunate subjects."

FURTHER READING: Carlson, Elof Axle, *The Unfit* (2001); Estabrook, Arthur H., and Davenport, Charles B., *The Nam Family* (1968); Haller, Mark H., *Eugenics* (1984); Kevles, Daniel J., *In the Name of Eugenics* (1985).

NATIVISM AND NATIVISTS

The conviction that the United States should be preserved for **Anglo-Saxon** Protestants with traditional American values is termed *nativism*. Nativist movements include all forms of animosity by native-born "pro-American" crusaders against perceived undesirable immigrants. This fear of "the other" and the desire to control them was an undercurrent of eugenics and **Clean Living** health-reform crusades of the **Jacksonian** (1830–1850) and **Progressive** (1890–1920) eras. Many social problems, including **alcoholism, prostitution, venereal diseases, crime**, and poverty, were blamed on foreigners. Publications on both sides of the Atlantic from the 1840s through the 1920s popularized the idea that all the good things about the United States, Britain, and northern Europe were rooted in the inborn "superiority" of the Anglo-Saxon or Germanic **Nordic race**. Middle-class reformers concluded that U.S. political, social, economic, and health problems were largely due to immigrant groups that did not conform to established ways or assimilate easily.

Nativism was rooted in antipathetic attitudes toward **Catholics** and fear of papal control, which was brought to the New World by early English **Protestant** colonists. These attitudes stemmed from struggles for political control in Britain between the Roman church and breakaway Protestants. Nativism emerged in the 1830s and reached its peak with the formation of the Know-Nothing party of the 1850s. Protestant health reformers were hostile toward the "alien menace" of Irish Catholic immigrants as they believed drunkenness, **tuberculosis**, and **pauperism** found among them would be passed down to succeeding generations based upon the theory of **Lamarckian inheritance** of acquired characteristics. As part of the religious **Great Awakening** of the Jacksonian Era, health and social crusades attempted to clean up society and achieve "perfectionism" in order to bring on the Millennium (Christ's reign on earth).

Nativism was pushed underground during the Civil War era. In the 1880s hostility toward immigrants reemerged as a wave of new immigrants arrived from eastern and southern Europe. Fear that the extinction of "real Americans" and **race suicide** were possible led to legislative acts in the 1890s to prevent **Asian** and **eastern and southern European** immigrants from entering the country. In addition, **Jewish** immigrants from eastern Europe were seen as anarchists and Bolsheviks (Communists) who wished to destroy U.S. democracy. Roman Catholics and the church hierarchy were thought to be undermining democracy by attempting to institute papal control over the nation. The desire for healthy offspring and for keeping the "race from being polluted" by **degenerate** immigrant groups led to eugenic reforms and the **immigration restriction movement** that reached a zenith immediately before and after **World War I**. **Eugenic sterilization laws** in numerous states resulted in the sterilization of thousands of people, of whom a high proportion were impoverished foreign-born. The Ku Klux Klan, historically hostile toward **African Americans**, reemerged in 1915 to fight for immigration restriction. This group, together with nativist eugenicists including **Harry H. Laughlin**, **Lothrop Stoddard**, and **Madison Grant**, fostered passage of the **Johnson-Reed Immigration Restriction Act of 1924**. With the relaxation of immigration quotas in the 1960s immigrants from **Asia** and other nations came in large numbers into the United States. Nativism reemerged in the 1980s with animosity toward immigrants with diseases such as the HIV virus, and in the twenty-first, against potential terrorists, following the terrorist attacks of September 11, 2001.

FURTHER READING: Billington, Ray Allen, *The Origins of Nativism in the United States, 1800–1844* (1974); Engs, Ruth Clifford, *Clean Living Movements* (2000); Haller, Mark H., *Eugenics* (1984).

NATURAL SELECTION

Also called *survival of the fittest* or **Darwinism**, the concept of natural selection is considered the process that drives the evolution of various species. It is also a mechanism that weeds out **unfit** or **degenerate** members of the species. The theory suggests that in the struggle for existence those organisms with the best traits for a particular environment will survive and reproduce. Over time these survival traits will become predominant, leading to changes in the species. In the late nineteenth century natural selection was applied to social, economic, and ethnic survival in the philosophy of **social Darwinism**. The theory influenced eugenic thought during the first third of the twentieth century, justified **pauperism** (inherited poverty), and became a tenet for improving the human race through **eugenic sterilization** and **segregation laws** in several countries.

The concept of survival of the fittest was implied by British scientist Herbert Spencer (1820–1903) as early as 1851; he used the term as the force behind biological evolution in his *Principles of Biology* (1864). British naturalists **Charles Darwin** and Alfred R. Wallace (1823–1913), in their independent explorations of the evolution of species, described the idea of natural selection in 1858 in essays published together introducing their theory. Darwin used the term *natural selection* as part of his title, and Wallace coined the phrase *struggle for existence*. Both naturalists had been influenced by Thomas Malthus (1766–1834), an economist who suggested that under favorable conditions, humans could produce far more offspring than could survive, and that overpopulation could lead to the demise of the species from famine, war, or disease. Under adverse environmental conditions, a struggle for survival allowed those individuals with the best traits for the environment to survive and reproduce. Those traits, passed to their offspring through inheritance, led them to become dominant in the population. The concept also fit the accepted theory of **Lamarckian inheritance of acquired characteristics**. With increased knowledge of **genetics** during the first third of the twentieth century, conflicts, especially in **Britain**, between geneticists regarding the importance of **nature-nurture** and the mechanism of inheritance emerged. By 1930 most scientists deferred to **Gregor Mendel**'s rules of inheritance (**Mendelism**), which provided a basic foundation for Darwin's theory of natural selection along with mutations, or changes, in the genes.

In the last half of the twentieth century, advances in molecular biology techniques suggested other theories regarding inherited changes in populations. For example, in a variety of organisms more genetic variation was detected than could be attributed to natural selection. As a result, a "neutral theory of evolution," or change governed by genetic drift (random genetic changes at the **DNA** level) was posited. Other proposals in the final decades of the century suggested that evolution in populations proceeds by a variety of mechanisms, and that it has been influenced by many factors.

FURTHER READING: Darwin, Charles, *On the Origin of Species* (1985); Dawkins, Richard, *The Extended Phenotype* (1999); Dover, Gabriel A., *Dear Mr. Darwin* (2000); Endler, John A., *Natural Selection in the Wild* (1986); Gould, Stephen Jay, *The Structure of Evolutionary Theory* (2002).

NATURE-NURTURE

Debate whether heredity or environment is the major factor in the development of human behaviors, characteristics, mental functions, and certain diseases has been fraught with controversy for many generations. Some scholars have contended that mental and behavioral characteristics are essentially biologically or genetically determined, while others have claimed they are primarily shaped by family, social, cultural, and physical environments. **Eugenics** tends to be in the ascendency when hereditarian beliefs are popular and loses favor when environmental factors are embraced. The heredity and environmental perspectives have been shaped by economic, political, social, religious, and other factors, and not necessarily by rigorous scientific study. In the mid-nineteenth century characteristics such as **alcoholism, pauperism, feeblemindedness, insanity**, and temperament (personality) were believed to be transmitted from generation to generation through heredity. In 1874 **Francis Galton**, who coined the term *nature versus nurture*, suggested that mental abilities and most human traits were inherited. This belief fueled the eugenics movement and peaked in the mid-1920s. In the late 1930s the genetic basis for mental functioning among different groups and the inheritability of certain behaviors began to be discounted. The human infant was seen as a blank slate on which society shaped behaviors. Positive or negative environmental influences, including education, home environment, and economic opportunity, were considered the primary factors that led to **intelligence**, social economic status, and other human conditions. However, in the 1980s arguments for a genetic base for behaviors including mental ability, alcoholism, aggression, mental illness, and sexual preferences emerged. The nature-nurture debate continues. In the early twenty-first century a prevailing argument based upon **twin studies** and other research suggests that an individual inherits genes for a predisposition to certain behaviors or characteristics, but that environmental factors promote or dampen the exhibition of these attributes. It is hypothesized that **genes** are "switched on or off" depending upon environmental conditions.

FURTHER READING: Clark, William R., and Michael Grunstein, *Are We Hardwired?* (2000); Ridley, Matt, *Nature via Nurture* (2003); Rosenberg, Charles E., *No Other Gods*

(1976); Scher, Steven J., and Frederick Rauscher, *Evolutionary Psychology* (2002); Steen, Grant R., *DNA and Destiny* (1996).

NAZI GERMANY (1933–1945)

The **German eugenics** movement, which originated in the German Empire and expanded in the Weimar Republic after **World War I**, was taken over by National Socialism (Nazism) in 1933 as a governmental function. Drawing upon the leading German eugenics text, *Human Heredity and Race Hygiene* (1921) and the writings of race theorist **Hans F. K. Günther**, Adolf Hitler (1889–1945), in *Mein Kampf* (*My Struggle*, 1925), formulated his "race purity" theories. While Hitler was German chancellor the **Aryan** or **Nordic** supremacy (Nordicism) movement merged with the **race hygiene**, or eugenics movement. Under the Nazi government, race hygiene continued to focus on increasing the birthrate of the "fitter" classes and preventing the reproduction of the **unfit**, as it had in the first three decades of the century. However, eugenicists and their leading professional organization, the **German Society for Race Hygiene**, were now mandated to focus exclusively on improving the "Aryan race," and not the entire population, because the "purity" of the Aryan or **master race** was an integral part of Nazi philosophy and belief.

The world economic depression of the 1930s led to drastic budget cuts for health and welfare programs. Eugenic sterilization laws to decrease breeding among the **degenerate** had been championed by most German race hygienists for a decade on the grounds that it would reduce the social welfare burden. In July 1933, within six months of the Nazi takeover, the Law for the Prevention of Genetically Diseased Offspring was passed. This sterilization law, based upon a 1932 Prussian proposal initiated by biologist and former Jesuit Hermann Muckermann (1877–1962) and adapted from the model sterilization law of the **United States**, allowed for mandatory sterilization of individuals with **feeblemindedness**, schizophrenia, manic depression, hereditary epilepsy, hereditary blindness, deafness, physical deformities, Huntington's disease, and serious **alcoholism** after a review by a genetic health court. At the end of the first year of the decree approximately 62,000 people had been sterilized. By the beginning of **World War II** over 360,000 had been sterilized, mostly ethnic Germans. This activity was claimed to have saved millions in welfare costs. Soon other laws were passed to improve the health and efficiency of ethnic Germans. In 1935 the Law for the Protection of the Health of the German People required couples to undergo a medical examination prior to marriage and prohibited the marriage of individuals with **venereal disease** or **genetic diseases**.

Because Nazi Aryan supremacists were attempting to create a new race of **supermen**, they took issue with interracial procreation between ethnic Germans and **Jews** or other racial groups. The Law for the Protection of German Blood and German Honor of September 1935 prohibited marriage or relationships outside of marriage between Jews and "nationals of German or kindred blood." In an effort to keep the **race** "pure," the Gestapo in 1937 secretly sterilized about 400 mixed-race children with **African** ancestry fathered by French colonial occupation troops after World War I. To encourage reproduction among fit ethnic Germans, the *Lebensborn* **Program** of maternity centers was created in 1935. In the 1940s it became a patriotic obligation for the genetically fit to reproduce. Legal **abortions** became difficult, and fit, unmarried soldiers were strongly encouraged to become engaged, marry, and have many children.

Under National Socialism, supporters of Nordicism, who included a few eugenic leaders such as physicians **Alfred Ploetz** and **Fritz Lenz**, anthropologist **Eugen Fischer**, Swiss-born psychiatric researcher Ernst Rüdin (1874–1952), and internationally known geneticist Otmar Freiherr von Verschuer (1896–1969), continued to hold leadership positions at the prestigious **Kaiser Wilhelm Institute** (KWI) and other research institutions. Non–Nordic supremacy eugenicists, who were concerned about increasing the health and efficiency of the entire German nation, not just Aryans, were asked to leave their positions; among them were Muckermann, director of the Eugenics Department at the KWI, and Artur Ostermann (1876–?) head of the Prussian Ministry of Welfare, as well as all Jewish researchers and professors. After passage of the 1933 sterilization law, eugenicists including Lenz, Fischer, and Rüdin participated in the sterilization review courts. Physicians were sent to the KWI for race hygiene instruction, and after passage of the 1935 race purity laws, to learn procedures for identifying different races.

The Nazi government embraced the claim, made by eugenicists in many countries, that **war** killed a nation's most genetically fit and valuable. They reasoned, therefore, that if productive members of society were called to sacrifice their lives, then the same sacrifice could be justified for mentally and physically handicapped individuals who were deemed a burden to society. At the outbreak of **World War II**, in order to conserve supplies and resources and to make room for the wounded at hospitals, Hitler secretly authorized the "mercy killing" of incurable mentally ill and handicapped adults and children. This decree was issued at the end of October 1939 but was backdated to September 1, the day the Nazis invaded Poland. It is unclear whether Lenz, who served on a committee designed to formulate a law permitting euthanasia, knew about this secret order. It is unlikely that other professional eugenicists had knowledge of the activity because German race hygienists, like eugenicists in other countries, including the **United States** and **Britain**, with a few exceptions, did not approve of euthanasia. Between 1939 and 1941 when the program was stopped in Germany, almost 100,000 mostly ethnic Germans with serious mental and physical disabilities were killed. After this date up to 150,000 with disabilities were killed in the occupied territories until the end of the war.

With the opening of concentration camps in the early 1940s other "undesirables" including **Jews**, Gypsies (Roma), members of religious orders, homosexuals, and opponents of the regime were killed. This became known as the Holocaust. For his research, Verschuer used specimens from dead Auschwitz prisoners given to him by his former student and assistant Josef Mengele (1911–1979?). Records from individuals and institutes such as the KWI were deliberately destroyed at the end of the war, so it is speculated but not known for certain whether Verschuer knew of the circumstances surrounding the specimens he obtained. In the postwar years, many researchers, including Verschuer, Lenz, and Rüdin, after going through a "de-Nazification" process, were allowed to return to scientific professorships at German universities. *Racial hygiene*, as a term and an area of research, was abandoned. Former eugenics organizations, publications, and institutions either ceased to exist or changed their names so as not to be associated with the twisting of the eugenics philosophy that had resulted in mass murder. However, the 1932 draft of the voluntary sterilization bill was accepted by the U.S. and British postwar occupation forces, as were restrictive birth-control and abortion laws introduced under National Socialism and not relaxed until the mid-1960s. The Nazi policies of genocide demonized the concept of—and even the term—*eugenics* into the twenty-first century, resulting in concerns by some regarding the possible negative use of **genetic testing and screening**, **genetic engineering**, and other techniques of the **new eugenics**.

FURTHER READING: Kühl, Stefan, *The Nazi Connection* (1994); Weindling, Paul, *Health, Race, and German Politics between National Unification and Nazism, 1870–1945* (1989); Weingart, Peter, "German Eugenics between Science and Politics" (1989); Weiss, Sheila Faith, *Race Hygiene and National Efficiency* (1987).

NEED FOR EUGENIC REFORM, THE (1926)

Written by British eugenicist **Leonard Darwin**, a son of naturalist **Charles Darwin**, *The Need for Eugenic Reform*, as stated by the author, was aimed at the "well-educated man or woman, without special scientific training, who is prepared not only to take racial problems seriously, but also to devote some considerable time and energy to the consideration of the eugenic method of attempting to benefit the human race. For the success in the eugenic campaign will finally depend on such as these." This classic British eugenics work was published in London by J. Murray and in New York by Appleton in 1926. The work was composed of many of Darwin's previous writings from his lectures, essays, and articles published in *Eugenics Review* and other journals. Darwin dedicated the book to his father and in the preface thanked **R. A. Fisher** for statistical help. The 529-page book has twenty-seven chapters with titles such as "Evolution," "Heredity and Environment," "Racial Poisons," "Inheritance of Acquired Differences," "Natural Selection," "The Lessons of the Stock Yard," "Feeblemindedness," "The Burden of the Less Fit," and the "Elimination of the Less Fit."

The author presents many arguments of the **British eugenics movement** during its peak years. He discusses the problem of **race degeneration** and the potential decline of civilization due to the differential birth rate. According to prevailing British attitudes, **paupers** and the **unfit** were reproducing faster than the upper and the productive middle classes, thus overtaxing the welfare system, reducing national vitality, and threatening the world status of the British Empire. The author makes recommendations for **positive eugenics**, such as child allowances, eugenic **family history** registries, national health insurance, school scholarships for children of good heredity, and encouragement of the fit to produce more children to prevent **race suicide**. He suggested eugenic segregation in colonies as a **negative eugenics** measure to prevent the unfit from reproducing. The work was reprinted in 1984 when study of the early-twentieth-century eugenics movement became popular and the "**new eugenics**" began to emerge.

FURTHER READING: Darwin, Leonard, *The Need for Eugenic Reform* (1984).

NEGATIVE EUGENICS

Preventing the **unfit** from reproducing was termed *negative eugenics* by **C. W. Saleeby**, a British physician, in 1907. He described it in detail in *Parenthood and Race Culture* (1909). Negative eugenics in the first half of the twentieth century reflected the belief that some people were physically, mentally, or socially **degenerate** by virtue of defective biology and that those **dysgenic** traits, passed on to offspring, would lead to **racial degeneracy** (weakening of the human race). It was the primary eugenics philosophy of the **United States, Germany**, and some Canadian provinces and Scandinavian countries. British naturalist **Francis Galton** coined the term *eugenics* in 1883 and recommended that those with high mental ability marry young and produce many children with financial help from the state. This concept became known as **positive eugenics** or, sometimes, *Galtonian* or *classical eugenics*. However, near the end of his life Galton suggested that those who are "seriously afflicted by lunacy, **'feeblemindedness,'** habitual **criminality**, and **pauperism**" (emphasis added) should not marry. By the middle of the second decade of the century both positive and negative eugenics concepts were found in eugenics publications.

Negative eugenics policies implemented to prevent the breeding of the unfit in the United States included mandatory **eugenic sterilization** or **eugenic segregation** in institutions during child-bearing years, marriage licensing laws, and **immigration restriction laws**. By definition, negative eugenics also included euthanasia, infanticide, and **abortion**. However, most eugenicists in the first four decades of the twentieth century rejected these measures. Beginning in 1939 and through 1945 in countries occupied by **Nazi Germany**, an estimated 20 million civilian adults and children, of many religious and ethnic groups, were killed or died of maltreatment in an effort to "purify the race." Of this number 6 million were **Jews**. This extremist form of negative eugenics, known as the Holocaust, became associated with all eugenic measures and demonized the entire concept in most Western cultures into the twenty-first century. Some groups, such as the Roman **Catholic Church**, consider **new eugenics** techniques of **assisted reproduction, in vitro fertilization, germinal choice** (selection of egg or sperm donors with specific characteristics), gender selection for **family balancing**, and abortion as negative eugenics.

FURTHER READING: Carlson, Elof Axel, *The Unfit* (2001); Lukas, Richard C., *The Forgotten Holocaust* (1986); Pernick, Martin S., *The Black Stork* (1996); Reilly, Philip R., *The Surgical Solution* (1991).

NEGROID

See Africans or African Americans

NEW EUGENICS

Practices and procedures to increase the probability of a healthy child or a child with specific characteristics, to identify an individual's genetic material, and to manipulate human **DNA** were called *the new eugenics* by the 1990s. New technological methods and an international effort to map the **human genome** renewed a discussion of eugenics ideas in the 1980s. British scientist **Francis Galton**'s 1904 definition of eugenics as anything used to "improve or impair the inborn qualities of human beings" covered a range of activities found at the turn of the twenty-first century and embraced the concepts of both heredity and environment (*nature-nurture*). In terms of environmental efforts, the new eugenics includes health education programs to persuade pregnant women to eat well-balanced diets and to abstain from **tobacco** and **alcohol** during pregnancy. It encompasses **assisted reproduction** techniques including **germinal choice** (selection of egg or sperm donors with specific characteristics),

artificial insemination, in vitro fertilization procedures, preimplantation diagnosis of embryos for **genetic diseases**, and **abortion**.

Some research results suggest that differences between individuals and/or groups in **intelligence** and artistic talent are largely inherited. Concern about the differential birthrate, in which the birthrate is increasing among the less intelligent and decreasing among the most intelligent, is part of new eugenics thought. Other aspects of the new eugenics include the creation of **sperm banks** such as the **Repository for Germinal Choice**, with sperm from selected or highly intelligent individuals. Biotechnology and **genetic engineering** techniques such as **genetic therapy** and therapeutic cloning to treat a person with his or her own cloned tissues, the potential of human **cloning, genetic screening and testing** for inherited conditions, and **genetic profiling** for identification are also part of the new eugenics. These and other rapidly developing techniques have caused some individuals to express concern that DNA profiles could lead to national data banks that could be used for discrimination in employment, educational opportunities, medical insurance, treatment, or other activities. Some evangelical **Protestants** and the Roman **Catholic Church** are concerned about the ethics and morality of the new procedures.

While the eugenics movement of the early twentieth century aimed to "improve the race," the new eugenics of the first decade of the twenty-first century has been consumer driven. Rather than promote governmental legislation to sterilize the **unfit** due to concerns about **racial degeneracy** and the decline of civilization, individuals and couples demand various techniques and medical procedures in order to increase the probability of having a healthy child, to avoid having a child with a debilitating **genetic disease**, to learn their family lineage, or to select sperm with either the X or the Y chromosome (preconception gender selection) for **family balancing** or "designer babies." Since the use of current and emerging techniques is likely to increase in developed nations and among the wealthy, the legal, privacy, civil-rights, ethical, and public-policy implications of many programs and procedures considered part of the new eugenics will likely provoke continued debate.

FURTHER READING: Carlson, Elof Axel, *The Unfit* (2001); Duster, Troy, *Backdoor to Eugenics* (2003); Hubbard, Ruth, and Elijah Wald, *Exploding the Gene Myth* (1993); McGee, Glenn, *The Perfect Baby* (1997); Reilly, Philip R., *Abraham Lincoln's DNA* (2000); Rifkin, Jeremy, *The Biotech Century* (1998); Stacey, Meg, (ed.), *Changing Human Reproduction* (1992); Stock, Gregory, *Redesigning Humans* (2002).

NORDICS OR NORDIC RACE

Individuals with ancestry from northwestern Europe, the British Isles, and Scandinavia, generally characterized by light skin, eyes, and hair; tall stature; and angular features, were classified as *Nordics* in the early twentieth century. The term was first used by French anthropologist Joseph Deniker (1852–1918) in 1900 and was popularized by German race theorist **Hans F. K. Günther**. The words *Teutonic, Germanic, Aryan*, and *Anglo-Saxon* were also used from the late nineteenth century onward. The preferred term under **Nazi Germany** in the 1930s and by some white supremacy groups in the United States and Europe during the late twentieth and early twenty-first centuries was Aryan. "Race theorists," anthropologists, and a few eugenicists of the late nineteenth and early twentieth centuries believed Nordics were a "superior race" and that they exhibited the characteristics of truthfulness, energy, good judgment, boldness, and strength necessary for the advancement of civilization. The promotion of this concept, known as *Nordicism*, was often fused with anti-Semitism. Fear of **race suicide** of Germanic peoples due to low reproduction, death of the fittest young men in **war**, overbreeding of non-Nordics, or interbreeding of Nordics with "inferior" ethnic and racial groups were a concern of many race theorists and **nativist** eugenicists.

Some late-nineteenth- and early-twentieth-century scholars believed that in antiquity Nordic Indo-European or Aryan tribes migrated in waves into southern Europe and Asia Minor from the north. As they acquired power, it was theorized, they caused the rise of Greek, Roman, and other civilizations. Eventual "denordization" (disappearance of Nordic blood) of the leadership was thought to have resulted in the downfall of these civilizations. From the late Roman Empire to the early eleventh century, waves of Germanic tribes migrated to, or invaded, the British Isles and central and southern Europe. This historic fact led to a belief that the Renaissance was due to the emerging leadership of people of "noble Nordic blood." French race theorist **Arthur de Gobineau** proposed in *Essay on the Inequality of Human Races* (1853) that "superior" northern Europeans advanced art, science, religion, politics, and the foundation of Western culture. In the 1920s Günther argued in ***Racial Elements of European History*** (1927) that notable individuals from antiquity to modern times had Nordic features and physical characteristics, giving apparent support to this theory. American nativist eugenicists **Madison Grant** and **Lothrop Stoddard** also presented similar ideas. The aim of race theorists in the 1920s, and during the Nazi regime was to prevent **racial degeneracy** and denordization among Germanic and Anglo-Saxon peoples in order to prevent the perceived downfall of Western civilization.

FURTHER READING: Coon, Carleton S., Stanley M. Garn, and Joseph B. Birdsell, *Races* (1981); Grant, Madison, *The Passing of the Great Race* (1970); Günther, Hans F. K., *The Racial Elements of European History* (1992).

NOYES, JOHN HUMPHREY (September 3, 1811–April 13, 1886)

A religious reformer, Noyes founded the **Oneida Community of Perfectionists**, the first group of the modern era to practice positive eugenic ideals over several decades. Born in Brattleboro, Vermont, he was one of nine children of a prominent businessman from an old-stock colonial **Anglo-Saxon** family. His mother, also from an old family, was devoted to the religious education of her children. When Noyes was ten, the family moved to Putney, Vermont. Noyes studied in Brattleboro and entered Dartmouth College (1826–1830), Hanover, New Hampshire from which he graduated with high honors. He spent a year studying law. Caught up in the religious fervor of the second **Great Awakening**, he decided to study theology and entered the Theological Seminary, Andover, Massachusetts (1831), but transferred a year later to Yale's seminary. He was licensed to preach in the Congregational Church in 1833. Noyes became a convert to Perfectionism, the belief that "man could achieve perfection on earth." In 1834 Noyes announced that a second evangelical conversion had brought him "complete release from sin," along with the revelation that Christ's second coming had occurred in 70 A.D. Because these beliefs were incongruent with mainstream Protestant theology he was forced to withdraw from theological studies and his license

John Humphrey Noyes, a religious reformer, founded the Oneida Community of Perfectionists. Under his guidance the community practiced positive eugenics, called "stirpiculture," in order to produce the best offspring. (Courtesy Oneida Community Mansion House.)

to preach was annulled. Noyes spent a few years visiting various Perfectionist groups in New York and Massachusetts and gathered a group of followers. He returned to his home in Putney in 1836. In 1841 the group formally organized and developed a community of "Bible or Christian Communists" whose aim was to spread the Perfectionist message. However, after being charged with adultery due to the sexual practice of the community, Noyes fled Vermont. In 1848 he and his followers reestablished their commune in Oneida, New York. Noyes also established communities in other places, but only those in Oneida and Wallingford, Connecticut, were successful. In his later years Noyes tended to reside at Wallingford.

Following a disappointing love affair, Noyes deemed that perfectionism could not be reached within a monogamous marriage, which he considered a "spiritual tyranny" and slavery. In 1837 he proclaimed his belief in "complex marriage," in which everyone in the commune was married to each other. Sexual intercourse was seen as having two forms, "amative" and "propagative." Complex marriage was first practiced in the Putney community in 1846. By 1869 the community practiced **stirpiculture**, or **positive eugenics**, in order to produce the best offspring. For this procedure, only the most "spiritually advanced" members were given permission by Noyes and a committee to have children. Sexual "interviews" for pleasure were allowed, but "male continence"—lack of ejaculation as a form of **birth control**—was required.

Noyes wrote a number of tracts concerning his perfectionist and eugenic theories including *Bible Communism* (1848), which explained Perfectionism, and *Male Continence* (1849), a widely condemned booklet that outlined a formula for allowing a broad range of pleasurable sexual experiences while at the same time retaining semen for the "sacred purpose" of propagation. Noyes expanded his eugenic concepts in *Essay on Scientific Propagation* (c. 1875). In this work he also criticized British naturalist **Francis Galton** for not presenting a plan of action for what would be later termed *positive eugenics*. After three decades of communal living and eugenic experimentation, growing dissatisfaction within the Oneida community and outside opposition to its practices forced Noyes to allow members to marry legally among themselves in 1879. In 1881 Noyes and a few followers fled to Canada to escape legal difficulties. The community disbanded and became a "joint-stock company." Noyes married Harriet A. Holton (1838) and fathered one surviving child with her and at least eight more with other women in the community. He died in Niagara Falls, Ontario, Canada, having created one of the longest-lasting positive eugenics experiments of modern times.

FURTHER READING: Kevles, Daniel J., *In the Name of Eugenics* (1985); Klaw, Spencer, *Without Sin* (1993); Walters, Ronald C., *American Reformers 1815–1860* (1978).

O

Eugenics, in asserting the uniqueness of the individual, supplements the American ideal of respect for the individual. Eugenics in a democracy seeks not to breed men to a single type, but to raise the average level of human variations, reducing variations tending toward poor health, low intelligence, and antisocial character, and increasing variations at the highest levels of activity.

Frederick Henry Osborn, *Preface to Eugenics* (1951)

OCCIDENTALS

See Anglo or Anglo-Saxon Americans; Euro or European Americans; Nordics or Nordic Race)

OLSON, HARRY (August 4, 1867–August 1, 1935)

A judge and eugenics supporter, Olson championed eugenic sterilization for criminals. Like most other American eugenicists, he was concerned that **racial degeneracy** in the United States would lead to the political and economic deterioration of the country from **unfit** immigrants and **criminals**, and to **race suicide** from inadequate reproduction among the healthy and fit. Born in Chicago, Olson was brought up in a Lutheran home on a Kansas farm until age thirteen, when his father died. He attended high school in small Illinois town and then taught school. He then attended Washburn College (1887–1888), Topeka, Kansas, and went to Chicago, where he earned a law degree from Union College Law School (1891). Olson was admitted to the bar in 1891 and became an assistant and, later, first assistant in the Office of State's Attorney for Cook County. In 1906 he was elected chief justice of the newly established Chicago Municipal Court, where he served until 1930. Olson pioneered the use of psychologists in criminal cases. He helped establish a "psychopathic laboratory" at his court in 1914 where inmates were tested using the newly developed **intelligence test**. Most criminals were found to be **feebleminded**, **morons**, or **insane**. These tests led Olson and others to conclude that a major cause of crime was "mental abnormality." He appointed **Harry H. Laughlin**, superintendent of the **Eugenics Record Office**, **Cold Spring Harbor**, New York, as the court's official eugenics authority.

Olson was one of the original incorporators of the **American Eugenics Society** and served on its advisory council and board of directors (1923–1930). He actively participated in the society and chaired its Committee on Crime Prevention. Olson was active in the **Eugenics Research Association** and served as president in the early 1920s. He believed that crime prevention was the first step in a eugenics program. He suggested "the weeding out of defective stocks" and supported **eugenic segregation** in state-run institutions and **eugenic sterilization laws**. When Laughlin was unable to find a publisher for his survey on eugenic sterilization laws, Olson helped facilitate its publication. For this work, *Eugenic Sterilization in the United States* (1922), Olson wrote the preface. Reflecting his concern over crime and undesirable immigrants, he suggested that "America, in particular, needs to protect herself against indiscriminate immigration, criminal degenerates, and race suicide" and favored **immigration restriction laws**. Over his lifetime he wrote numerous legal papers, many on the subject of criminality and eugenics. He received several honorary degrees. He married Bernice Miller (1902), with whom he had three children. He died at his home in Oak Park, a Chicago, Illinois, suburb, of a heart attack.

FURTHER READING: Haller, Mark H., *Eugenics* (1984); Reilly, Philip R., *The Surgical Solution* (1991).

ONEIDA COMMUNITY OF PERFECTIONISTS (1848–1881)

Founded by reformer **John Humphrey Noyes**, the Oneida Community was a successful religious commune and one of the longest-lasting **positive eugenic** experiments of modern times. It arose out of the ferment of the Second **Great Awakening** and the **Jacksonian Era** (1830–1860) reform movements and predates the eugenics concept of British naturalist **Francis Galton**. Noyes underwent a religious conversion and gathered a group of followers who embraced *perfectionism*, the belief in obtaining a sinless life on earth. This small group organized more formally in 1841 and pooled their resources to create a primitive form of "Bible or Christian Communism," in which property was shared. In 1846 "complex marriage," meaning that each man and each woman were married in common to one another, was instituted. To avoid prosecution for adultery due to this practice, Noyes had to flee Putney, Vermont, but he and his followers reestablished the community in Oneida, New York, in 1848. In its new location it prospered after establishing an animal-trap business and other endeavors, some based on inventions of the commune members such as a mop wringer. Noyes made several attempts to establish similar communes, but only two were successful. In 1874 the Oneida group had 235 members, and one in Wallingford, Connecticut, numbered 40.

Community members considered themselves "saints" purified by religious experience, and they were committed to absolute fellowship among themselves. They saw

In the Oneida Community, women and men were equal in business and social life. Women wore their hair and skirts shorter than the custom of the day for comfort and ease of work. (Courtesy Oneida Community Mansion House.)

incompatibility between communal relationships and the exclusive legal and physical bonds of conventional marriage, which they viewed as "slavery" for women, who were often overburdened with childbirth. Women and men were equal in business and social life; women wore both their hair and skirts shorter than the custom of the day. Commune members lived together in one house, a "unity house," which by the late 1860s had become a 300-room mansion with each adult member having his or her own small, narrow bedroom. For complex marriage, rigorous rules were established. Individuals were discouraged from becoming attached to one other person. Sexual intercourse was encouraged as long as the woman consented and the man did not reach orgasm when sex was solely for "amative," not reproductive, purposes. The contraceptive practice of "male continence," by which the man refrained from ejaculation, was required.

Until around 1860 mutual consent between two people was generally all that was required for sexual interaction in the Oneida Community. However, by 1860 all requests for a sexual meeting had to be made through a third party and were recorded in a ledger. This led to a positive eugenic program with planned reproduction that Noyes termed **stirpiculture**. In 1869 a committee was formed to suggest scientific combinations of community members to become parents. Drawing upon **Lamarckian inheritance of acquired characteristics**, Noyes believed that moral traits were passed on to children, and that the men and women selected should be the most spiritually advanced in the community, as well as physically healthy.

When only certain individuals were allowed to have children, the resulting resentments polarized the group and led to the disintegration of the Oneida communal marriage system. Discontent along with outside pressures against the community's sexual practices persuaded Noyes in 1879 to abandon complex marriage, and many community members married conventionally. This in turn led to abandonment of communal property ownership the following year and formation of a "joint-stock company," the Oneida Community Ltd. In 1881 Noyes and a few followers went to Canada to escape legal difficulties, and the Perfectionist community disintegrated within a few years. Many community members continued to work in the industries. Some residents lived in the mansion in apartments created from several bedrooms, which are still rented by some mostly elderly descendants of Stripicult children. The Oneida Company flourished in the late twentieth century, manufacturing silver-plate and stainless steel tableware. As of 2005 the company was still manufacturing kitchenware, although it has encountered economic cutbacks like other American manufacturing operations.

FURTHER READING: Klaw, Spencer, *Without Sin* (1993); Sokolow, Jayme A., *Eros and Modernization* (1983); Walters, Ronald C., *American Reformers 1815–1860* (1978).

ORIENTALS
See Asians or Asian Americans

OSBORN, FREDERICK HENRY (March 21, 1889–January 5, 1981)
During the 1930s Osborn played a key role in shifting the emphasis of eugenics in the United States to the study of population and social biology through his leadership in the **American Eugenics Society**. Born in New York City to a wealthy, old stock **Anglo-Saxon** colonial family, he was the eldest of four siblings. His father was an attorney and philanthropist, and his uncle was museum director and pioneer eugenics leader

Henry Fairfield Osborn. After attending a private school, he entered Princeton University and graduated Phi Beta Kappa with an A.B. (1910). He attended Trinity College at Cambridge University (1911–1912) and then entered the business world, becoming president of a Detroit railroad company, which he sold to car manufacturer Henry Ford (1863–1947) in 1920. Osborn then went into investment banking in New York City, but he retired in 1928 and became active in eugenics, environmental, and population issues. He became a research associate at the American Museum of Natural History and studied anthropology, sociology, and psychology. At the beginning of **World War II** he was given a temporary appointment as a general in the U.S. Army and named director of the Moral Branch, later called the Information and Education Division (1941–1945). After the war he was deputy representative from the United States to the United Nations Atomic Energy Commission (1947–1950). He was involved with numerous organizations the rest of his life.

Like many others who supported eugenic from the late nineteenth century onward, Osborn was concerned that new medical and **public health** measures were weakening the **genetic** pool and leading to **racial degeneracy**, based upon the premise that those who normally would have died early though the process of **natural selection** (survival of the fittest) were now maturing to bear children. However, he refuted some of the more racist opinions held by his uncle and **nativist** eugenicists such as **Lothrop Stoddard** and **Madison Grant** and distanced himself from the **Eugenics Record Office** in **Cold Spring Harbor**, New York. Osborn promoted **positive eugenics**, suggested financial aid to mothers, and recommended that "privileged" individuals produce more children. Osborn became actively involved with fostering the eugenics movement in the **United States**. He was a member of the **Galton Society**. He joined the American Eugenics Society soon after its inception and was on its advisory council until his death (1928–1981). He became secretary in 1931, a role he continued for many years. He was secretary-treasurer (1936–1945; 1959–1970) and president (1946–1952). When the old eugenics movement shifted direction in the 1930s he was influential in refocusing the outward goals of the society from hereditarian toward environmental concerns and population control. In the post–World War II era, he helped shift the society to an organization with more scientific goals that examined both genetic and environmental aspects of human behavior and populations.

As the early-twentieth-century eugenics movement dissipated and evolved into population concerns and **social biology** over the second half of the century, Osborn helped found the **Pioneer Fund** to fund research on population differences and was its president (1947–1956). Unlike many earlier eugenicists, he supported **birth control** and belonged to birth control organizations. He was involved with eugenics on the international level, helped fund the **Third International Congress of Eugenics** (1932), and was a member of the London-based **Eugenics Society**. Osborn wrote numerous eugenics and population articles and books including *Heredity and Environment* (1933), *Preface to Eugenics* (1940), and *Possible Effects of Differential Fertility on Genetic Endowment* (1952). Over his lifetime he received several honorary degrees. He married Margaret L. Schieffelin (1914) and, following his own advice for positive eugenics, the two had six children—four girls and two boys. He died in Garrison, New York, after a long, active life.

FURTHER READING: Haller, Mark H., *Eugenics* (1984); Kevles, Daniel J., *In the Name of Eugenics* (1985).

OSBORN, HENRY FAIRFIELD (August 8, 1857–November 6, 1935)

An eminent paleontologist, anthropologist, and director of the American Museum of Natural History in New York City, Osborn was a leading pioneer of the eugenics movement. He had **nativist** leanings and championed immigration restriction. Born in Fairfield, Connecticut, Osborn was the second of four children of a wealthy Presbyterian family of "old colonial" **Anglo-Saxon** stock. Educated at the private Lyons Collegiate Institute of New York, he graduated from Princeton University (1877) with an A.B. in archaeology and geology. Following graduation he participated in archeological expeditions to the western states (1877–1878), studied anatomy and physiology at the College of Physicians and Surgeons of New York (1878–1879), spent a year in Europe studying embryology (1879–1880), and received a Sc.D. degree from Princeton (1881), where he taught natural science for ten years (1881–1891). In 1891 Osborn accepted a joint appointment at Columbia University and the American Museum of Natural History. At the museum he quickly created a world center of vertebrate paleontology. Financial backing through family connections supported expeditions around the world to collect fossil displays and establish educational programs. In 1908 he became president of the museum, where he remained until his retirement.

Osborn's status as a scientist lent respectability to both the eugenics and the **immigration restriction** movements in the **United States**. Disturbed by the nation's growing urbanization and ethnic diversity, Osborn viewed unrestricted immigration as a threat to **public health** and the survival of old-stock **Anglo-Saxon** Americans and their assumed "superior" values and heredity. In *Men of the Old Stone Age* (1915) he implied that humanity **degenerated** when "inferior" **eastern and southern European** "**races**" interbred with the more intelligent and artistic "Cro-Magnons," or "**Nordics**." Osborn also wrote the preface to **Madison Grant's** *The Passing of the Great Race* (1916), which had a similar theme. Along with Grant and **Charles Davenport**, Osborn was a founder of the **Galton Society** and was active in other eugenic organizations including the **Eugenics Record Office's** Committee on Eugenics. He was an organizer of the **First International Eugenics Congress** (1912) and president of the Second Congress, which was postponed due to **World War I** until 1921. Under his presidency the American Museum of

Anthropologist Henry Fairfield Osborn, director of the American Museum of Natural History, New York, was a major leader of the eugenics movement in the United States. (From: *Eugenics in Race and State* [1923], Main Library, Indiana University, Bloomington, Indiana. Image courtesy of the Digital Library Program.)

Natural History sponsored both the **Second** and **Third International Congresses of Eugenics** (1932).

Osborn considered **birth control** detrimental to eugenic principles. While the middle and upper classes were limiting their number of children, the lower classes, who were not using contraceptives, were producing many **unfit** offspring. When his views and scientific interpretations became less acceptable to the scientific community Osborn was pressured, in 1933, to resign as president of the museum. Over his career, his large staff engaged in numerous projects and wrote many of his 900 publications, for which they received little credit. He received numerous honorary awards and was considered an eminent man of science during the peak of the **Progressive Era**. Osborn married Lucretia Perry (1881), with whom he had five children. Two years after retiring he died of a heart attack in his study overlooking the Hudson River in Garrison, New York, while writing a tome on the evolution of the elephant.

FURTHER READING: Haller, Mark H., *Eugenics* (1984); Kevles, Daniel J., *In the Name of Eugenics* (1985); Reilly, Philip R., *The Surgical Solution* (1991).

P

Positive eugenics seeks to improve the race by encouraging greater reproductivity among the racially fitter, the civically more worthy, stocks. Negative eugenics aims to prevent contamination and degeneration by prohibition of parenthood to the obviously and grossly unfit. The peculiar means employed by positive eugenics are mainly educational, by negative eugenics legislative.

Harvey Ernest Jordan, *Eugenics: Twelve University Lectures* (1914)

PARENTHOOD AND RACE CULTURE: AN OUTLINE OF EUGENICS
(1909)

Written by British physician and eugenics supporter **C. W. Saleeby**, *Parenthood and Race Culture* became the most popular eugenics text of the **British eugenics** movement. It was also popular in the **United States** and was reprinted several times through 1917. Initially published by Cassell and Company in New York and London, it contained 531 pages and seventeen chapters divided into two parts, "The Theory of Eugenics" and "The Practice of Eugenics." Titles of chapters included "Natural Selection and the Law of Love," "Heredity and Race-culture," "The Supremacy of Motherhood," "Selection through Marriage," "The Racial Poisons," and "The Promise of Race-Culture." The book was dedicated to "**Francis Galton**, the August Master of All Eugenists." The author explained in the preface that the work was a general introduction to eugenics that aimed to define the scope of eugenics to the middle and working classes. In this work Saleeby coined the terms **positive** and **negative eugenics**. He argued that positive eugenics was the original eugenics, as defined by Galton, which encouraged "parenthood by the worthy." Negative eugenics discouraged "parenthood of the unworthy." The author also coined the term **racial poison** to include substances and diseases such as **alcohol** and syphilis. He recommended eugenic programs including financial help to parents likely to have "worthy children"; education for parenthood; eugenic marriage; and **eugenic segregation** (institutional care) of the **feebleminded**, **insane**, and **criminals**, but he was neutral about **eugenic sterilization**. The book received positive reviews on both sides of the Atlantic.

PASSING OF THE GREAT RACE, THE: THE RACIAL BASIS OF EUROPEAN HISTORY (1916)

This best-selling popular work, written by American **nativist** eugenics leader **Madison Grant**, was a major influence on the **immigration restriction movement** and the nativist branch of the eugenics movement in the United States. The book had gone through six printings by 1921 and was acclaimed as an important work by the popular press. It influenced the direction of the eugenics movement in regard to the "social worth" of various **races** in the United States and Europe. The book was published by Charles Scribner's Sons in New York. The first edition had 245 pages in two parts and fourteen chapters. Part 1, "Race, Language and Nationality," included titles such as "Race and Democracy" and "The Competition of Races." Part 2, "European Races in History," contained titles such as "The Mediterranean Race" and "The Expansion of the Nordics." The preface was written by anthropologist **Henry Fairfield Osborn**. The basic premise of the book was based upon **social Darwinism** and the racial classification theories of **Arthur de Gobineau** and others.

The publication approached European history in terms of heredity as manifested through three so-called European **races**: Nordic, Alpine, and Mediterranean. Moral, intellectual, and spiritual attributes were believed to be transmitted unchanged from generation to generation in the same manner as physical traits, resulting in three races with different characteristics. Grant argued that the **Nordic** (northern European) race was "racially superior" to the other groups due to its achievements in science, religion, economics, and government throughout Western history. He proposed that such attainments made those with northern European ancestry a **master race**. He expressed concern about **race suicide** among this group because they were not reproducing as rapidly as "lower" or "inferior" races. To deal with this situation he recommended sterilization of "the criminal, the diseased and the insane" and, eventually, "worthless race types." The book was praised by race theorists in other countries, including **Hans F. K. Günther** in Germany. Many of Grant's prescriptions were later found in the racial policies of the **Nazi German** government of the 1930s.

In the **United States** the work was widely read by academics, politicians, reformers, and the educated public. Politicians such as President Theodore Roosevelt (1858–1919), respected scientists, and other prominent individuals wrote generally favorable reviews in both popular and scientific journals. The book fueled the **immigration restriction movement**, which culminated in the **Johnson-Reed Immigration Restriction Act of 1924** that led to quotas for immigrants from many nations who were considered inferior, especially **eastern and southern Europeans**. Although Grant represented the extreme nativist branch of the eugenics movement, his views were considered respectable by the majority of educated Americans at the time.

FURTHER READING: Grant, Madison, *The Passing of the Great Race* (1970); Paul, Diane B., *Controlling Human Heredity* (1995).

PAUPERISM AND PAUPERS

A nineteenth-century term, *pauperism* was considered inherited poverty because poverty and other "**degenerate**" traits, such as **feeblemindedness** and **alcoholism**, were often found in certain families over several generations. Although the poor have been a social phenomenon since antiquity, they were generally cared for by religious groups, relatives, and local charities until the late nineteenth century. By the 1890s public institutions and welfare programs (often termed "poor relief") had become common in Western cultures. As social welfare programs became institutionalized, paupers were considered a major social and economic drain on society. In the United States the problem increased as poor immigrants from **eastern and southern Europe** filled urban slums, leading to **nativist** fears of the demise of the middle-class work ethic, the **degeneration** of society, and the decline of civilization.

In the last decade of the nineteenth century, social welfare reformer Henry Boies (1837–1903), in *Prisoners and Paupers* (1893), classified paupers into three groups. The first category was composed of the physically, mentally, and morally defective—the feebleminded, **prostitutes**, the **insane**, epileptics, hardened criminals, "imbeciles," and "cripples." The second group included "tramps, idlers, beggars, pickpockets, and thieves." These two groups were classified as "hereditary paupers." The third class, sometimes called the "deserving poor," included the elderly, the sick, accident victims, and orphans and widows who were forced into poverty and who deserved charity. This group was believed capable of resuming productive lives under better circumstances. To eliminate the indigent and their increasing economic drain on society, some social reformers argued that hereditary paupers should be prevented from breeding. This philosophy manifested itself as **eugenic sterilization** and other laws beginning in the first decade of the twentieth century.

During the Depression of the 1930s, when numerous working- and middle-class men were without employment, hereditarian views of pauperism began to wane. In addition, the developing field of **genetics** was discovering that human behavior was far more complex than simple **Mendelian inheritance**. Environmental problems and lack of opportunity, rather than heredity, were now seen as the root cause of the lack of social and economic success. This remained the prevalent view throughout the rest of the century. Although the reasons for social and economic failure are complex, a few researchers found in the 1990s that in some families certain genetic traits appeared to be associated with certain behaviors, or mental dysfunctions, that might contribute to social problems and lack of success. It was also postulated that heredity influenced behavior and that environment influenced the "switching" of **genes** on or off for the manifestation, or lack, of certain characteristics. However, the possibility that the expression of certain genetic traits could lead to low achievement and poverty is fraught with controversy.

FURTHER READING: Broder, Sherri, *Tramps, Unfit Mothers, and Neglected Children* (2002); Carlson, Elof Axle, *The Unfit* (2001); Haller, Mark H., *Eugenics* (1984); Kevles, Daniel J., *In the Name of Eugenics* (1985).

PEARSON, KARL (March 27, 1857–April 27, 1936)

An internationally known British statistics pioneer, Pearson developed mathematical measurements to support **Charles Darwin**'s theory of evolution and inheritance and to establish eugenics as a scientific field. Born in London the son of a Quaker lawyer, with the given name Carl, Pearson later changed his first name to Karl after living in Germany. As a child he exhibited mathematical ability while being educated at home. He attended the University College School but withdrew in 1873 because of poor health. After a year of private tutoring, Pearson was awarded a scholarship to King's College, Cambridge University (1875), where he graduated with honors in mathematics (1879). Upon graduation he went to Germany to study. He returned to England, studied law, and received law degrees in 1881 and 1882, but never practiced. He then taught mathematics part time at Cambridge. In 1884 Pearson was appointed Goldsmid Professor of Applied Mathematics and Mechanics at University College London. He was also a lecturer in geometry at Gresham College, London (1891–1894). In 1906 Pearson became director of the Eugenics Records Office founded by **Francis Galton** at University College. Upon Galton's death in 1911, Pearson was appointed head of the new Department of Applied Statistics at the laboratory and became the first Galton Professor, a position he retained until his retirement in 1933.

British statistician and eugenicist Karl Pearson developed statistics to support Charles Darwin's theory of evolution and to establish eugenics as a scientific field. (From: *Journal of Heredity* [1916], Main Library, Indiana University, Bloomington, Indiana. Image courtesy of the Digital Library Program.)

Pearson developed mathematical measures based upon correlations Galton had introduced in 1889 to study living organisms and to test the Darwinian theory of evolution. This use of mathematics with biology was called *biometrics*. Between 1893 and 1912 Pearson published a series of papers under the title Mathematical Contributions to the Theory of Evolution. Between 1905 and 1925 he improved his statistical measures, such as the Pearson r (or product moment) correlation, which are integral to many fields of research including psychology, sociology, and physiology today. Like most eugenics supporters in the early twentieth century, Pearson embraced **social Darwinism**, was concerned about the decline in the birthrate among the middle and upper classes, and promoted **positive eugenics**. He accepted the concept of a **hierarchy of races** in terms of ability and **intelligence**, as suggested by the racial theories of **Arthur de Gobineau** and others. Pearson was not active in eugenics organizations and did not join the **Eugenics Society**.

Unlike most scientists who accepted **Mendelian inheritance** as the primary mechanism of heredity, Pearson championed **Darwinism** and **natural selection**, a vew that led to conflicts with other researchers in Britain. British **genetics** pioneer William Bateson (1861–1926), who considered statistics useless for biology and Pearson's support for Darwinism misguided, was instrumental in rejecting one of Pearson's papers for publication in 1900. This rejection triggered bitter dispute between biometricians and supporters of **Gregor Mendel**'s laws of inheritance. The conflict was not prevalent in **Germany** and the **United States**, where most scientists embraced Mendelism. Pearson and others formed their own journal, *Biometrika*, in 1901, as a venue to publish their papers. **Charles Davenport**, pivotal leader of the U.S. eugenics movement, was a coeditor. A misunderstanding between upcoming statistician and eugenics supporter **R. A. Fisher** and Pearson, concerning a paper Fisher submitted to *Biometrika*, created lifelong hostility between the two men.

Pearson's laboratory was among the groups that discredited **Lamarckian inheritance of acquired characteristics**. In 1925 he founded the scientific journal *Annals of Eugenics* to publish studies using biometry to investigate human inheritance and remained its editor until his retirement. His coeditor was **Ethel Elderton**, a researcher in his laboratory. Over his lifetime Pearson published more than 300 books and papers including the three-volume *Life, Letters and Labours of Francis Galton* (1914, 1924, 1930). His eugenics works included *The Scope and Importance to the State of the Science of National Eugenics* (1909) and *The Problem of Practical Eugenics* (1912). He received many honors but, as an advocate of socialism, he refused the offer to be knighted. He married Maria Sharpe (1890) and the couple had three children. After her death (1928) he married Margaret V. Child (1929), a colleague at University College. He died in Coldharbour, Surrey, in southeastern England, of heart disease shortly after his retirement.

FURTHER READING: Farrall, Lyndsay Andrew, *The Origins and Growth of the English Eugenics Movement, 1865–1925* (1985); Mazumdar, Pauline M. H., *Eugenics, Human Genetics, and Human Failings* (1992); Porter, Theodore M., *Karl Pearson* (2004); Searle, G. R., *Eugenics and Politics in Britain, 1900–1914* (1976); Soloway, Richard A., *Demography and Degeneration* (1990).

PERFECT BABY

See Designer Baby

PERKINS, HENRY FARNHAM (May 10, 1877–November 24, 1956)

A zoologist and eugenicist, Perkins was most noted for his **Eugenics Survey of Vermont** (1925–1936). Born in Burlington, Vermont, he was the only son of a distinguished professor and dean at the University of Vermont (UVM). He was from a prominent old-stock **Anglo-Saxon** New England family that were devoted Congregationalists. Perkins followed in his father's footsteps and also became an academic. He graduated Phi Beta Kappa from UVM (1898), received a Ph.D. in zoology from Johns Hopkins University (1902), Baltimore, Maryland, and returned to the University of Vermont that year to begin teaching. During the summer months, early in his career, he was also a research assistant at a marine biology laboratory in Florida (1902–1906). Perkins worked his way up the academic ladder and became chair of the zoology department at UVM. In 1922 he reorganized the zoology curriculum and began teaching elective courses on heredity and evolution. He was curator of the Fleming Museum in Burlington from 1931 until his retirement in 1945. He obtained WPA funds to organize, preserve, and cataloged the materials from his eugenic studies for the use of further generations.

In 1922 Perkins contacted geneticist **Charles Davenport**, director of the **Eugenics Record Office** (ERO) for ideas concerning student research. Davenport suggested the collection of family histories, which led to the Eugenics Survey of Vermont, carried out between 1925 and 1936 and modeled on the studies of Arthur Estabrook of the ERO. Based upon the study of sixty-two **degenerate** Vermont families, Perkins argued for **eugenic sterilization laws** for the "socially inadequate," which included people with **alcoholism**, **pauperism**, **tuberculosis**, and syphilis. A sterilization law was passed in 1931, which Perkins called, near the end of his life, "one of the most important and Progressive measures on the statute books." Perkins shouldered leadership roles in some eugenic organizations. He was president of the **American Eugenics Society** (1931–1934) and director until 1937. He edited the proceedings for the **Third International Congress of Eugenics** (1932). By the end of **World War II**, due to advances in genetic research and changes in political and social thought, eugenics was no longer an accepted field of study and his work was considered antiquated. Perkins married Mary Edmunds (1903), from a prominent Baltimore family, and fathered two daughters. When his mother died in 1904 the couple moved in with his father in the large house where he had been born and raised, and lived there the rest of his life.

FURTHER READING: Dann, Kevin, "From Degeneration to Regeneration: The Eugenics Survey of Vermont, 1925–1936" (1991); Gallagher, Nancy L., *Breeding Better Vermonters* (1999).

PHRENOLOGY

A mid-nineteenth-century field of study to determine the personality and character traits of an individual by measuring the bumps, indentations, shape, and size of the head is termed *phrenology*. It was used as a method for determining the most compatible marriage partner for producing healthy offspring. Phrenology was developed by Austrian physician Franz Joseph Gall (1758–1828) and had five principles: (1) the brain is the "organ" of the mind; (2) the mind is made up of thirty-seven separate "organs," each corresponding to a distinct mental "faculty" or personality trait; (3) each faculty is found on a certain part of the brain's surface; (4) the strength of any personality trait is determined by the size of that faculty; and (5) the size of the outer surface of the skull matches the organ on the brain surface; thus a trained phrenologist could determine the temperament and the character of a person. Phrenology was brought to the United States by lecturer and physician John Gasper Spurzheim (1775–1832) in the early 1830s. American phrenologists, in the spirit of perfection and self-improvement of the **Jacksonian Era**, added several more principles to Gall's: (1) each and every faculty is susceptible to improvement; (2) the developmental direction of the different faculties depends on education and circumstances; and (3) perfection of mind and character depends upon proper training and direction.

Phrenology, hereditary concerns, and health reform during the 1840s and 1850s became intertwined. A protoeugenics message, or **inherited realities**, was brought to the middle class by phrenologists and other health advocates. Health reformers in the **Clean Living Movement** such as **William Alcott** endorsed phrenology, while phrenology publicists **Orson Fowler**, his brother Lorenzo (1811–1896), and his sister-in-law **Lydia Folger Fowler** urged the need for hereditary improvement through proper diet and "hygienic marriages." Both **positive** and **negative eugenics** concepts were advised. Based upon phrenology "head readings" and the theory of **Lamarckian inheritance of acquired characteristics**, phrenologists encouraged marriage only between the most healthy and compatible individuals to produce the best offspring. They did not recommend marriage for the very young, "delicate females," individuals disposed to hereditary diseases, "partners too nearly allied in blood," or couples with great differences in age because they were all likely to produce imperfect offspring. By the late

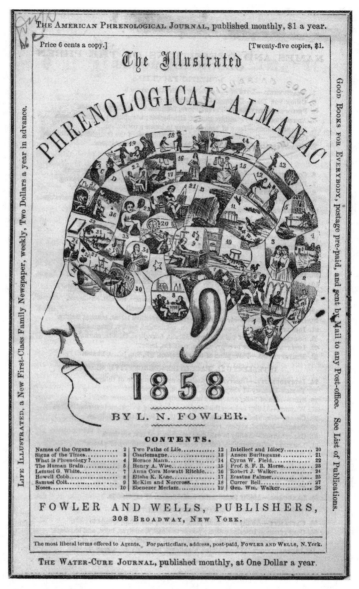

The study of human character through the shape and attributes of the head is called *phrenology*. It was a protoeugenics method to select the most compatible partner in order to produce the most intelligent and healthy offspring. (From: *Phrenological Almanac* [1858]. Courtesy of Lilly Library, Indiana University, Bloomington, Indiana.)

nineteenth century academics discounted phrenology as a science, but its measurement techniques became a component of physical anthropology.

FURTHER READING: Nissenbaum, Stephen, *Sex, Diet, and Debility in Jacksonian America* (1988); Stern, Madeline B., *Heads and Headlines* (1971); Walters, Ronald C., *American Reformers 1815–1860* (1978).

PHYSICAL CULTURE OR PHYSICAL EDUCATION

Physical activity to improve the physical fitness and health of the individual and the human race has been termed *physical culture, physical education,* and *physical exercise* over the past 150 years. It was championed by reformers in the **Clean Living Movements** of the **Jacksonian** (1830–1960), **Progressive** (1890–1920), and **Millennium** (1970–   ) social reform eras. The first woman physician in North America, **Elizabeth Blackwell**, advocated physical exercise as a **law of health** to improve the health of women and their offspring, based upon the theory of **Lamarckian inheritance of acquired characteristics**. At the turn of the twentieth century the terms *physical culture* and *physical education* were use interchangeably in the popular press. Physical education, however, tended to focus on physical activity and sports in the schools in addition to actions for overall health, including abstinence from **alcohol, tobacco**, and sexual activities outside of marriage. Its leaders included school health and physical education professionals such as Luther Gulick (1865–1918) and Dudley Sargent (1849–1924). Physical culture was associated with Bernarr Macfadden (1868–1955), a self-styled physical culturist, and included vegetarianism along with physical fitness. Physician and eugenics supporter **John Harvey Kellogg** combined the two philosophies. The concept of the "whole man," which included a sound, or healthy, mind in a sound body (*mens sana in corpore sano*), was an underlying theme of the early-twentieth-century eugenics movement. In both **Fitter Family** campaigns and **Race Betterment Foundation** contests, physical fitness along with **intelligence** was essential to becoming a prizewinner. Being physically fit was also a necessary component of the **British** and **German** eugenics movements.

FURTHER READING: Engs, Ruth Clifford, *Clean Living Movements* (2001); Greene, Harvey, *Fit for America* (1986); Whorton, James C., *Crusaders for Fitness* (1982).

PIONEER FUND (1937–present)

A not-for-profit organization that gives grants to academic researchers studying variations among human population groups, the Pioneer Fund was founded and endowed by Wickliffe Preston Draper (1891–1972), the heir to a New England textile manufacturer. Eugenics researcher **Harry H. Laughlin**, superintendent of the **Eugenics Record Office** in **Cold Spring Harbor**, New York, and **Frederick Osborn**, the most significant figure in the later years of the **American Eugenics Society**, were founding members and presidents of the organization. The fund, incorporated on March 11, 1937, in New York, had two objectives: to provide financial assistance to parents of children who were "likely to become socially valuable citizens who would make important contributions to their society," and to provide grants for research into the "study of human nature, heredity, and eugenics." Over its years of existence the fund has focused primarily upon the second objective. In 1958 Osborn resigned from the board when population **genetics** and environmental interpretations of social problems became prevalent, in direct opposition to Draper's strictly hereditarian philosophy. The Pioneer Fund has a five-member voluntary board that makes funding decisions. Areas of research the fund has supported include behavioral and medical **genetics**, cognitive ability, demographic characteristics, and population variation. It has often financed controversial projects that were unlikely to receive funding from mainstream grants agencies. The largest single area of research support has been **twin studies**, which have yielded information about the heritability of **intelligence**, personality traits, physical traits, diseases, and disabilities. The organization, particularly in the 1990s, became controversial because of funding to researchers whose findings suggested that variations in **intelligence test** scores among racial, ethnic, gender, and socioeconomic groups had a strong genetic component.

FURTHER READING: Kenny, Michael G., "Toward a Racial Abyss: Eugenics, Wickliffe Draper, and the Origins of the Pioneer Fund" (2002); Lynn, Richard, *The Science of Human Diversity* (2001).

PLANNED PARENTHOOD
See American Birth Control League

PLOETZ, ALFRED (August 22, 1860–March 20, 1940)

Founding father and pivotal leader of the **German eugenics** movement in the pre–**World War I** era, Ploetz established the world's first eugenics society and journal. Like other German eugenicists he supported programs for improving the nation's health and efficiency, and population growth among the fit. Unlike most others, he leaned toward Nordicism ("superiority" of the **Nordic race** and its improvement). He coined the term *Rassenhygiene* (**race hygiene**) for hereditary improvement of the human race and throughout his life preferred this ambiguous term to *eugenics* because it also implied keeping the Nordic race "pure." Born to a wealthy manager of a soap company in Swinemünde, near the Baltic Sea, Ploetz spent his youth in the ethnically diverse Breslau (now Wroclaw). As a youth he became obsessed with the "glories of the Germanic Teutonic past," and along with a small circle of friends he established a "league to invigorate the race" to bring Germany back to its "heroic past." Ploetz embraced socialism and began to study social economics at the University of Breslau in 1884. While there he formed a society with the goal of establishing a Teutonic socialist utopian community. This led him to travel to the **United States** to work on a socialist commune. Disillusioned with the experience, he went to Zurich, Switzerland, and completed a medical degree (1890). He worked intermittently as a physician and journalist until 1898, when he had obtained the financial means through a fortuitous marriage to devote himself full time to the promotion of race hygiene. By 1907 he moved to Munich, where he remained for most of his life.

After receiving his medical degree Ploetz interned at a Swiss mental hospital. From this experience he perceived **alcohol** as the major cause of social problems and became active in temperance causes. After his internship he went to the United States and practiced medicine in Springfield, Massachusetts, and then in Connecticut for a few years. He returned to Berlin and finished *Die Tüchtigkeit unser Rasse und der Schutz der Schwachen* (Fitness of Our Race and the Protection of the Weak, 1895). This work gave a theoretical basis for eugenics, which he termed *Rassenhygiene*. Ploetz rejected the theory of **Lamarckian inheritance of acquired characteristics** and supported August Weismann's (1834–1914) theory of the **germ-plasm** as the basic unit of heredity. Like most other eugenics supporters on both sides of the Atlantic, he accepted **social Darwinism** and, upon its rediscovery, **Mendelian inheritance**. Ploetz advocated methods of **positive eugenics**, including selection of genetically healthy eggs and sperm when scientific advances made this techniques possible.

In order to promote eugenics and research in the field, Ploetz founded the *Archiv für Rassen- und Gesellschafts-Biologie* (*Archive for Racial and Social Biology*) in 1904 in Berlin and remained its editor until around 1939. Along with others, in 1905 he founded the **German Society for Race Hygiene** to further advance the eugenics cause among the elite and was its secretary until 1914; he then became its vice chairman. To encourage interest in the society on an international level, Ploetz in 1910 visited England, where he spent time with **Francis Galton**, statistician **Karl Pearson**, and members of the **Eugenics** (Education) **Society**, to which he was named a vice president in 1916. However, due to the international political climate he found little international interest in the society. Ploetz presented a paper at the **First International Eugenics Congress** (1912) in London. He formed groups to promote Nordicism and improvement of the **Aryan race** in Germany. His protégé, geneticist **Fritz Lenz**, became a major leader of the German eugenics movement in the post–World War I era. After the onset of the war Ploetz increasingly devoted his time to **family history** research and pulled back from active involvement in the eugenics movement. He was, however, one of the few German eugenicists who supported race theorist **Hans F. K. Günther**'s work. In **Nazi Germany** in the 1930s, Ploetz, who was now elderly, was a member of Nazi Expert Committee for Sterilization and was given honorary awards and titles. He also spoke out against **war** as eugenically disastrous for the Western world. For this position, the Nazi government unsuccessfully nominated him as a candidate for the 1936 Nobel Peace Prize. Ploetz married Pauline Rüdin (1890), the sister of psychiatrist and eugenics supporter Ernst Rüdin (1874–1952). He divorced her and married Anita Nordenholz (1898); they had three children. Ploetz died at his estate in Herrsching near Munich.

FURTHER READING: Kühl, Stefan, *The Nazi Connection* (1994); Weindling, Paul, *Health, Race and German Politics between National Unification and Nazism, 1870–1945* (1989); Weingart, Peter, "German Eugenics Between Science and Politics" (1989); Weiss, Sheila Faith, "The Race Hygiene Movement in Germany" (1987).

POPENOE, PAUL BOWMAN (October 16, 1888–June 19, 1979)

A proponent of eugenic sterilization, Popenoe founded the first marriage counseling institution in the United States and helped popularize eugenics through his writings. Born in Topeka, Kansas, he was the son of a businessman and newspaper owner whose French Huguenot ancestors came to the United States in 1696. After graduating from Washburn Academy (1905), Popenoe moved to the Los Angeles area with his parents. He attended Occidental College (1905–1907) and Stanford University (1907–1908). He soon left Stanford due to his father's illness and became city editor of the *Pasadena Star*. He then became an agricultural explorer in North Africa and subsequently moved to Washington, D.C., to become editor of the *Journal of Heredity* (1913–1918). During **World War I** he served on the surgeon general's staff as director of the **venereal diseases** control section. After the war he became executive secretary of the **American Social Hygiene Association** (1919–1920) and was involved with anti–venereal disease and anti**prostitution** campaigns. In 1920 Popenoe moved to a ranch near Los Angeles and grew dates, wrote *Modern Marriage* (1925), and became involved with various eugenics causes and organizations. He established the Institute of Family Relations (1930), the first such organization in the country that provided premarital examination in addition to heredity and marriage counseling. He remained as director until 1960, when he semiretired and became chairman of the board.

Popenoe was involved with organizations associated with eugenics. He was a delegate to the **First National Conference on Race Betterment** (1914) where he met educator and eugenics supporter **Roswell H. Johnson**. Four years later they coauthored *Applied Eugenics* (1918), which helped popularize eugenics. Popenoe was on the advisory council of the **American Eugenics Society** and was active in the **Eugenics Research Association**. In 1926 he moved to Altadena, California, to become secretary and director of research of the **Human Betterment Foundation**, founded by philanthropist **Ezra Gosney** to promote eugenic practices. Under the auspices of this organization he investigated the effectiveness of **eugenic sterilization** and with Gosney wrote *Sterilization for Human Betterment* (1929). In his later years he wrote publications on marriage and sexuality for popular magazines. He married Betty Stankovitch (1920) and fathered four sons. He died in Miami, Florida, a year after his wife's death.

FURTHER READING: Kline, Wendy, *Building a Better Race* (2001); Pickens, Donald, *Eugenics and the Progressives* (1968).

POSITIVE EUGENICS

The original eugenics philosophy of **Francis Galton** that encouraged parenthood among the fit was termed *positive eugenics* in the first decade of the twentieth century. The "fit" were the "best stock," and "the worthy," and included healthy, educated, talented, or financially well-off individuals without any mental or physical impairments. They were deemed to have "civic worth," or the ability to make positive contributions to society. Positive eugenics was the primary eugenics philosophy of **Britain** and its **Eugenics Society**. The term *positive eugenics* was generally used in contrast with **negative eugenics**, which discouraged reproduction among the **unfit** and was primarily found in the **United States**, **Germany**, and the Scandinavian countries. The two terms were coined by **C. W. Saleeby**, a British physician, in 1909. As the two concepts began to be discussed, Galton, who championed positive eugenics in 1901, wrote, "The possibility of improving the race of a nation depends on the power of increasing the productivity of the best stock. This is far more important than that of repressing the productivity of the worse." Positive eugenics included educational programs to encourage parenthood by healthy and fit parents. The programs gave advice for appropriate mate selection, parenting, child-rearing, and actions individuals could take to produce better and healthier babies. Young people were advised to avoid marriage to those with a family history of **insanity**, **feeblemindedness**, epilepsy, or **criminality**. **Better Babies** and **Fitter Families** contests were held in the United States, and families were encouraged to complete and file **family history and pedigree** forms with the **Eugenics Registry**. Programs to eliminate **racial poisons** such as **venereal disease**, **tuberculosis**, and **alcohol** were recommended.

In the United States reformers advocated granting bonuses to healthy young couples for producing children, requiring a medical exam before marriage, and granting marriage certificates only to healthy couples. The idea of compulsory mating as a positive eugenics method was generally rejected, although the late-nineteenth-century **Oneida Community** carried out a practice of selective mating in a procedure called **stirpiculture**. The Mormon community (**Church of Jesus Christ of Latter-day Saints**) also encouraged marriage only among the physically, mentally, and morally fit. Positive eugenics concepts were championed by some, including biologist **Hermann Muller** in the post–**World War II** era. By the beginning of the twenty-first century positive eugenics was found in medicine in the techniques of **assisted reproduction** such as **in vitro fertilization** and **germinal choice** (choosing a desirable egg or sperm donor). Positive eugenics came from the demands of prospective parents wanting to assure a healthy, "perfect," or "**designer baby**," or to have a child of a specific gender to ensure **family balancing**.

FURTHER READING: Rifkin Jeremy, *The Biotech Century* (1998); Stacey, Meg, (ed.), *Changing Human Reproduction* (1992); Stock, Gregory, *Redesigning Humans* (2002)

Positive eugenics encouraged the intellectually, physically, economically, morally, and socially fit to produce many children in order to improve the race. (From: *Progress of Eugenics* [1914], Main Library, Indiana University, Bloomington, Indiana. Image courtesy of the Digital Library Program.)

POVERTY

See Pauperism and Paupers

PREGNANCY TERMINATION

See Abortion

PROGRESSIVE ERA (1890–1920)

The period of time in the United States during the last decade of the nineteenth and first decades of the twentieth centuries, when numerous health and social welfare laws were enacted, is termed the *Progressive Era*. Eugenics was one of many middle-class progressive reform movement to "clean up America" and its health and social problems. Several interpretations have been given for this reform era. Some suggest that the old-stock **Protestant** middle class feared that their historic rural traditions were being destroyed by a combination of economic and political power from **Catholic** and **Jewish** immigrants with different values and cultures. Concerned that the country was degenerating, they longed to bring the United States back to a mythical golden age, and at the same time desired a forward-looking philosophy, or progressivism. Others argue that the era was a reaction against corporate dominance of American life and corruption. These populist sentiments resulted in a flurry of legislation including Prohibition, pure food and drug laws, mandatory immunization, **eugenic sterilization** and **marriage-restriction laws**, narcotics controls, and **immigration restriction**, as well as anti**prostitution** initiatives. Emphasis was placed on hygiene, **physical culture**, good diet, sexual purity, and the "whole man," who embodied physical, mental, and moral fitness. Countering purity and chastity campaigns, however, a crusade for **birth control**, sex education, and the elimination of

venereal disease through prevention and education emerged. New religions that stressed mental healing became popular, alongside fundamentalism that strove to bring back biblical teachings and "family values." Less controversial sanitation and **public health** measures, campaigns against **racial poisons** such as **tuberculosis** and **tobacco**, and improved medicine helped increase the life span of the U.S. population. However, near the end of this **clean living**, or health reform aspect, of the era, when the new health measures were forecast to bring America to a healthy new future, **World War I** broke out, and in 1918 and 1919 influenza killed large numbers of young Americans.

FURTHER READING: Burt, Elizabeth V., *The Progressive Era* (2004); Engs, Ruth Clifford, *Clean Living Movements* (2000); Engs, *The Progressive Era's Health Reform Movement* (2003); Hofstadter, Richard, (ed.), *The Progressive Movement* (1986)

PROSTITUTION AND PROSTITUTES

The sale of sexual services, mostly by women to men, prostitution was called the *ancient profession, social vice, white slavery,* or the *social evil* during the **Progressive Era** in the United States. It was considered a moral, social, and health problem by reformers in the **Clean Living Movement**, the health reform aspect of the era, who viewed it as the main cause of **venereal diseases**. Laws were passed in the 1880s and 1890s to raise the age of sexual consent (before the new laws it was ten in many states) in an effort to prevent young girls, many of whom were sold by impoverished parents, from being drawn into the prostitution lifestyle. Some reformers considered any nonmonogamous female sexual activity as prostitution. Turn of the century surveys of prostitutes reported a high proportion of immigrant women engaged in this occupation. Popularization of these reports fostered **nativist** and **immigration restriction** sentiments among eugenicists and reformers. However, the survey results were largely artifactual inasmuch as most of the surveys were conducted in eastern urban cities with a high percentage of immigrants. Low wages and limited job opportunities for girls migrating from farms to larger towns were also factors in many areas.

Social hygiene, temperance, **public health**, and eugenics reformers campaigned against prostitution. In the first decade of the twentieth century attitudes changed from containment of prostitutes in "red-light districts" to its abatement, or elimination. Prostitutes began to be seen as **feebleminded**. **Eugenic sterilization** and **segregation** in state-run facilities were recommended. The Mann Act was passed by Congress in 1910 to prohibit interstate traffic in prostitution. Social reformer Abraham Flexner's *Prostitution in Europe* (1914) suggested that European state regulation of the "social evil" had failed. Subsequently, repressive measures were instituted against prostitutes during **World War I** in an attempt to eliminate them from military base communities. Legal prostitution became prohibited, and it was still illegal in most states in the early twenty-first century.

FURTHER READING: Connelly, Mark Thomas, *The Response to Prostitution in the Progressive Era* (1980); Grittner, Frederick K., *White Slavery* (1990); Pivar, David J., *Purity Crusade* (1973); Pivar, *Purity and Hygiene* (2002); Rosen, Ruth, *The Lost Sisterhood* (1982).

PROTESTANTS

Most health, hereditarian, or eugenics reformers of the **Jacksonian** (1830–1860) and **Progressive** (1890–1920) social reform eras in the United States were from mainstream Protestant churches including Congregational, Quaker, Unitarian, Methodist, and Episcopalian denominations. In the reform period of the late-twentieth-century **Millennium Era** (1970–), **Catholics** joined with evangelical Protestants to oppose the techniques of the emerging **new eugenics**. For all three reform eras, health crusades were often an extension of religious beliefs. In the early nineteenth century a wave of Protestant revivalism, the Second **Great Awakening**, swept the United States. Out of this ferment arose strongly devout individuals with a passion for perfectionism and the elimination of society's problems in order to achieve the "reign of Christ on earth." Fear of Irish Catholic immigrants, in particular, led to anti-Catholic **nativist** sentiments. This fear had origins in the colonial period, stemming from political conflicts between British Catholic and Protestants powers for political control. Unlike in Britain, where religious antipathies were sublimated over time and more accepting attitudes were adopted, the isolated American colonists deepened in their "no-popery prejudices." **Anglo-Saxon** Protestant Americans, proud of their old colonial heritage, believed that God had ordained the New World as a special place and led temperance, sanitation, diet, and exercise crusades to keep established middle-class Protestants values intact and to bring on Christ's reign.

The Social Gospel Movement, a liberal Protestant social reform movement, gave impetus to the turn-of-the-twentieth-century health and social reform crusades of the Progressive Era, including those for pure food and drugs, prohibition, sexual purity, and eugenics. As thousands of Catholics and **Jews** from **eastern and southern Europe** settled in eastern urban slums, disease and societal problems were blamed on these non-Protestant immigrants. Poor families with many children were perceived as leading to **race suicide** of the Anglo-Saxon **race**, **racial degeneracy**, and the decline of the nation. **Eugenic sterilization** and **immigration restriction laws**, in part, were efforts to control the "immigrant menace" and keep established, rural, Protestant mores intact. In order to popularize the eugenics concept, the **American Eugenics Society** sponsored sermon contests open to all clergy, including Jewish and Catholic. One sermon topic was "Religion and Eugenics: Does the Church Have Any Responsibility for Improving the Human Stock?"

In the post–**World War II** era a "baby boom" swelled the population. By the mid-1970s evangelical Protestants and older Americans began to react against what they perceived as immoral and unpatriotic behaviors on the part of youth. Their concerns coalesced into political action, termed the *religious right*, that emerged out of the Fourth Great Awakening, They campaigned against the use of drugs, **alcohol**, and sexuality outside marriage, and for "family values." Simultaneously, a more secular health-reform movement, that to some became a "religion," surged out of the youthful generation that often rejected the religious values of their elders. Fitness and physical exercise, diet, alternative religions and medicine, consumers' rights, and smoke-free environment crusades surfaced. By the mid-1980s consumer demand for **in vitro fertilization** and other **assisted reproduction** techniques of the new eugenics emerged. Catholics and Evangelical Protestants joined to fight against many of the aspects of the new eugenics, including **abortion**, human **cloning**, the manipulation of the human **genome**, embryonic "stem-cell research," and some forms of therapeutic cloning. These issues remain controversial in the first decade of the twenty-first century.

FURTHER READING: Engs, Ruth Clifford, *Clean Living Movements* (2000); McLoughlin, William G., *Revivals, Awakenings, and Reform* (1978); Rosen, Christine, *Preaching Eugenics* (2004); Rosenberg, Charles E., *No Other Gods* (1976).

PUBLIC HEALTH

The applied field of science that includes sanitation, preventive medicine, the study and control of epidemics, personal hygiene education, organization of health services, and the development of public policies and governmental organizations to ensure a healthy population is *public health*. At the turn of the twentieth century the terms *hygiene* and *sanitarian* were also used. The public health movement that began in the last half of the nineteenth century by the 1920s had resulted in proper sewage treatment, clean water, elimination of insect-borne diseases, instruction in hand-washing, immunization against infectious diseases, and improvement in nutrition. The efforts dramatically improved the health and longevity of the population in westernized cultures. Not everyone, however, supported the public health movement. In the United States, by the second decade of the twentieth century organizations such as the National League for Medical Freedom arose to protest mandatory immunizations. Some fundamentalist **Protestant** religious groups believed that prevention and medical treatment were "against God's will." Based upon the theory of **natural selection**, or survival of the fittest, some physicians and social welfare professionals were afraid that the decreasing death rates among infants and children was weakening the human race as less healthy children were living to reproduce, leading to **racial degeneracy**. By the first decade of the twentieth century, in **Britain**, **Germany**, and the **United States** eugenics and public health measures had became intertwined. Public health workers were advised by health leaders, such as **Adolphus Knopf**, that they needed to stop the propagation of those with "defective genes" just as they had fought germs. Many public health programs became indistinguishable from eugenics programs.

In the United States the popularization of **Darwinism**, along with health and social reforms of the **Progressive Era** (1890–1920), led to public health legislation that owed much to nativist and eugenics sentiments. Immigrants were blamed for several epidemics in New York City, which added to rising **nativism** in the 1890s. The nativist attitude was reflected in the National Quarantine Act of 1893, which gave powers to New York City's health board to prevent immigrants from entering the nation if suspected of carrying infectious diseases. The germ theory of disease intensified social pressure for various kinds of cleanliness, both social and personal. This was manifested in the broad-based **Clean Living Movement**, including Prohibition, crusades for "sexual purity," personal hygiene, anti-**tobacco** and anti-**prostitution** initiatives, and the early identification and treatment of **tuberculosis** and **venereal diseases**. Other public health measures included **eugenic sterilization** and **segregation** (custodial care), and **eugenic marriage-restriction laws** that focused on premarital tests to screen for syphilis. Many of these efforts resulted in a decrease of infectious diseases.

Public health campaigns of the late-twentieth-century's Clean Living Movement also had eugenic implications in terms of mothers producing healthy offspring. These included anti**alcohol** campaigns to prevent fetal alcohol syndrome, antismoking crusades to prevent various conditions among infants whose parents smoked, the "War on

Drugs" to prevent fetuses from becoming damaged, abstinence-based sex education programs to prevent sexually transmitted diseases and pregnancies, and **genetic testing** to determine carrier status for genetic diseases. Modern medicine, new drugs, and public health programs saved the lives of people who would have died in previous generations, many with **genetic diseases**. By the late twentieth century the question whether the survival of those with serious diseases who chose to reproduce was leading to the weakening of the "gene pool" had became an aspect of the **new eugenics**.

FURTHER READING: Duffy, John, *The Sanitarians* (1992); Engs, Ruth Clifford, *Clean Living Movements* (2000); Jones, Helen, *Health and Society in Twentieth-Century Britain* (1994); Pernick, Martin S., *The Black Stork* (1996); Weindling, Paul, *Health, Race, and German Politics between National Unification and Nazism, 1870–1945* (1989).

R

The Race Betterment movement resulted from the recognition of the rapid increase of race degeneracy, especially in recent times. The fact of race degeneracy is evident from the alarming increase of the insane and other mental defectives, who now constitute one percent of the whole population of the United States.

"The Race Betterment Foundation," *Official Proceedings of the Second National Conference on Race Betterment August 4, 5, 6, 7 and 8, 1915* (1915)

RACE BETTERMENT

The term *race betterment* embodies methods for healthy living and eugenics. It was first used around 1906 when health reformer **John Harvey Kellogg** founded the **Race Betterment Foundation** at Battle Creek, Michigan, to address concerns about **race degeneration**. Kellogg was convinced that the human race was deteriorating physically, mentally, and morally from unhealthy lifestyles and behaviors that included poor diet, lack of exercise, and the use of **alcohol** and **tobacco**. At the **Second National Conference on Race Betterment**, in 1915, he listed ten methods to increase health and longevity, or race betterment. These included "Simple and Natural Habits of Life, Total Abstinence from the Use of Alcohol and Other Drugs, Eugenic Marriage, Medical Certificate before Marriage, Vigorous Campaign of Education in Health and Eugenics, Eugenic Registry, and Sterilization or Isolation of Defectives."

FURTHER READING: Carson, Gerald, *Cornflake Crusade* (1957); Money, John, *The Destroying Angel* (1985); Schwarz, Richard William, *John Harvey Kellogg, M.D.* (1981).

RACE BETTERMENT FOUNDATION (1906–c. 1955)

A eugenics and hygiene organization, the Race Betterment Foundation supported conferences, publications, and a **eugenics registry** and funded a college. Reformer **John Harvey Kellogg** established the foundation in 1906 at Battle Creek, Michigan, to combat **racial degeneration**. Kellogg believed that race deterioration could be reversed through a combination of heeding his principles of healthy living and selective mating of individuals to perpetuate desirable traits (traits reflecting good health). The term *race betterment* was congenial to Progressive middle-class intellectuals who comprised the leadership of the eugenics and other health-reform movements of the **Progressive Era**. The foundation enabled Kellogg to introduce his "biologic living" ideals to important molders of public opinion through conferences and publications. It consisted of several units including a Eugenics Department. Race betterment became so great a cause for Kellogg that in 1920 he rented space at the Battle Creek sanatorium to house foundation offices. During the peak of its influence the foundation subsidized the **First**, **Second**, and **Third National Conferences on Race Betterment**, held in 1914, 1915, and 1928. After the 1928 conference, the foundation took over sponsorship of the **Fitter Family** campaign from the **American Eugenics Society**. After Kellogg's death the Race Betterment Foundation lost its major source of leadership and funding. It limited its later activities to public lectures and the publication of *Good Health* and became inactive after 1955.

FURTHER READING: Boyle, T. Coraghessan, *The Road to Wellville* (1993); Carson, Gerald, *Cornflake Crusade* (1957); Kellogg, J. H., *Plain Facts for Old and Young* (1974); Money, John, *The Destroying Angel* (1985); Schwarz, Richard William, *John Harvey Kellogg, M.D.* (1981).

RACE IMPROVEMENTS IN THE UNITED STATES (1909)

At its thirteenth annual meeting in 1909 the American Academy of Political and Social Science concentrated on eugenics, or *race improvement*. This was one of the first academic conferences with a eugenics focus and it led to publication of *Race Improvements in the United States*, published by the academy in Philadelphia. The 171-page volume is divided into five parts: "Heredity and Environment in Race Improvement," "Influence of City Environment on National Life and Vigor," "Obstacles to Race Progress in the United States," "Relation of Immigration to Race Improvement," and "Clinical Study and Treatment of Normal and Abnormal Development." Prominent reformers from varying fields of interest including **public health**, social welfare, **physical education**, public recreation, "slum" problems, and immigration presented papers that were included in this work. Presenters and authors included geneticist and eugenics leader **Charles Davenport**. Problems of **racial degeneracy** and solutions of both **positive** and **negative eugenics** were addressed. The conference theme and resulting publication were symbolic of the rising importance of eugenics to social science and welfare professionals.

RACE OR RACIAL CLASSIFICATIONS

The scientific term used until the late twentieth century to designate people with different physical characteristics or from different ethnic groups was *race*. Three basic racial subgroups of the human species, first classified in the eighteenth century (although known from antiquity), were *Caucasoid, Negroid,* and *Mongoloid.* Through the early twentieth century national and ethnic identities, such as Germans, Irish, and Italians, were also considered biologically different races with particular characteristics. **Hans F. K. Günther**, a German anthropologist and race theorist, classified five European races: *Nordic,* found in northern Europe; *Mediterranean,* comprising southern Europe; *Dinaric,* corresponding with Bavaria, Austria, and Hungary; *Alpine,* meaning of Czech, Slovak, Austrian, and French origins; and *East Baltic,* consisting of Russia, Finland, and Croatia. Race in some contexts from the mid-nineteenth through the early twentieth centuries referred to those with northern European ancestry, as in the phrase **race suicide**, for the decline in population of people with this heritage. Race was also used in phrases such as "her race," referring to gender. Race was important to many early-twentieth-century eugenicists who assumed that those with northern European or **Anglo-Saxon** backgrounds were superior to all other groups. This belief was based upon French writer **Arthur de Gobineau**'s theory of a **hierarchy of races**. The most "superior" race in his classification, the **Aryans**, had created advanced civilizations and cultures. The belief that some racial groups were inferior in terms of **intelligence** and other attributes led to the **immigration restriction movement** in the

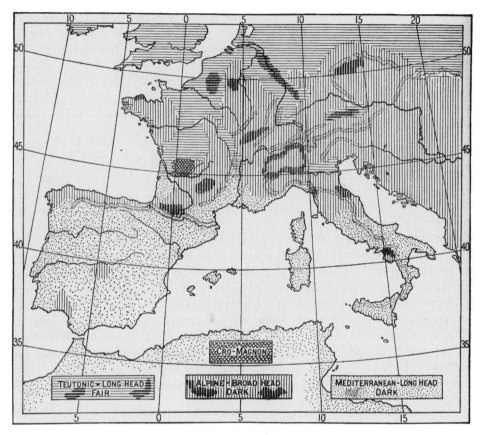

Race theorists and some eugenicists of the early twentieth century divided Europe into several "races" including Teutonic, Alpine, and Mediterranean. (From: *Journal of Heredity* [1917], Main Library, Indiana University, Bloomington, Indiana. Image courtesy of the Digital Library Program.)

United States to deny **eastern and southern Europeans** and **Asians** entry into the country. A major precept of **Nazi Germany** was to create a **master race** of blond, blue-eyed Aryan **Nordics**. By the late twentieth century many anthropologists and sociologists considered the classification of races to be a social construct without much meaning. While the word *race* remained in popular use, the term *population* began to be used in scientific circles to describe human subgroups who had different frequencies of specific genes (genetic markers) and were from different continents or geographic areas. Although populations were seen as being on an extremely complex and everchanging continuum, certain genetic disorders, diseases, and drug reactions were more common in some groups than others.

Around 1919 ethnic differences in the frequency of blood antigens A and B were found. This discovery set the stage for the use of biochemical or genetic markers to study human diversity. By the mid-twentieth century, cultural changes along with new scientific information caused some scientists to began to question the concept of race. In 1963 Carleton Coon (1904–1981), in *The Origin of Races*, documented patterns of variation among populations similar to variations that had led to earlier classifications.

Coon suggested that people belonged to five groups corresponding to the major continents or geographical regions of the world: Africa, Asia, Europe, Melanesia, and the Americas. Controversy surrounding the work led to the new field of *population studies*, which took into account the geographically gradual nature of human diversity and patterns of variation across the species.

Based upon archaeology, language, and **genetics**, new ideas concerning populations were advanced by the late twentieth century. Several studies, including a substudy using genetic material from the **Human Genome Project**, investigated human diversity. By using gene frequency analysis, Italian researcher Luigi Luca Cavalli-Sforza (b. 1922) and his team in 1994 reported population patterns similar to those described by Coon and traditional physical anthropologists. Statistical analysis of genetics and language suggested that early humanoids broke into two separate populations: *Africans* and *non-Africans*. From the non-African group, Asian and European peoples developed. The Melanesia populations evolved from southeastern Asians, and the people of the Americas developed from northeast Asians. A larger genetic distance in terms of gene frequency was found between Europeans and sub-Saharan Africans and also between these Africans and Australian Aborigines, compared with other populations. Small differences were found among Europeans, Middle Easterners, and Asians. Genetic variations among populations were thought to be due to random mutations in isolated groups, **natural selection** for desirable traits in a particular climate, and migration and interbreeding with neighboring tribes.

Studies showed that most human variation is found within, and not between, population groups. About 94 to 95 percent of the genetic variation in **DNA** is due to differences among individuals within a continent. However, the remaining 5 percent of variation can be used as a genetic marker to indicate the geographic region from which an individual's recent ancestors came. All genetic traits have been found in all population groups, although some, like blood type, certain diseases, and reactions to medications, are more common in some groups. This led many researchers to argue that an individual's ethnicity or population identity is important in terms of health and medical diagnosis and treatment. Others, however, contend that **genetic screening and testing** for potential health problems common in certain populations could set the stage for discrimination in health insurance, employment, and other aspects of individual freedom.

Further Reading: Cavalli-Sforza, L. L., *Genes, Peoples, and Languages* (2000); Coon, Carleton Stevens, *The Origin of Races* (1971); Marks, Jonathan, *Human Biodiversity* (1995); Regal, Brian, *Henry Fairfield Osborn* (2002); Renfrew, Colin, *Archaeology and Language* (1988).

RACE OR RACIAL HYGIENE (*RASSEN-HYGIENE*)

A German concept that fused **public health** and sanitation, maternal and child care, bacteriology, hygiene, physical culture, and race improvement was coined *race hygiene* by **Alfred Ploetz** in 1895. The term had a broader scope than *eugenics*, used in the **United States** and **Britain**. Race hygiene in the **German eugenics** movement included the promotion of programs to improve the **genetic** quality and the "national efficiency" of the nation, and to increase in the nation's population to prevent **race suicide**. Under **Nazi Germany** the term *race hygiene* also connoted the "superiority" of the **Nordic** or **Aryan** Germanic **race** and efforts for "racial purity." Although the Germanized term

Eugenik was introduced by pioneer eugenicist **Wilhelm Schallmayer** in 1907, German eugenic leaders with Nordic sympathies, such as physician Alfred Ploetz and anthropologists **Fritz Lenz**, found the double connotation of *Rassenhygiene* desirable as it embraced both Nordic supremacy views and the idea of increasing the health and reproduction of the "fit."

FURTHER READING: Weiss, Sheila Faith, *Race Hygiene and National Efficiency* (1987); Weingart, Peter, "German Eugenics between Science and Politics" (1989).

RACE REGENERATION

Another name for *eugenics* in the early twentieth century, *race regeneration* was used by some British eugenicists. It was mentioned in 1911 in two tracts published by the National Council for Public Morals, entitled "New Tracts for the Times" by **British** eugenicists **C. W. Saleeby** in *Methods of Race-Regeneration*, and Havelock Ellis (1859–1939) in *The Problem of Race-Regeneration*. These tracts discussed basic concepts of **positive** and **negative eugenics**. Race regeneration included social reform that encouraged the fit to reproduce (positive) and employ measures to keep the **unfit** from reproducing, by means of segregation, **eugenic sterilization**, or **eugenic marriage-restriction laws** (negative). The term had been replaced by *eugenics* by the mid-1920s.

FURTHER READING: Searle, G. R., *Eugenics and Politics in Britain, 1900–1914* (1976).

RACE SUICIDE

The decline of the birthrate among the educated middle classes or **Anglo-Saxon** or **Nordic** populations, and the increase in the birthrate of **unfit** or "inferior" populations was termed *race suicide* in the early twentieth century. This differential birthrate became a theme of the eugenics movements in **Britain**, **Germany**, and the **United States**. The term was coined by sociologist Edward A. Ross (1866–1951) in 1901 and popularized by U.S. president Theodore Roosevelt (1858–1919). Ross warned that undesirable immigrants, especially **eastern and southern Europeans**, were reproducing much faster than "more valuable" old-stock Americans. **Nativists** and eugenicists were concerned that higher birthrates among inferior groups would lead to **racial degeneracy** and the loss of traditional American values. Some eugenicists opposed **birth control**, fearing it contributed to race suicide, while others were hopeful that if birth control were available to all social classes the problem of high fertility among the poor and the unfit would be eliminated. A major effort of the eugenics movement was to encourage educated and intelligent people to bear more children in order to prevent race suicide.

FURTHER READING: Carlson, Elof Axel, *The Unfit* (2001); Kevles, Daniel J., *In the Name of Eugenics* (1985); Kline, Wendy, *Building a Better Race* (2002).

RACIAL DEGENERACY

The deterioration of the human race due to an increase in people with mental, physical, medical, and moral weaknesses and infirmities and the decline of the capable and healthy was termed *racial degeneracy*. The term **degeneracy** was also used. Blamed on prolific breeding by the **unfit** and lack of offspring among the "fit," race degeneracy was a core concern of the early-twentieth-century eugenics movement. Both **positive** and **negative eugenics**, including the encouragement of reproduction among the middle class and the prevention of reproduction among the poor and diseased,

To prevent race suicide, a positive eugenics measure encouraged the fit to have large families. (From: *Journal of Heredity* [1923], Main Library, Indiana University, Bloomington, Indiana. Image courtesy of the Digital Library Program.)

were advocated to counteract the trend. Although the concept can be traced to antiquity, the philosophy solidified in Europe during the late nineteenth century as an offshoot of **social Darwinism** and degeneracy theory. The idea was also discussed in the United States as **inherited realities** earlier in the nineteenth century during the **Clean Living Movement** (1830–1860) of the **Jacksonian Era**. The birthrate among the middle and upper classes declined in many industrialized nations in the late nineteenth century, leading to fears of **race suicide** among this group. It was observed that **paupers, alcoholics**, the **feebleminded**, and other unfit individuals were producing many children. Because of belief in **Lamarckian inheritance of acquired characteristics**, it was proposed that the higher reproductive rate among these **degenerates** would lead to the moral, intellectual, physical, political, and social decline of North American and European cultures. Factors contributing to racial degeneracy included **racial poisons** such as **alcohol, venereal disease**, and **war**, and **public health** and medical interventions that saved the weak allowing them to reproduce. Interbreeding with "inferior" races was also thought to lead to degeneration. These factors and fear of unfit immigrants led to **immigration restriction** and other eugenics laws in the United States and other countries.

FURTHER READING: Carlson, Elof Axel, *The Unfit* (2001); Money, John, *The Destroying Angel* (1985); Rafter, Nicole Hahn, (ed.), *White Trash* (1988).

RACIAL ELEMENTS OF EUROPEAN HISTORY (1927; *Rassenkunde Europas*, 1925)

Written by German anthropologist and "race-theorist" **Hans F. K. Günther**, *Rassenkunde Europas* describes the history of various European *races* and lauds the "superior" **Nordics** as the developers of Western civilization. It supports the thesis of *Nordicism,* or the "superiority of the **Aryan** race," as termed by the regime of **Nazi Germany**. Originally published in 1925, this 296-page work was translated by G. C. Wheeler into English in 1927 and published by Methuen in London. The publication is divided into twelve chapters with titles such as "The Bodily Characteristics of the European Races," "The Nordic Race in Prehistory and History," and "The Present Day from the Racial Point of View." The author classified five European races (ethnic groups) based upon skull and other physical measurements and "mental characteristics." The book contains several hundred photographs illustrating differences in various races. Although Günther suggested that most of Europe is a "mingling of the five European races," he contended that the tall, blond, and blue-eyed Nordic people were the most intelligent, industrious, truthful, and attractive of all. He suggested that in antiquity these nomadic Teutonic Indo-Europeans established ancient civilizations including Greece and Rome. When they interbred with "inferior" peoples, the civilizations collapsed. He argued for keeping the Nordic race "pure" on the grounds that mixing Nordic with non-Nordic types would result in **racial degeneration** and the decline of Western civilization. Günther noted that American eugenicists **Madison Grant** and **Lothrop Stoddard** also presented similar ideas. He suggested that social welfare and **public health** were saving the **unfit** and those with inferior hereditary qualities, "the weak, the unstable, the work-shy, the harlot, the tramp, the drunkard, the weak-minded," whose presence and progeny would contribute to the downfall of civiliza-

tion. This book was popular in Britain and North America among eugenicists and the educated, and gave an ideological basis for eugenic programs in nations with **Anglo-Saxon** and Germanic populations.

FURTHER READING: Günther, Hans F.K., *The Racial Elements of European History* (1992).

RACIAL POISONS

The term *racial poisons* was coined in 1907 by **C.W. Saleeby**, a **British** eugenicist whose works were internationally distributed among academics and the middle classes. Racial poisons were toxic substances and conditions that caused congenital defects. They were thought to injure the **germ-plasm**—egg and sperm cells—preventing "healthy, effective and intelligent offspring" and thus leading to the decline of the human race, or **racial degeneracy**. Racial poisons included **alcohol**, **tobacco**, lead, **venereal diseases**, **tuberculosis**, dissipated lifestyles, and **war**. The term was primarily used in British eugenics literature but was an underlying concept in the eugenics and other health reform

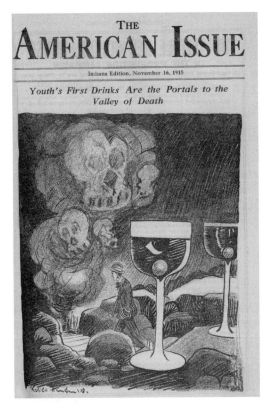

THE
AMERICAN ISSUE

Indiana Edition, November 16, 1915

Youth's First Drinks Are the Portals to the Valley of Death

Health reformers of the early twentieth century considered toxic substances and diseases, such as alcohol and tuberculosis, as racial poisons, harmful to genetic material and offspring. (From: *The American Issue* [1915], Main Library, Indiana University, Bloomington, Indiana. Image courtesy of the Digital Library Program.)

movements of the era in the **United States** and **Germany**. By the second decade of the twentieth century, most health experts considered problems in offspring from toxic materials to be environmentally induced, in the womb or in early development. However, a few reformers, including American physician **John Harvey Kellogg**, still considered the resulting **degeneracy** caused by alcohol, tobacco, and venereal disease to be passed on to future offspring through **Lamarckian inheritance of acquired characteristics**.

FURTHER READING: Engs, Ruth Clifford, *Clean Living Movements* (2000); Engs, *The Progressive Era's Health Reform Movement* (2003).

RED PLAGUE

Coined by educator and eugenics supporter **David Starr Jordan**, the term *Red Plague* was an euphemism for **venereal diseases**, including gonorrhea and syphilis. These sexually transmitted diseases were considered major **racial poisons** leading to sterile mothers or deformed infants and, ultimately, the decline of the health and fitness of the nation and **racial degeneracy**. Jordan clarified the derivation of the term in his autobiography, *The Days of a Man, Vol. 2* (1922). He contended that it was a natural derivative from "red light district," and analogous to "white" plague (**tuberculosis**), "yellow" fever, and "black" (bubonic) plague.

REPOSITORY FOR GERMINAL CHOICE (1979–1999)

A storage facility for frozen sperm of notable men, the Repository for Germinal Choice was established as "a means of breeding higher intelligence." Founded by Robert K. Graham (1906–1997), the wealthy inventor of plastic, shatterproof eyeglass lenses, it was based upon the **positive eugenics** philosophy of American geneticist **Hermann Muller**. This "high **IQ**" sperm bank opened in 1980 near San Diego, California. Although initially established to store sperm from Nobel laureates, it later expanded to accept genetic material from men with athletic, artistic, and other "superior" abilities. As early as 1935 Muller had advocated a repository for sperm from distinguished men to be used in **artificial insemination**. Concerned about the dysgenic effect upon the human gene pool, as individuals with **genetic diseases** were now reaching reproductive age, of advances in medicine and **public health**, Muller in 1959 encouraged eugenic human evolution by voluntary "**germ cell** choice" for sterile couples who wanted children. Rather than using sperm from just any sperm donor, Muller argued that sperm from highly intelligent and altruistic individuals could lead to improvement of the human race. Graham shared Muller's ideas, but differences in their concepts of the sperm bank resulted in its establishment only after Muller's death. The repository claimed several hundred successful pregnancies, but it was considered highly controversial. Lack of funding, a rule allowing only married women to use its sperm, and competition from numerous other sperm banks diminished the influence of the repository and it closed in 1999, within two years of Graham's death.

FURTHER READING: Carlson, Elof Axel, *Genes, Radiation, and Society* (1981).

RISING TIDE OF COLOR AGAINST WHITE WORLD-SUPREMACY (1920)

Written by **nativist** writer **Lothrop Stoddard**, this book influenced the passage of the **Johnson-Reed Immigration Restriction Act of 1924**, which legislated minimal quotas for immigrants from **Asian**, **eastern and southern European**, and Middle Eastern nations. Nativist eugenics supporter **Madison Grant** wrote the introduction. This 320-page work, first published in New York by Scribner, is divided into three parts and has twelve chapters. Part 1, "The Rising Tide of Color," includes chapter titles such as "The World of Color," "The Yellow Man's Land," and "The Brown Man's Land. Part 2, "The Ebbing Tide of White," includes chapters titled "The White Flood" and "The Beginning of the Ebb." Among the chapters in Part 3 are "The Deluge on the Dikes," "The Inner Dikes," and "The Crisis of the Ages." This widely read, popular book was also read by the educated middle class who formed the **British** and **German eugenic movements.**

Stoddard argued that the worldwide domination of the "white race" was due to its intellectual, economic, physical, religious, and moral "superiority." However, he expressed concern that the "yellow races" from eastern Asia, particularly China, would be the greatest "menace" to the white race. He claimed that cheap labor, many technical schools, and inexpensive manufacturing in China and other Asian countries, would enable those countries to wrest control of the world's markets from the west. Stoddard warned that **Anglo-Saxon** and **Nordic** Americans, who built the nation but were experiencing a declining birthrate, were committing **race suicide** because the multitudes of immigrants pouring into the United States had an increasing birthrate. He argued that immigration is "filling our own land with the sweepings of the European east and south." Stoddard expressed fear that race-mixing of these immigrants with northern European stock would lead to "mongrelization," resulting in a "walking chaos" of "worthless" types of individuals, leading to further **racial degeneracy** and the demise of Western civilization. To prevent these problems Stoddard recommended **immigration restriction laws** to keep out undesirable **races** and encouraged fit northern European Americans to reproduce.

FURTHER READING: Stoddard, Lothrop, *Rising Tide of Color against White World-Supremacy* (1971).

S

Sterilization . . . has immediate and vital bearing on human life: on our personal happiness, on the welfare of our families, on the individual and community pocket-book, on the quality of our race in the long run.

Leon F. Whitney, *The Case for Sterilization* (1934)

SALEEBY, C(ALEB) W(ILLIAMS) (May 3, 1878–December 9, 1940)
British physician, writer, and health crusader Saleeby popularized eugenics through numerous publications and lectures. He coined the term **racial poisons**, was the first to use the terms **positive eugenics** and **negative eugenics**, and often referred to eugenics as **race regeneration**. Born in Worthing, England, his father was a founder and headmaster of a private school. He was educated by his mother and at the Royal High School in Edinburgh, Scotland. Saleeby studied medicine at the University of Edinburgh, where he received an M.B., Ch.B. (1901), and M.D. (1904). He was also a Scott Scholar in gynecology and worked as an anatomy demonstrator during his medical training. For a short time Saleeby was also resident physician in the Maternity Hospital and Royal Infirmary and worked at other Edinburgh hospitals and briefly in York, England. In 1904 he was influenced by a presentation given by **Francis Galton**, a British naturalist and the father of eugenics. Saleeby spoke of Galton as "my master" and argued that eugenics education was necessary to prevent **race degeneracy** and to "save the world." After receiving a substantial inheritance Saleeby quit his gynecological practice, moved to London, and became involved full time in eugenics and other health causes for the rest of his career. For a while he was science editor of the *Pall Mall Gazette* and Methuen Medical Science's editor. From 1907 through 1923 he worked intermittently as a eugenics lecturer at the Royal Institution. After 1924 Saleeby was a BBC radio broadcaster on eugenics and health subjects.

Like most eugenicists in **Britain**, **Germany**, the **United States** and other countries, Saleeby embraced **social Darwinism** and was concerned with the declining birthrate among the "fit." He leaned toward socialism and **Arthur de Gobineau**'s theory of the hierarchy of human **races**. He also asserted that the primary role for women was that of "child bearer." Saleeby accepted **Charles Darwin**'s theory of evolution, **Mendelian inheritance**, and August Weismann's (1834–1914) theory of **germ-plasm** by which genetic material is passed unchanged to offspring. Although Saleeby suggested that a predisposition to a problem or disease "ran in families," he contended that **degeneracy** found in the offspring of **unfit** parents was largely due to **racial poisons** (a term he coined in 1907), of which the three major ones were **alcohol**, **tuberculosis**, and **venereal diseases**, which all caused congenital defects. He suggested that degeneracy was also caused by rearing infants and children in unhealthy environmental conditions. Thus he argued for educational programs, sanitation and **public health** efforts, and postnatal care to prevent disease and reduce infant mortality. He also coined the term *smog* (the combination of smoke and fog).

Saleeby's writings arguing for an interplay between **nature** and **nurture** in producing all characteristics resulted in conflict with statistician **Karl Pearson** at the **Galton Laboratory** and its biometric school, which discounted environmental influences. Saleeby was active in many organizations, in particular those dealing with alcoholism, including the National Temperance League and the Society for the Study of Inebriety. He was on the governing councils of the National Council for Combating Venereal Diseases, the Sociological Society, and the National League for **Physical Education**. In the immediate post–**World War I** years he was a member of the National Birthrate Commission, which sought to create a ministry of health and raise the birthrate of the "fitter classes." In addition, Saleeby was chairman of the Sunlight League he founded in 1924, an organization that advocated fresh air and unclothed exposure to sunlight as a method of improving health and preventing disease.

Saleeby was influential in eugenics organizations but often clashed with other eugenicists. Galton considered his enthusiasm for eugenics reform as "obnoxious to the cause." However, Saleeby was instrumental in the formation of the **Eugenics Society** (first called the Eugenics Education Society) and was a member of its governing council in 1909. He was not reelected due to conflicts with other eugenicists concerning the need for prohibition. He was an active participant in the **First International Eugenics Congress** held in London (1912) and was known for several books that popularized eugenics for the middle classes. These included ***Parenthood and Race Culture*** (1909), which became the major British eugenics text, *The Methods of Race Regeneration* (1911), *The Progress of Eugenics* (1914), and *The Eugenic Prospect* (1921). These books were also popular in other countries. Saleeby also wrote the eugenics section in Arthur Mee's *Children's Encyclopedia*, in which he presented the basic theory of **genetics**, considered controversial at the time by some scientists. Over his lifetime Saleeby published over twenty books and numerous articles. His first wife was Monica Mary Meynell, with whom he had a daughter and at least one other child; his second wife was Muriel Gordon Billings (1929). Saleeby died in Apple Tree, Aldbury, a rural area north of London, from heart failure.

FURTHER READING: Farrall, Lyndsay Andrew, *The Origins and Growth of the English Eugenics Movement, 1865–1925* (1985); Rodwell, Grant, "Dr. Caleb Williams Saleeby: The Complete Eugenicist" (1997); Searle, G. R., *Eugenics and Politics in Britain, 1900–1914* (1976); Soloway, Richard A., *Demography and Degeneration* (1990).

SANGER, MARGARET HIGGINS (September 14, 1879–September 6, 1966)

The most prominent leader of the early-twentieth century birth control movement, Sanger also supported **birth control** as a eugenics technique. Although she endorsed eugenics concepts, as did most social welfare and health professionals of the time, she was not a major leader of the eugenics movement. One of eleven children, Sanger was born in Corning, New York, the daughter of Irish Catholic immigrant parents. Her father was a socialist stonemason and her mother a devout Catholic. Sanger attended Claverack College and Hudson Institute near Hudson, New York (c. 1896–1899) and later studied nursing in White Plains, New York. Her education was cut short when she married William Sanger (1873?–1961), an architect and aspiring painter, in 1902. She lived in several communities in upstate New York and in 1912 moved to New York City, where she worked as a visiting nurse on Manhattan's Lower East Side. There she encountered many women whose health had been destroyed by excessive childbearing or botched abortions, and she decided to do something about the situation.

In 1913 Sanger traveled to Glasgow, Scotland, and Paris to obtain contraceptive information she claimed was unavailable in the United States. Upon her return she launched *Woman Rebel* in March 1914, a radical sheet that discussed controversial subjects including **venereal diseases** and *birth control*, a term she helped coin in 1914. She argued that birth control was a fundamental tenet of women's rights. Sanger published pamphlets concerning contraceptives and was arrested for distributing this information through the U.S. mail. Distribution of such material was considered illegal under the Comstock laws on the grounds that it

Margaret Sanger, the primary leader of the birth control movement in the United States, like most other health and social reformers, supported eugenics. (Image courtesy of the Library of Congress.)

was obscene. To avoid prosecution, she fled to Europe in October 1914. In Britain she briefly interacted with **Marie Stopes**, a British birth control and eugenics crusader. In September 1915 she was lured back to the United States when Anthony Comstock (1844–1915), head of the New York Committee for the Suppression of Vice, entrapped her husband William into giving him a copy of the banned contraceptive publication *Family Limitations* and arrested him. Comstock died within a few weeks of this ploy. Sanger returned to New York, and the federal government dropped the charges against her in 1916. Due to conflicts with other birth control leaders, such as Mary Ware Dennett (1872–1947), Sanger began to champion the dispensing of birth control devices by physicians and worked to get support from the medical community for their legal distribution. She also clashed with physician **Lydia De Vilbiss**, whom she also saw as a competitor. In Brooklyn, New York, in October 1916, Sanger and her sister opened the first birth control clinic in the country. Within weeks they were arrested. Publicity from this and other arrests helped bring the birth control movement public attention.

By 1917 Sanger was giving public support to the eugenics movement and the eugenic benefits of contraception. In *Birth Control Review*, which she launched in January 1917, she discussed **race betterment**, or eugenics, along with birth control and other sexual topics. In November 1921 she added a subtitle to the review's masthead: *Birth Control: To Create a Race of Thoroughbreds*. In her writings Sanger addressed typical eugenic concerns about "overbreeding" among the poor as the source of many social problems and encouraged **negative eugenics** to prevent the **unfit** from breeding. However, she did not endorse the idea of promoting fertility, or **positive eugenics**, among the fit. Sanger argued that if women had the choice to control the size of their families through contraception, most would do so in an effort to improve their lives.

In November 1921 Sanger organized the first American Birth Control Conference, held in New York City, and formed the **American Birth Control League**, a lobby and educational group. After divorcing William Sanger in 1921 she married James Noah Henry Slee (1860–1943), a rich industrialist who underwrote her family-planning efforts. In 1923 Sanger opened the Clinical Research Bureau in New York, the first legal birth-control clinic in the United States. She also tried unsuccessfully, in cooperation with gynecologist **Robert Dickinson**, to get a dispensary licence for the clinic. Due to politics at the state level, opposition from the **Catholic Church**, and resistance among physicians over the controversial nature of birth control, it was not granted. In June 1928 Sanger resigned as president of the league and a few months later, as editor of *Birth Control Review* due to conflicts with the board. She retained control of the clinic, which was renamed the Birth Control Clinical Research Bureau, until she semiretired in 1939.

In the 1930s Sanger lobbied for the repeal of various federal anti–birth control laws, established international birth control and population control organizations, and lectured internationally on the cause of family planning as a basic human right. She was a prolific writer. Besides numerous articles, Sanger published several books including *Margaret Sanger: An Autobiography* (1938). However, material in her autobiography sometimes conflicts with biographical and other sources as to dates, experiences, and interpretations. Throughout her later life she was honored by many organizations and countries. With William Sanger she had three children, one who died in infancy. She retired to Tucson, Arizona, where she died of arteriosclerosis in a nursing home.

FURTHER READING: Chesler, Ellen, *Woman of Valor* (1992); Gray, Madeline, *Margaret Sanger* (1978); Gordon, Linda, *Women's Body Women's Right* (1990); Reed, James, *From Private Vice to Public Virtue* (1978); Sanger, Margaret, *My Fight for Birth Control* (1969); Sanger, *The Selected Papers of Margaret Sanger* (2003).

SCHALLMAYER, FRIEDRICH WILHELM (February 10, 1857–April 10, 1919)

The first German eugenics advocate, Schallmayer established the theoretical base for eugenics in Germany. He helped launch the **German eugenics** movement with a prize-winning essay (1900) that became the standard German eugenics textbook until the early 1920s. Schallmayer was against Nordicism (superiority of the **Nordic race**) and criticized **Arthur de Gobineau**'s theory of the **hierarchy of races** as unscientific. However, he considered social-class differences important and promoted reproduction among the socially productive middle class (**positive eugenics**). Schallmayer was born in Mindelheim, a small town in Bavaria near Augsburg; his father had a lucrative wagon and carriage business. After graduation from secondary school in Augsburg (1876), Schallmayer enlisted in the army as a one-year volunteer in order to fulfill his military obligations but was found to have a "lack of fitness for military duty." He took classes in history, philosophy, and geography at Würzburg and then studied law at the University of Munich. He left Munich in 1879 and went to Leipzig, where he studied philosophy, anatomy, national economy, and sociology and embraced socialism. In 1881 Schallmayer returned to Munich to study medicine and complete his degree (1883). After practicing medicine in several situations and specializing in urology, he gave it up in 1897 to devote his full time to scholarship and the promotion of eugenics.

To collect information for a dissertation after completing his medical degree, Schallmayer worked for a year as an intern in a Munich psychiatric clinic. Afterward he traveled for a year and in 1886 worked as a physician on a ship, where he wrote the draft of his first eugenics treatise. This small tract, *The Menacing Physical Deterioration of Civilized Man* published in 1891, focused on **social Darwinism** and **degeneracy** theory. It did not attract much attention. In 1887 Schallmayer began working as a general practitioner in Kaufbeuren, near Augsburg. He soon began to view his practice as counterproductive for "race improvement" inasmuch as he believed that **natural selection** was the primary agent of all social progress and was concerned that **public health** was keeping the **unfit** artificially alive, thus enabling them to reproduce. In order to prevent and treat diseases and conditions, such as venereal disease, that were leading to **racial degeneracy**, he left Kaufbeuren and trained in urology and gynecology in Vienna, Leipzig, and Dresden. Schallmayer then began a medical practice in Düsseldorf and worked again as a ship's physician. By 1897 he had acquired the financial means from inheritance and other sources to enable him to give up his practice.

In 1899 Schallmayer wrote an essay encouraging the creation of a "Ministry of Medicine" to be responsible for the health and efficiency of the nation, with eugenics as a branch of medicine. In 1900 he entered a writing contest on evolution and its relationship to politics, sponsored secretly by Friedrich Krupp, a member of a wealthy industrialist family. He was awarded first prize for his essay promoting a "hygienic-sociological" approach to race degeneracy. In it he argued that the real political lesson to be learned from **Darwinism** was that the long-term power of the state depended on the biological vitality of the nation. He believed that eugenic reforms must be implemented for the survival of Germany. His essay was revised and published in 1903 as *Heredity and Selection in the Life History of Nations* (*Vererbung und Auslese im Lebenslauf der Völker*) and became the standard German eugenics textbook until his death. The publicity surrounding this book helped to launch the German eugenics movement. Schallmayer, like other reformers and physicians of the era, campaigned for **social hygiene** programs and the prevention of **alcoholism** due to concern that **venereal disease** and **alcohol** were **racial poisons** that caused congenital defects and damaged the **germ-plasm** (hereditary material) and led to race degeneracy. Around 1907, after becoming familiar with British naturalist **Francis Galton**'s work on eugenics, Schallmayer used the Germanized term *Eugenik* in his writings rather than "race hygiene" (first used by German physician **Alfred Ploetz** in 1895), on the grounds that he did not like its "Nordic purity" connotations. He was involved with several organizations engaged in preventing social and health problems and increasing population. He married the daughter of the mayor of Kaufbeuren, Elise Bachschmied, in 1888, and, after his first wife's death in 1909, he married Gertrud Fritze (1911) with whom he had two children. Schallmayer developed an international reputation and had published forty-two articles in prestigious medical, social science, and eugenics journals at the time of his death. He died in Krailling from a heart attack after being ill for several years from heart disease and asthma.

FURTHER READING: Weindling Paul, *Health, Race, and German Politics between National Unification and Nazism, 1870–1945* (1989); Weingart, Peter, "German Eugenics between Science and Politics," (1989); Weiss, Sheila Faith, *Race Hygiene and National Efficiency* (1987); Weiss, "The Race Hygiene Movement in Germany" (1987); Weiss, "Wilhelm Schallmayer and the Logic of German Eugenics" (1986).

SECOND INTERNATIONAL CONGRESS OF EUGENICS (1921)

At the successful 1912 **First International Eugenics Congress**, a planning committee formed to plan the next congress for 1913. However, due to **World War I**, the Second International Congress did not take place until September 22–28, 1921. It was held at the American Museum of Natural History in New York. Its purpose was to bring together geneticists and eugenicists from around the world to "discuss results of their research and their application to race improvement." Most noted eugenicists attended. Educator **Clarence C. Little** was elected general secretary; anthropologist **Henry Fairfield Osborn**, president; **Alexander Graham Bell**, honorary president; and eugenicist **Madison Grant**, treasurer. **Harry H. Laughlin** headed exhibits. **Charles Davenport**, the pivotal leader of the American eugenics movement, and Osborn did most of the organizational work. It was funded by philanthropist **Mary Harriman**, the **Carnegie Institution of Washington**, and other groups.

The congress was divided into four sections: "Human and Comparative Heredity," "Eugenics and the Family," "Human Racial Differences," and "Eugenics and the State." Proceedings and presentations were published in two volumes and printed in Baltimore by Williams and Wilkins Company (1923) with a total of 96 papers. The first volume, *Eugenics, Genetics and the Family*, included 439 pages plus photographs. Titles of papers included "Darwinian Evolution by Mutation," "Inheritance of Mental Disorders," "The Mayflower Pilgrims," and "The Oneida Community Experiment in Stirpiculture." The second volume, *Eugenics in Race and State*, contained 471 pages. Papers included "Notes on the Body-Form of Man," "Harmonic and Disharmonic Race Crossings," "The Present Status of Eugenical Sterilization in the United States," and "Eugenics in Relations to the Tuberculosis Problem."

The congress registered 393 participants. Those from the United States included **Arthur H. Estabrook**, **Roswell H. Johnson**, **Wilhelmine Key**, **Hermann Muller**, and **Samuel J. Holmes**. Presenters from Britain included **Leonard Darwin**, **R. A. Fisher**, and **C. W. Saleeby**. Australia, Belgium, Denmark, France, and Norway were also represented; due to political tensions from World War I, no Germans attended. Participants were given tours of the **Eugenics Record Office** and the Station for Experimental Evolution at **Cold Spring Harbor**, New York. Out of this meeting American economist **Irving Fisher** formed an ad interim committee that evolved into the **American Eugenics Society**. A committee was formed to plan another meeting, the **Third International Congress of Eugenics**, which did not take place for almost another decade.

At the Second International Congress of Eugenics a certificate illustrating that eugenics derived from many disciplines was presented to exhibitors. (From: *Journal of Heredity* [1922], Main Library, Indiana University, Bloomington, Indiana. Image courtesy of the Digital Library Program.)

FURTHER READING: Engs, Ruth Clifford, *The Progressive Era's Health Reform Movement* (2003); Haller, Mark H., *Eugenics* (1984); Perkins, Harry F., (ed.), *A Decade of Progress in Eugenics* (1934).

SECOND NATIONAL CONFERENCE ON RACE BETTERMENT (1915)

Sponsored by the **Race Betterment Foundation** headed by health reformer and eugenics supporter **John Harvey Kellogg**, the Second National Conference on Race Betterment was held in connection with the Panama Pacific Exposition in San Francisco

August 4–8, 1915. The **First National Conference** in 1914 had received much public attention, leading Kellogg and his associates to sponsor another gathering the following year. The purpose of the second conference, as stated on the cover of the proceedings, was to "assemble and discuss the evidence of race deterioration and to promote race betterment." The president of the conference was Stephen Smith, vice president of State Board of Charities for New York City. Conference vice presidents included eugenics supporters economist **Irving Fisher** and J. N. Hurty, head of the Indiana State Board of Public Health. Geneticist and eugenics leader **Charles Davenport** and educator **Charles Eliot** were on the planning committee. Although the second conference had fewer delegates and papers than the first, several leaders of the eugenics, **public health**, **social hygiene**, and **birth control** movements, including **Wilhelmine Key**, **David Starr Jordan**, **Luther Burbank**, and **Paul Popenoe** presented papers.

A major outcome of the conference was formation of the **Eugenics Registry**, suggested by Kellogg at the first conference, to encourage "intelligent persons to consider more fully the importance of hereditary traits in planning their marriages." The registry was founded as a "partnership" of the Race Betterment Foundation and the **Eugenic Records Office** (ERO) in **Cold Spring Harbor**, New York. The presentations and exhibits were published as the *Official Proceedings of the Second National Conference on Race Betterment, August 4, 5, 6, 7 and 8, 1915; Held in San Francisco, California* (1915). The publication contained 173 pages. Papers included "Eugenics and War," "Natural Selection in Man," "The Human Life Cycle," "The Eugenics Registry," "Heritable Factors in Human Fitness and Their Social Control," "Alcohol Prohibition," "Some Practical Methods in the Social Hygiene Movement in the United States," and "The Commitment of the Insane." The publication also showed photographs of exhibits.

FURTHER READING: Schwarz, Richard William, *John Harvey Kellogg, M.D.* (1981).

SEXUALLY TRANSMITTED DISEASES
See Venereal Diseases

SHARP, HARRY CLAY (November 1872–October 31, 1940)
An Indiana physician, Sharp promoted vasectomy for **eugenic sterilization** of the **unfit** in the United States during the early-twentieth-century eugenics movement. His advocacy of this form of **negative eugenics** became the foundation of sterilizations laws, not only in the **United States** but also in **Germany** and other countries. Like many of the era, Sharp considered **degeneracy** inherited and subscribed to **Lamarckian inheritance of acquired characteristics**. He believed that people who **masturbated**, who were "sodomites" (homosexuals), or had syphilis were unfit to procreate because these conditions had tainted their **germ-plasm**. Sharp was born in Charleston, Indiana; however, little is known about his background or childhood. He studied medicine at Ohio State Medical School and received his M.D. from the University of Louisville (1893). In 1895 Sharp became a physician at the Indiana State Reformatory in Jeffersonville (1895–1908), and he was on the board of directors for the reformatory from 1899 to 1914. He left to become state hospital superintendent (1910–1913). Sharp and colleagues also established a hospital at West Baden, Indiana. During **World War I** he conducted surgery in the U.S. Army Medical Corps in France. After the war he en-

gaged briefly in private practice in Indianapolis but soon became a surgeon for the U.S. Public Health Service. He ended his career as head of a veterans hospital.

In 1899 Sharp was influenced by a paper by physician Albert John Ochsner (1858–1925), who suggested using vasectomies as a sterilization procedure with criminals, and as an alternative to imprisonment and castration. Without legal authority Sharp sterilized inmates at the reformatory between 1899 and 1907 to "reduce sexual excitation in delinquent boys." Soon Sharp saw it as a tax-saving measure to counter the rapid increase in the number of institution-

Physician Harry Sharp performed the first eugenic sterilizations at the Indiana Reformatory Hospital without legal authority. His enthusiasm for the procedure led to Indiana's being the first state in the United States to pass a sterilization law. (From: *Biennial Report, Indiana Reformatory, Jeffersonville, Indiana [1897–1898]*, Main Library, Indiana University, Bloomington, Indiana. Image courtesy of the Digital Library Program.)

alized **degenerates** and published a pamphlet, *Vasectomy: A Means of Preventing Defective Procreation* (1908). He used his influence as vice president of the Indiana State Medical Society (1905–1906) to persuade the Indiana Legislature to enact the first state eugenical sterilization laws (1907) authorizing compulsory sterilization "to prevent procreation of confirmed criminals, idiots, imbeciles and rapists." Though the legislation passed, many physicians and the governor opposed more sterilizations. It is estimated that some 450 male inmates may have been sterilized until the law was repealed in 1921. The legislation, known as the Indiana Plan, became the model for compulsory sterilization in other states. Eugenicist **Harry H. Laughlin** incorporated it as part of his model sterilization law. Sharp promoted vasectomy nationwide and consulted with legislative bodies to enact eugenic sterilization laws. He married and had one son. Sharp died a month before he planned to retire as chief medical officer of the United States Veterans Administration Hospital in Lyons, New Jersey.

FURTHER READING: Carlson, Elof Axel, *The Unfit* (2001); Gugliotta, Angela, " 'Dr. Sharp with His Little Knife" (1998); Reilly, Philip R., *The Surgical Solution* (1991).

SOCIAL BIOLOGY OR SOCIOBIOLOGY

The study of the biological, genetic, and evolutionary basis of animal and human social systems and behaviors is termed *social biology, sociobiology,* or *behavioral ecology.* It embraces the biological and sociocultural forces that affect the structure and composition of human populations. The concept was initially formulated by British naturalist **Charles Darwin** and others who studied the influence of inheritance and evolution on behavior. The term was popularized by American biologist Edward O. Wilson (b. 1929) in *Sociobiology: The New Synthesis* (1975). As an academic field of study, social biology is interdisciplinary, combining biology and the social sciences to make the case that animal behaviors can be explained by **Darwinian** evolutionary theory. It suggests that animals will behave in ways that maximize their chances of transmitting copies of their **genes** (hereditary material) to succeeding generations. The evolutionary process of **natural selection**, or survival of the fittest, fosters behavior and traits that increase an individual's chances of surviving and reproducing.

Wilson's concepts were considered controversial when they were introduced in the 1970s. Some people interpreted his theory as suggesting that all human behavior and negative social conditions such as poverty, lack of education, **criminality**, and low **intelligence** were due to inheritance, bringing to mind early-twentieth-century ideas of **degenerate** families and **races**. Wilson suggested, however, that both hereditary and environmental factors (**nature-nurture**) were important for shaping human behavior, which by the turn of the twenty-first century was accepted by most scholars. In the latter decades of the twentieth century, when the term *eugenics* fell out of favor as an academic topic, eugenics organizations on both sides of the Atlantic adopted the term *sociobiology* as best representing their field of interest or research. In 1969 the American journal *Eugenics Quarterly* changed its name to *Social Biology*. That same year the British **Eugenics Review** became the *Journal of Biosocial Science*. The organizations sponsoring these journals also changed their names. The **American Eugenics Society** was renamed the Society for the Study of Social Biology (1973) and the British **Eugenics Society** became the Galton Institute (1989).

FURTHER READING: Krebs, John R., and Nicholas B. Davies, (eds.), *Behavioural Ecology* (1991); Wilson, Edward O., *On Human Nature* (1978).

SOCIAL DARWINISM

This philosophical and sociological theory, which emerged during the second half of the nineteenth century, drew analogies between **Darwinism** (evolution found in nature) and human socioeconomic development. It became a basic element of **eugenics.** Social Darwinism posits that individuals, groups, and societies are subject to the same laws of **natural selection** as found among plants and animals. This process resulted in the "survival of the fittest," a phrase coined by British intellectual and sociologist Herbert Spencer (1820–1903), who popularized social Darwinism. Social Darwinists believed that the process of natural selection, acting on a human population, would lead to continued improvement of the human race and survival of the best social, economic, and political systems. As a justification for class structure, social Darwinism in Britain suggested that successful socioeconomic classes were composed of the most biologically superior. This ideology was also embraced by German eugenicists.

As part of social Darwinian philosophy, different **races**, or ethnic groups, were considered at different stages of development in the "upward ascent of man." Based upon **Arthur de Gobineau**'s theory of **hierarchy of races**, some groups such as the **Nordics** (northern Europeans) were considered biologically and culturally superior, while others, such as southern Europeans, were deemed inferior. In the United States the philosophy gained much support in the late nineteenth century among industrialists, such as Andrew Carnegie (1835–1919), and intellectuals. Some eugenics supporters argued that newly developed immunizations against infectious disease and life-saving medical treatment would lead to **racial degeneration** because the **unfit** and the poor would survive to reproduce. Disease was viewed as a natural process to eliminate the inferior through natural selection. **Paul Popenoe**, a leading California eugenicist, for example, viewed the high infant morality rate among the poor as an example of natural selection at work. Likewise, eugenicists advocated **eugenic sterilization laws** to keep the unfit from breeding; some were opposed to **birth control** on the grounds that it would lead to **race suicide** among the "fit" middle class. Social Darwinism declined after the third decade of the twentieth century as increased knowledge of environmental and genetic mechanisms undermined many of its basic tenets.

FURTHER READING: Bannister, Robert C., *Social Darwinism* (1988); Hawkins, Mike, *Social Darwinism in European and American Thought, 1860–1945* (1997); Hofstadter, Richard, *Social Darwinism in American Thought* (1959).

SOCIAL HYGIENE

A euphemism for many sex-related subjects including sex education, prevention of venereal disease, elimination of prostitution, and morality education, the phrase *social hygiene* was first used around 1907 in the United States by a Chicago newspaper to indicate the scope of activities proposed by the developing social hygiene movement and remained in common use until the mid-twentieth century. The term has roots from France in the 1840s. In Europe the meaning became broader than in the United States as it embraced social medicine. Social hygiene was an aspect of the eugenics movement on both sides of the Atlantic inasmuch as an aim was to prevent **racial degeneration** (deterioration of the human race) from the **racial poison** of **venereal disease**. Because the term suggested broad interests, many new health reform organizations of the **Progressive Era** (1890–1920) adopted the expression, including the **American Social Hygiene Association**, founded in 1913. Social hygiene was also linked with **birth control** and the use of condoms to prevent the spread of sexually transmitted diseases. The concept was also a philosophy and program of action in regard to all aspects of sexuality. Sexually related issues in this era were not discussed openly in the press or in polite conversation, and the Comstock laws of 1873 made it illegal to publish explicit material. However, eugenics and social hygiene reformers considered open discussion of sexuality important. By the 1920s social hygiene involved four aspects: education to instruct the public concerning venereal disease and a single standard of sexuality to avoid venereal diseases; recreation programs and physical education aimed at minimizing sexual desire and giving "wholesome activities" to youth; medical programs to detect and treat syphilis and gonorrhea; and law-enforcement measures that included arrest of **prostitutes**. These ideas are still considered important in the early twenty-first century.

FURTHER READING: Brandt, Allan M., *No Magic Bullet* (1985); Clarke, Charles Walter, *Taboo* (1961).

SPERM BANKS

A repository where sperm is kept frozen in liquid nitrogen for later use in **artificial insemination** is called a *sperm bank*. By the late 1990s some of these repositories also stored eggs and embryos. The phrase "banks for freezing sperm" was suggested by American geneticist **Hermann Muller** in 1959 for storage of sperm from highly intelligent men. The term *sperm-bank* was first used by British biologist Julian Huxley (1887–1975) a year later. Artificial insemination using sperm from notable men had originally been discussed in 1935 by Muller, along with the concept of repositories for their semen, in his book *Out of the Night*. Artificial insemination of humans, done secretly until the mid-1960s, was generally accomplished with fresh semen. Although attempts had been made to freeze sperm for the procedure, it was not until 1949 that a method was developed using glycerol, a sugary, viscous substance, to protect semen from injury during deep-freezing. Freezing enabled sperm to be stored for years or shipped to other locations, and by the mid-1970s several sperm banks had been established. In the United States the **Repository for Germinal Choice** for Nobel laureates and, later, men with superior talents and abilities, based upon Muller's **eugenics** philosophy, opened for business in 1980. In the mid-1980s, after AIDS and the HIV virus had emerged, most sperm banks had begun rigorous testing procedures. These included infectious disease testing, detailed semen analysis, family and medical histories, and **genetic screening and testing** of prospective donors. By the turn of the twenty-first century many westernized countries had sperm banks. Donors tended to be healthy individuals under age thirty, often students. In addition, military personnel and men undergoing radiation and other medical procedures began to store sperm in repositories for possible future use. In the United States donors are offered anonymity; however, in many European countries donor identity can be obtained when the child has become an adult. A few repositories also began to collect and store eggs for **in vitro fertilization** procedures. Over 300 centers for **assisted reproduction** operated in the United States in 2003, and some also stored sperm, eggs, and embryos. Because repositories seek genetic material from healthy individuals without genetic diseases to be used by infertile couples and single women desiring children, they foster **positive eugenics** and are a facet of the **new eugenics**.

FURTHER READING: Henig, Robin Marantz, *Pandora's Baby* (2004); Stacey, Meg, (ed.), *Changing Human Reproduction* (1992).

STERILIZATION

See Eugenic Sterilization

STERILIZATION FOR HUMAN BETTERMENT: A SUMMARY OF RESULTS OF 6,000 OPERATIONS IN CALIFORNIA, 1909–1929 (1929)

Written by **eugenics sterilization** champions **E. S. Gosney** and **Paul Popenoe**, of the **Human Betterment Foundation**, the purpose of *Sterilization for Human Betterment* was to support sterilization laws that granted a state the right to sterilize the **unfit** that had been upheld by the U.S. Supreme Court in the **Buck v. Bell** case. The "unfit" included the **feebleminded, insane**, and **criminals**. The publication also aimed to present the results of sterilization in California as an effort to promote the procedure as an efficient eugenics measure to prevent **dysgenic** individuals from reproducing and passing their "defective **genes**" to offspring. Published in New York by Macmillan, the book contained 202 pages. It was divided into two parts, "The Facts" and "The Conclusions." Within these two sections were twelve chapters, with titles that included "The History of Sterilization," "Effects on the Patient's Behavior," "Voluntary Sterilization," "The Operation," and "Sterilization for Eugenic Reasons." The history of sterilization and its need as a **negative eugenics** measure to prevent **racial degeneration** was discussed, along with the vasectomy procedure. The authors reported that between 1910 and January 1, 1929, 6,255 sterilizations had been performed in California state institutions, about three times as many conducted in the rest of the United States. Of the sterilizations in California state hospitals, 601 more males than females had had the surgery. Of the total 1,488 sterilizations for feeblemindedness, 330 more females than males had been sterilized. Feebleminded patients were sterilized before they were allowed to leave the institution. One in twelve of all mentally ill individuals and one in five new admissions to state hospitals were sterilized. The authors stated that consent of the nearest relative was always obtained before the surgery and that most patients were satisfied with the operation.

FURTHER READING: Carlson, Elof Axel, *The Unfit* (2001); Reilly, Philip R., *The Surgical Solution* (1991).

STIRPICULTURE

The **positive eugenics** practice of selective breeding to improve the human race used in the **Oneida Community** was called *stirpiculture*. The term was coined by **John Humphrey Noyes** in 1869, founder of the community. The term had roots in agricultural breeding from the word *stirp* meaning to produce the best offspring or stock. The community's "improvement of human beings through scientific breeding" originated in the **Jacksonian Era**'s health reform and hereditarian movement. It was undertaken over a ten-year period (1868–1879) to produce morally, mentally, and physically healthy children. This experiment in human breeding produced fifty-eight live children from eighty-one parents. About 100 prospective parents participated in the venture. A central committee approved or disapproved an application from two individuals to produce a child based upon their spiritual, intellectual, mental, and physical characteristics. However, if a couple was not approved, the committee attempted to match the individuals with someone more suitable if they were considered "fit." Noyes and the community's leaders, like many health reformers of the era, subscribed to **Lamarckian inheritance of acquired characteristics** and attempted to facilitate the best matings. In some cases the committee requested that two specific individuals mate for the purpose of child-bearing. Noyes himself fathered at least nine children.

In 1921 Hilda Herrick Noyes and George Wallingford Noyes, descendants of the community, presented a paper on the stirpiculture experiment at the **Second International Congress of Eugenics**. They reported that although Noyes first published his views regarding selected human breeding in the first annual report of the Oneida Community in 1849, well before British naturalist **Francis Galton** had discussed the concept, this positive eugenics experiment did not start until the foundation of community beliefs, practices, and financial resources had been fully established. The Noyeses reported that the mortality rate of stirpiculture children was lower than expected and the children were healthier than other children in rural areas. Parents of the stirpiculture children were reported to have been unusually healthy, with about a half surviving past age eighty-five. Of the forty men participating, eight were college graduates. Over time resentment by those not chosen to be parents led to internal problems within the community. Stirpiculture was halted and the community disbanded in 1881.

FURTHER READING: Klaw, Spencer, *Without Sin* (1993); Sokolow, Jayme A., *Eros and Modernization* (1983); Walters, Ronald C., *American Reformers, 1815–1860* (1978).

STODDARD, LOTHROP THEODORE (June 29, 1883–May 1, 1950)

Stoddard, a political analyst and eugenicist, helped shape the **nativist** segment of the eugenics movement in the United States. He championed **immigration restric-**

tion laws and **birth control** as eugenics measures. Born in Brookline, Massachusetts, to an old New England family that took pride in its pedigree, Stoddard grew up in a privileged environment, the only child of a noted lecturer and travel writer. After private school he attended Harvard University, where he graduated *magna cum laude* (1905). He then studied law at Boston University and was admitted to the bar in 1908. Upon graduation Stoddard made a grand tour of Europe and became convinced the world was soon headed for war. This led him to study international relations, in which he received a M.A. (1901) and Ph.D. (1914). Stoddard wrote several books concerning European political complexities and in 1918 took over the foreign affairs department of the popular periodical *Worlds Work*. Over his lifetime he wrote twenty-two books and numerous magazine articles focused on political, social, and racial theory. In the early 1930s he moved to Washington, D.C., and in 1939 became a special correspondent for the North American Newspaper Alliance in Germany. On his return to the United States, he served as a foreign policy expert for the *Washington Evening Star* (1940–1944).

In the immediate post–**World War I** era, Stoddard became increasingly concerned about the demise of **Anglo-Saxon** stock in the United States. He advocated immigration restriction and popularized eugenics through a series of articles and books including **The Rising Tide of Color against White World-Supremacy** (1920), and *Revolt against Civilization: The Menace of the Under Man* (1922), in which he proposed that unrestricted immigration of unfit "under men" was "a grim peril to civilization" because the "biologically unfit" were taking power away from the "eugenically elite." Stoddard was active in several eugenics organizations. He was a member of the Advisory Council of the **American Eugenics Society**, on the organizing committee and head of publicity for the **Second International Congress of Eugenics** (1921), and an early member of the **Galton Society**. He testified at immigration restriction hearings that led to passage of the **Johnson-Reed Immigration Restriction Act of 1924**, and was on the board of directors of the **American Birth Control League** (1921–1928). While in Germany during the 1930s he interviewed race theorist **Hans F. K. Günther** and prominent German eugenicists including **Eugen Fischer** and **Fritz Lenz**, and wrote sympathetically of the early **Nazi Germany** regime in *Into the Darkness* (1940). In the aftermath of **World War II** and the Holocaust, his reputation as a theorist was ruined. Stoddard married Elizabeth Guildford Bates (1926) and fathered two children. His wife died in 1940 and he married Zoya Klementinovskaya (1944). Stoddard died of cancer in Washington, D.C., almost a forgotten man. His ashes were buried at West Dennis on Cape Cod, Massachusetts, where he spent his summers.

FURTHER READING: Haller, Mark H., *Eugenics* (1984); Paul, Diane B., *Controlling Human Heredity* (1995).

STOPES, MARIE CHARLOTTE CARMICHAEL (October 15, 1880–October 2 1958)

The leading British **birth control** and sex education reformer in the early twentieth century, Stopes supported eugenics and established the first birth control clinic in Britain. Born in Edinburgh, Scotland, she was the oldest daughter of well-educated parents. The family moved to London after her birth. She was educated at home by her mother until age twelve, attended private schools, and graduated with a B.S. (1902)

from University College, London, with honors in botany, geology, and physical geography. She received a Ph.D. (1904) from the Botanical Institute of Munich University in Germany. Upon returning to Britain she became the first woman science faculty member at Manchester University and earned a D.Sc. in 1905 from London University. Stopes was well respected as a researcher and was awarded a grant from the British Royal Society to conduct research in Japan (1907–1908). She briefly returned to Manchester and then joined the faculty at University College (1913–1920) to lecture, conduct research, and publish in paleobotany. In 1911 she married botanist Reginald Gates, but the marriage was annulled due to "non-consummation" in 1916. She turned from scientific research and teaching, and later wrote that problems in her marriage were the reason for her crusade for sexual enlightenment.

Through her writings and lectures Stopes broke the taboo against openly discussing sexuality and birth control. With the financial help of her wealthy second husband, Humphrey Verdon Roe (1878–1949), an aircraft manufacturer whom she married in 1918, Stopes published *Married Love* (1918), which was extremely popular. With his finances she also founded the first British birth control clinic, The Mother's Clinic, in London (1921). She published several other books including *Wise Parenthood* (1918), *A Letter to Working Mothers* (1919), and *Radiant Motherhood* (1920). In the 1920s both the medical community and the **Catholic Church** condemned Stopes's writings, her advocacy of contraception, and her establishment of the birth control clinic. Stopes viewed birth control as a eugenics method and was a member of the British **Eugenics Society**. She encouraged both **positive** and **negative eugenics** and advanced, "more children from the fit, less from the unfit." Her marriage to Roe produced one living child from whom she became alienated because he did not marry a "eugenically suitable" mate. She separated from her husband in 1938 and moved to Norbury Park. In her later life, Stopes engaged in literary ventures and supported the arts. Distrusting physicians because her first son had been stillborn, she failed to immediately seek medical attention and died of advanced breast cancer at her home. She bequeathed her birth control clinic, library, and many other assets to the Eugenics Society.

FURTHER READING: Hall, Ruth E., *Passionate Crusader* (1977); Rose, June, *Marie Stopes and the Sexual Revolution* (1992); Soloway, Richard A., *Demography and Degeneration* (1990).

SUPERMAN, SUPER RACE, OR *ÜBERMENSCH*

The pinnacle of human moral, intellectual, and physical development and attainment was termed the *superman*, or *Übermensch*, sometimes referred to as the *higherman* or *overman*. This late-nineteenth-century concept became entwined with **social Darwinism**. Some early-twentieth-century eugenicists used the philosophy as a basis for **positive** and **negative eugenic** programs to promote reproduction among the fit and prevent breeding among the **unfit**. Philosopher Friedrich Wilhelm Nietzsche (1844–1900) crystalized the superman ideology in *Also Sprach Zarathustra* (Thus Spoke Zarathustra) (1883), the same year as British naturalist **Francis Galton** published his *Human Faculty* in which he coined the term eugenics. The idea of a Germanic super race had been introduced by composer Richard Wagner (1813–1883) in *Der Ring des Nibelungen* (The Ring of the Nibelung) (1848–1876), the cycle of four music dramas, based upon Germanic mythology. Nietzsche was influenced by Wagner, with whom he had a close friendship from 1868 until a bitter falling out in the late 1870s. Nietzsche, drawing upon Germanic **Nordic** myths, **Darwinism**, and mid-nineteenth-century **hierarchy of races** theory introduced in 1853 by **Arthur de Gobineau**, conceptualized humans as consisting of two separate **races**, one superior and the other inferior (masters and slaves). The superman represented the superior noble aristocracy. After Nietzsche's death, his sister Elizabeth altered the philosophy. The new ideology was used by **Nazi Germany** as a justification for building a **master race** of **Aryans**—northern European stock—through the extermination of "inferior races," including **Jews** and Gypsies. This mass killing became known as the Holocaust.

FURTHER READING: Cate, Curtis, *Friedrich Nietzsche* (2002); Clay, Catrine, and Michael Leapman, *Master Race* (1995); Stone, Dan, *Breeding Superman* (2002).

SURVIVAL OF THE FITTEST
See Natural Selection

T

The Eugenic ideal—the belief that we owe a paramount duty to posterity dependent on the laws of heredity.

Leonard Darwin, "Preface," *Problems in Eugenics, Vol. 2* (1912)

TEST-TUBE BABIES

The term was first used for **artificial insemination** in the mid-1930s but changed in meaning in the late 1970s to denote **in vitro fertilization** (IVF). "Test tube" is a misnomer inasmuch as the egg and sperm are mixed in a glass petri dish for the procedure. *Test-tube baby* was coined around 1934 to describe a child conceived by the clinical transfer of sperm into the vagina or uterus to treat infertility. The phrase alludes to British authors Aldous Huxley's *Brave New World* (1932) and George Bernard Shaw's *Back to Methuselah* (1921), in which humans are created outside the womb. By the early 1950s the term had entered into popular culture. A low-budget film, *Test Tube Baby* (1953), for example, had a plot involving artificial insemination. The first successful in vitro fertilization (collection, fertilization, selection, and implantation of embryos into the uterus resulting in a live birth) occurred in Britain July 1978 with the birth of **Louise Brown**. The method has **positive eugenic** implications because embryos are selected to eliminate **genetic diseases** and in some cases preimplantation gender selection is accomplished for **family balancing**.

FURTHER READING: Henig, Robin Marantz, *Pandora's Baby* (2004); Kerr, Anne, and Tom Shakespeare, *Genetic Politics* (2002); McGee, Glenn, *The Perfect Baby* (1997).

THIRD INTERNATIONAL CONGRESS OF EUGENICS (1932)

Due to the worldwide Depression, the Third International Congress of Eugenics, held at the American Museum of Natural History in New York City, August 21–23, 1932, was smaller (with 267 delegates) than the two previous congresses. Although **British eugenics** leader **Leonard Darwin** did not attend, he submitted a paper. The aim of the conference was to "mark the advance made in the field of eugenics, both as a pure and as an applied science" since the **Second International Congress of Eugenics** in 1921. Geneticist and pivotal American eugenics leader **Charles Davenport** served as president of the congress. Others on the planning committee included noted American eugenics supporters Clarence G. Campbell, **Irving Fisher**, **Madison Grant**, **Frederick Osborn**, **Leon Whitney**, and **Harry H. Laughlin**, who chaired the exhibits committee. In addition to meeting at the museum, congress delegates took a field trip to the **Eugenics Record Office** in **Cold Spring Harbor**, New York.

Biologist **Henry F. Perkins** chaired the publication committee for the proceedings, *A Decade of Progress in Eugenics* (1934), which also included Campbell, Grant, Osborn, Laughlin, and **Paul Popenoe**. The volume was dedicated to Mrs. E. H. **(Mary) Harriman**, "founder of the **Eugenics Record Office**." The **Carnegie Institution of Washington** financed the publication and it was printed by the Williams & Wilkins Company, Baltimore, Maryland. The 531-page proceedings was divided into two parts. Part 1, "Scientific Papers," contained eight sections devoted to eugenic research. Part 2, "Exhibits," described the numerous exhibits shown at the conference. Listed in scientific papers, the section topics included research on "Anthropometric Methods and Tests," "Race Amalgamation," "Education and Eugenics," "Positive and Negative Eugenics," "Disease and Infertility," "Differential Fecundity," and "Human Genetics." A total of sixty-five articles comprise the volume, many by American eugenics researchers or promoters including **Henry Fairfield Osborn**, **Samuel J. Holmes**, **Wilhelmine Key**, **Hermann Muller**, **Michael F. Guyer**, **E. S. Gosney**, and **Roswell H. Johnson**. Other papers were presented by researchers from Russia, Italy, Holland, Norway, and other countries. The publication was positively reviewed by biologists and considered a "genuine contribution to a scientific eugenics."

FURTHER READING: Engs, Ruth Clifford, *The Progressive Era's Health Reform Movement* (2003); Haller, Mark H., *Eugenics* (1984); Perkins, Harry F., (ed.), *A Decade of Progress in Eugenics* (1934).

THIRD NATIONAL CONFERENCE ON RACE BETTERMENT (1928)

After the successful **First** and **Second National Conferences on Race Betterment**, a third was planned but was delayed by **World War I**. It was not held until January 2–6, 1928. The Battle Creek Sanatorium, a combination spa and hospital, hosted the event. It was sponsored by the **Race Betterment Foundation** of Battle Creek, Michigan, under the directorship of physician and eugenics advocate **John Harvey Kellogg**. A 747-page publication, *Proceedings of the Third Race Betterment Conference, Jaunary 2–6, 1928* was published by the foundation. The purpose of the conference, as stated on the cover of the *Proceedings* was to "bring together a group of leading scientists, educators and others for the purpose of discussing ways and means of applying science to human living . . . in the promotion of longer life, increased efficiency and well-being and of race improvement." **Clarence C. Little**, president of the University of Michigan, planned and presided over the proceedings. Unlike the previous two conferences, it was an academic-oriented program with presentations of scholarly works. The central planning committee included a number of eugenicists including **Henry Fairchild**, **Michael F. Guyer**, **Harry H. Laughlin**, and **Albert E. Wiggam**, along with physicians, statisticians, and professionals engaged in **public health**, bacteriology, chemistry, education, physical education, and nutrition. Presenters included eugenics leaders **Charles Davensport**, **Irving Fisher**, and **Paul Popenoe**.

The major sessions of the conference bore the titles "Heredity and Eugenics," "Fitter Family Contests," "Crime and Sterilization," "Environment," "Factors in Living Long," "The Biologic and Physiologic Life," "Nutrition," "Chemistry in the Service of Health," "The Physics and Therapeutic Uses of Sunlight," "The New Education," and "Physical Education." The conference received wide publicity and alerted Americans to eugenics and social problems. In conjunction with it, a **Fitter Family** contest was held to determine the most mentally and physically healthy families. A fourth congress was planned, but the Depression, **World War II**, and Kellogg's death intervened. After the war, race betterment was no longer an acceptable concept for academic discussion. As a result, no more conferences were held.

FURTHER READING: Haller, Mark H., *Eugenics* (1984); *Proceedings of the Third Race Betterment Conference, January 2–6, 1928*; Schwarz, Richard William, *John Harvey Kellogg, M.D.* (1981).

TOBACCO

A plant indigenous to the Americas, tobacco contains nicotine, a stimulant. Cured and processed tobacco is generally ingested by chewing, sniffing, or smoking to get the stimulating effect. Since the early nineteenth century tobacco has been considered addictive and harmful to health. Early-twentieth-century eugenicists viewed it as a **racial poison** that could damage offspring and lead to **racial degeneracy** (deterioration of the human race). Late-twentieth-century **public health** professionals deemed tobacco, and especially cigarettes, a major contributor to chronic disease and early deaths. Tobacco was first used by American Indians, who smoked it in pipes for medicinal and ceremonial purposes. European explorers to the New World introduced tobacco around the world at the beginning of the sixteenth century. Even though many Old World cultures prohibited tobacco, its use continued. In the United States tobacco consumption increased in the late 1840s when the cigar was popularized during the Mexican War (1846–1847). Cigarette consumption rose after introduction of the machine-rolled cigarette in 1894 and became the dominant tobacco product from 1921 into the twenty-first century. Health reformers of the **Progressive Era**'s **Clean Living Movement** (1890–1920) believed tobacco, particularly cigarettes, led to **alcoholism**, **insanity**, **crime**, and vice. Although antitobacco laws were passed in a number of cities during this era, they generally were not enforced. In 1964 the U.S. Surgeon General's *Report on Smoking and Health* proposed that, based upon research, tobacco was a major cause of chronic diseases and early preventable deaths. An antitobacco campaign emerged that became the initial spark of the health reform crusade of the **Millennium Era's Clean Living Movement**. From tobacco's peak consumption around 1980, when almost 50 percent of the U.S. population smoked, use declined to about half that by the first decade of the twenty-first century.

FURTHER READING: Ferrence, Roberta G., *Deadly Fashion* (1989); Kluger, Richard, *Ashes to Ashes* (1996).

TREND OF THE RACE, THE: A STUDY OF PRESENT TENDENCIES IN THE BIOLOGICAL DEVELOPMENT OF CIVILIZED MANKIND (1921)

Written by zoologist and eugenics supporter **Samuel J. Holmes**, *Trends of the Race* helped to popularize eugenics. It was aimed at sociologists, physicians, and the educated middle class to apprise them of the latest scientific information regarding **genetics** and eugenics. The book, published in New York City by Harcourt, Brace and Company, contains 396 pages and sixteen chapters. It was based upon lectures given by Holmes and presents basic information concerning the **Mendelian inheritance** mechanism. The major focus of the book is on various influences on the human race that lead to **racial degeneracy**, the decline of civilization. These included the differential birthrate, by which the "fit" produced fewer children compared with the **unfit**, and environmental factors such as the **racial poisons** of **war**, **venereal disease**, and **alcohol**. The author argued that "everywhere the nemesis of degeneracy hangs threateningly over the organic world." Chapter titles include "The Inheritance of Mental Defects and Diseases," "The Heritable Basis of Crime and Delinquency," "The Inheritance of Mental Ability," "The Decline of the Birth Rate," "Natural Selection in Man," "The Selective Influence of War," "Consanguineous Marriages and Miscegenation," "The Possible Role of Alcohol," and "Disease in Causing Hereditary Defects."

FURTHER READING: Carlson, Elof Axel, *The Unfit* (2001); Kevles, Daniel J., *In the Name of Eugenics* (1985).

TRIBE OF ISHMAEL, THE: A STUDY IN SOCIAL DEGRADATION
(1888)

One of the first studies to explore the possibility of "inherited **pauperism**" was *The Tribe of Ishmael* by Oscar McCulloch (1843–1891), an Indianapolis, Indiana, minister and social reformer. Influenced by *The Jukes*, the first published exploration of an impoverished nomadic kinship, McCulloch traced a "ganglion of pauper families" for ten years in the Indianapolis area. The tribe, a collection of around 250 families interrelated by marriage. He identified 1,692 individuals from thirty families of this kinship from six generations and traced their ancestors to indentured servants and criminals shipped from England in the late seventeenth century. The tribe was considered white, although intermarriage with escaped and freed blacks and nomadic Indians had occurred. Descendants of the original immigrants moved to the Midwest in the post–Revolutionary War western migration. The nomads left Indianapolis in the spring and migrated to Champaign-Urbana, Illinois. In the summer they moved to Decatur, Illinois, and returned in the fall to Indianapolis. McCulloch reported that the tribe lived by "petty stealing, begging, ash-gathering and gypsying." He suggested they were "unusually licentious, 121 being prostitutes, but as a rule they are not intemperate." The results of this study were presented in 1888 at the National Conference of Charities and Corrections at Buffalo and published as an 8-page pamphlet. Between 1915 and 1922 **Arthur H. Estabrook**, a researcher at the **Eugenics Record Office**, resumed the study. The kinship was then estimated to number around 10,000. He reported his findings at the **Second International Congress of Eugenics** in New York in 1921. Estabrook, in his update, suggested that "the men were shiftless; the women immoral, and the children, ill-fed and clothed, the typical feeble-minded people who are so easily recognized today." This **family history and pedigree study** of a **degenerate** family was one of several that gave impetus to **eugenic sterilization laws** and other measures to reduce reproduction among the "socially undesirable" and **unfit** in the first third of the twentieth century.

FURTHER READING: Carlson, Elof Axel, *The Unfit* (2001); Rafter, Nicole Hahn, (ed.), *White Trash* (1988).

TUBERCULOSIS (TB)

An ancient affliction also known as *TB*, the *white plague*, *phthisis*, and *consumption*, tuberculosis was the major cause of death in many countries during late nineteenth century. Because it ran in families, the disease was thought to be inherited. In 1882 Robert Koch (1843–1910) identified the tuberculosis bacillus, a rod-shaped bacterium.

Tuberculosis's familial occurrence was then deemed to be caused by both hereditary and environmental (*nature-nurture*) factors. Since individuals in tubercular families were thought to have inherited susceptibility to becoming infected when exposed to the bacteria, they were advised to marry individuals from families who did not have the disease. Eugenicists during the first decades of the twentieth century considered TB a **racial poison** that led to **racial degeneracy**. In the United States during this era, tuberculosis was prevalent among immigrants living in crowed East Coast slums. In New York City Irish immigrants had a tuberculosis rate almost

In the first decades of the twentieth century tuberculosis was often found in several generations of poor families, leading to the belief that a trait for its susceptibility was genetically transmitted to offspring. (From: *A Social Study of Mental Defectives* [1923], Main Library, Indiana University, Bloomington, Indiana. Image courtesy of the Digital Library Program.)

three times that of native-born whites and about twice that of other European-born immigrants. The "colored component" (**African** and **Asian Americans**) had a rate nearly five times that of native-born whites in the city. The high rate among the Irish was blamed on their "frequent intemperance and generations of poverty." Italians, Russians, Hungarians, and Poles were considered much less likely to die from tuberculosis even though they also lived in the tenements. Temperance advocates noted that these particular immigrants were "comparatively free from drunkenness" and considered their more moderate drinking patterns the reason for lower tuberculosis rates. A tuberculosis movement emerged in the **Progressive Era** (1890–1920) to eliminate the disease through education and **public health** measures. By the mid-twentieth century the disease was rare. However, in the late 1980s, in the wake of the AIDS epidemic, the incidence of tuberculosis began to increase.

FURTHER READING: Bates, Barbara, *Bargaining for Life* (1992); Dubos, Rene, and Jean Dubos, *The White Plague* (1987); Teller, Michael E., *The Tuberculosis Movement* (1988).

TWIN STUDIES

Studies of twins to determine human characteristics and traits that are influenced by heredity and/or by environment (**nature-nurture**), have been termed *twin studies* since the early-twentieth-century eugenics movement. These studies have taken place since the late nineteenth century, when introduced by **Francis Galton**, and have been integral to eugenics and the field of **genetics**. Since identical or "look-alike"(monozygote) twins have the same genetic material (both from the same egg), psychosocial characteristics that are similar among a pair of twins, especially if they are reared apart, are regarded as evidence for inheritance. On the other hand, characteristics that are different are considered to be caused by life experiences. Fraternal (two eggs) twins

Identical or "look-alike" (monozygote) twins have the same genetic material (both are from the same egg). (From: *Journal of Heredity* [1919], Main Library, Indiana University, Bloomington, Indiana. Image courtesy of the Digital Library Program.)

tend to be no different than other siblings. Studies over the past century have suggested that many psychosocial traits including **intelligence**, personality, and abilities, and health problems including mental illness, **alcoholism**, depression, and addictive behaviors are to some extent inherited. The studies showed that when tested for a particular characteristic, at least 50 to 90 percent of identical twins were "concordant" (having the condition). This led most researchers to conclude that human traits are influenced by both nature and nurture. By the turn of the twenty-first century, research suggested that **genes**, or hereditary material, may "switch on or off" depending upon environmental influences. The environment may include slight differences prenatally, or different life experiences during critical periods in infancy and childhood or trauma or stress in adulthood.

Twin studies have provoked controversy since the mid-twentieth century. In Germany medical geneticist Otmar von Verschuer (1896–1969) established the first "twin registry," which contained information on thousand of twins. His student, Josef Mengele (1911–1979?) of **Nazi Germany**, in the early 1940s did unethical research on twins in the Auschwitz concentration camp. Mengele's experiments dampened all twin research in the aftermath of the Holocaust and **World War II**. In the postwar years researchers shied away from studies that investigated heredity's influence on human differences. Some twin research, however, was conducted in Britain; results were mixed on the effect of heredity and environment on human characteristics. A resurgence in the United States of twin research began when a twin registry, the Minnesota Study of Twins Reared Apart, was founded in 1979 by psychologist Thomas Bouchard at the University of Minnesota. The center investigates various physical, mental, and sociological aspects of twins reared both together and apart. As genetic techniques became more sophisticated in the early 1990s it became possible to identify human genes and **DNA** segments linked to certain charac-

Fraternal (two eggs) twins tend to be no different than other siblings. They can be of either gender, while identical twins are always the same gender. (From: *Journal of Heredity* [1916], Main Library, Indiana University, Bloomington, Indiana. Image courtesy of the Digital Library Program.)

teristics. These techniques are now being used in twin studies to determine the nature of heredity and environment, not only in human traits, but also **genetic disease**. In 2002 a European twin registry, the GenomEUtwin, a collation of national twin registries from many European countries and Australia, was established to do similar studies.

FURTHER READING: Ridley, Matt, *Nature via Nurture* (2003); Segal, Nancy, *Entwined Lives* (2000); Stewart, Elizabeth A., *Exploring Twins* (2003).

U

The permanent basis of civilization is the quality of the germ plasm of the people. Social caste, economic station, these things are will-o'-wisps in the long march of time. They are quickly acquired and quickly lost, but the permanent basis goes on until the last man dies.

Leon F. Whitney, "The American Eugenics Society:
A Survey of Its Work," *Eugenics* 3 (July 1930)

The survival of the unfittest is the primal cause of the downfall of nations.

David Starr Jordan, *The Blood of the Nation* (1902)

UNFIT

A term commonly used from the late nineteenth century through the first half of the twentieth century to describe people who were genetically, socially, mentally, economically, or medically inferior was *unfit*. The terms *defective*, **dysgenic**, or **degenerate** were also used. The unfit in industrialized countries included the **insane**, the **feebleminded**, and those with **genetic diseases** or other illnesses such as **venereal disease** or **tuberculosis**. They included **alcoholics**, **criminals**, **prostitutes**, **"masturbators,"** and **paupers**. In the **United States** the unfit also included impoverished immigrants from **eastern and southern Europe**. The unfit were thought to contribute to **racial degeneracy** and **race suicide** as they "reproduced their own kind," leading to the decline of Western society. In **war** the "fit" were often killed in battle, while the weak and those unfit for military duty remained at home to reproduce. **Eugenic sterilization laws** in the United States, **Germany**, and some other countries were enacted to reduce the propagation of the unfit during the early part of the twentieth century.

FURTHER READING: Carlson, Elof Axel, *The Unfit* (2001); Rafter, Nicole Hahn, (ed.), *White Trash* (1988).

UNITED STATES

Focus on hereditarian, or genetic, causes of human behavior has emerged in the United States over the past 150 years as an aspect of the **Jacksonian** (1820–1860), **Progressive** (1890–1920), and **Millennium** (1970–) health and social reform eras. In the early nineteenth century there emerged out of the second **Great Awakening** a perfectionist movement to bring on the "millennium," or the reign of Christ on earth. Reformers preached that to have healthy children without negative inherited characteristics people needed to obey the **laws of health** such as **physical exercise** and **alcohol** abstention and be aware of **inherited realities** or traits that were thought to be inherited based upon **Lamarckian inheritance of acquired characteristics**. The early-twentieth-century eugenics movement grew out of these late-nineteenth-century beliefs in the inheritability of poverty and moral behaviors, concepts of **social Darwinism**, and concerns over the population decline of old-stock **Anglo Americans**. Similar to the **British** and **German eugenics** movements, it was supported by the educated middle class and considered a vital social program to prevent **race suicide** and **racial degeneracy** and the decline of Western civilization. It was led by men, although many women were involved with the movement. Its members were concerned about the differential birthrate, wherein the birthrate was decreasing among the middle classes and the "fit" and increasing among the poor and the "unfit," and worried about the health and vitality of the nation. When in 1900 the rules of **Mendelian inheritance** were rediscovered, American researchers quickly embraced the theory.

The U.S. eugenics movement is divided into three phases. During the first phase, from about 1870 to 1905, hereditarian attitudes took root among professionals and reformers. Early proponents who became pioneers of the movement included inventor **Alexander Graham Bell** and physician and health reformer **John Harvey Kellogg**. Physician **Harry Sharp** of Indiana began to perform eugenic sterilizations in 1899, and in the first decade of the new century the **American Breeders Association** was founded to advance the fields of **genetics** and eugenics. By 1905 three factions of the eugenics movement with differing agendas and activities had evolved. Social welfare, health professionals, and institutional care-givers were concerned about preventing **feeblemindedness**, **crime**, disease, and poverty. They advocated **eugenic sterilization**, **segregation** (custodial care), and **marriage-restriction laws**. Geneticists and other academicians carried out **family history and pedigree studies**, genetics research, and mental ability and **intelligence testing** to determine what diseases and human characteristics were inherited. **Nativists**, alarmed by the massive influx of impoverished **Catholic** and **Jewish** immigrants from **eastern and southern Europe**, whom they considered to be of "inferior racial stocks" compared with the **Protestant Anglo-Saxon races**, advocated immigration-restriction legislation to prevent race suicide.

The second phase of the U.S. movement, when it had its greatest impact, lasted from 1905 through the late 1920s. During the second decade of the century, the **Eugenics Record Office** (ERO), directed by geneticist **Charles Davenport** with **Harry H. Laughlin** as superintendent, became the hub of the movement. The ERO, in addition to the **Eugenics Research Association**, the **Eugenics Registry**, and the **Race Betterment Foundation**, further promoted the cause with the **First** and **Second National Conference on Race Betterment**. Family pedigree studies, including those by **Henry Goddard** and **Arthur H. Estabrook**, advanced the case for eugenic sterilization and segregation laws. By 1914 eugenics principles were taught in forty-four colleges. **Positive eugenics** educational programs, including **Better Babies** and **Fitter Families** contests and popular lectures, attempted to promote the movement. In the 1920s organizations such as the **American Eugenics Society** and the **Human Betterment Foundation** were established to further educate the public. The **First**, **Second**, **and Third International Congresses of Eugenics** linked eugenicists from all over the world. Support from reputable researchers, including **Henry Fairfield Osborn** and **Irving Fisher** and university presidents **Charles Eliot** and **David Starr Jordan**, lent the movement prestige. Throughout this phase many states passed eugenic sterilization laws. Although such laws were often challenged and eventually repealed, Virginia's involuntary sterilization law was upheld by the 1927 Supreme Court decision *Buck v. Bell*, which led to more sterilizations during the 1930s.

The third phase of the U.S. movement began in the late 1920s—the decline of its prestige and influence. By then eugenics had begun to lose scientific support based upon new information from the fields of genetics and psychology. It shifted direction to population studies because it had outlived its political usefulness; sterilization laws, premarital health exam laws, and immigration-restriction laws had been passed. In addition the movement's nativistic faction, including **Madison Grant**, **Lothrop Stoddard**, and Harry H. Laughlin, moved to the forefront. Their support of **Nazi Germany**'s eugenic sterilization program in the mid-1930s became an embarrassment to the movement's funding sources, such as the **Carnegie Institution of Washington**. Funds were withdrawn and several eugenic organizations were forced to close in the late 1930s. Older leaders died and few successors emerged to replace them. By the 1940s the movement had only a few adherents. Nazi Germany's program of genocide to eliminate "inferior races" and foster a "superior **Ayran** race" became identified with eugenics. This association discredited eugenics throughout the rest of the twentieth century. However, by the 1980s a new **eugenics**, based upon consumer demand, began to appear in **assisted reproduction** techniques such as **in vitro fertilization**.

FURTHER READING: Allen, Garland E., "Science Misapplied: The Eugenics Age Revisited" (1996); Haller, Mark H., *Eugenics* (1984); Kevles, Daniel J., *In the Name of Eugenics* (1985); Paul, Diane B., *Controlling Human Heredity* (1995); Pickens, Donald, *Eugenics and the Progressives* (1968); Reilly, Philip R., *The Surgical Solution* (1991).

In some cases individuals suffering from venereal disease can only give birth to defective children. The use of Birth Control in all cases of venereal disease would eliminate many of the defectives who are a burden to themselves and society.

"Birth Control Primer," *Birth Control Review* 11 (April 1927)

VENEREAL DISEASES

The sexually transmitted diseases gonorrhea and syphilis were called *venereal diseases (VD)*, *social diseases*, or the **red plague** from the early to the late twentieth century. By the 1980s the term *sexually transmitted disease* became common and was applied to other infections transmitted by sexual contact, such as Acquired Immune Deficiency Syndrome (AIDS). During the **Progressive Era** (1890–1920) prevention and elimination of VD was a factor in several health reform campaigns, including **birth control**, eugenics, and **social hygiene** (anti-VD) movements on both sides of the Atlantic. In the United States publications by eugenics supporters and other reformers like physician **Robert Dickinson** and educators **Charles Eliot** and **David Starr Jordan** helped erase the "conspiracy of silence" regarding the mention of these diseases in print or public. Syphilis transmitted from mother to infant caused death or permanent damage of the child. In later years it could lead to mental illness or heart conditions. Gonorrhea caused sterility among adults and blindness among infants. These diseases, because of their damage to "innocent wives and offspring" when spread by unfaithful husbands, were considered **racial poisons** by eugenicists. The newly formed **American Social Hygiene Association** began to disseminate factual information in the early 1910s concerning the prevention of VD. Campaigns emerged to test for syphilis before marriage, seen as both a eugenic and a preventive **public health** measure. By 1912 physical examinations to rule out venereal disease were required in Connecticut, Washington, Utah, Michigan, and Colorado before a marriage licence was granted. Effective cures for both gonorrhea and syphilis were not readily available until the introduction of antibiotics in the **World War II** era.

FURTHER READING: Brandt, Allan M., *No Magic Bullet* (1985).

The sound personal health of the parents, combined with the sound germ plasm which they may carry, is from four to five times as important in the future health of the children, as pure milk, good doctors, open air, physical culture, and hygiene all put together.

Albert Edward Wiggam, *The Fruit of the Family Tree* (1924)

WAR

Darwin's theory of **natural selection**, proposed in the latter half of the nineteenth century, suggested that war was necessary for the development and progression of the human race. The conquest of one primitive tribe by another allowed the group with the strongest and cleverest warriors to pass on their heredity to future generations. However, by the end of the nineteenth century new warfare techniques led to mass casualties among combatants, and war began to be seen as **dysgenic**. The "strong and brave" volunteered for military service or were drafted, while the "cowardly and weak" and those unfit for military duty remained at home to reproduce. Eugenicists on both sides of the Atlantic discussed war as a **racial poison** that would lead to **racial degeneration**. **World War I**, **World War II**, and other conflicts through the rest of the twentieth century produced increasingly more lethal and destructive weapons, including poison gases, thermonuclear bombs, biochemical and biological weapons, and the potentiality of "ethnic chemical weapons" designed to attack naturally occurring differences in vulnerabilities among specific population groups.

FURTHER READING: Cashman, Greg, *What Causes War?* (2000); Copeland, Dale C., *The Origins of Major War* (2000).

WATSON, JAMES DEWEY (April 6, 1928–)

A molecular biologist and a codiscover of the structure of **DNA** (the genetic code of most organisms), Watson for many years directed the Cold Spring Harbor Laboratory in **Cold Spring Harbor**, New York. That community was a center of the early-twentieth-century eugenics movement. Watson was the central figure in shaping the **Human Genome Project**, which determined the "genetic blueprint" of humans and whose results laid the foundation for the **new eugenics**. Born in Chicago, Illinois, Watson was educated in Chicago's public schools. He was an extremely bright child and entered the University of Chicago at age fifteen in 1943, graduating with a B.S. in zoology (1947). He proceeded to Indiana University for graduate work and received a Ph.D. (1950), with core studies in virology. **Hermann Muller**, who had won a Nobel Prize for demonstrating that X-rays cause mutation, was on the faculty. Watson was awarded a National Research Council fellowship grant to study the molecular structure of proteins in

Copenhagen, Denmark. Subsequently, he was awarded a fellowship at the Cavendish Laboratory in Cambridge, England, from 1951–1953, where he met British molecular biologist Francis Crick (1916–2004). The two frequently discussed the nature of DNA and jointly explored a model of its structure. In 1953 the twosome identified DNA's three-dimensional structure after viewing X-ray diffraction images taken by Rosalind Franklin (1920–1958) at Kings College, London, and shown to them by her colleague Maurice Wilkins (1916–2004) without her knowledge. The three men won the 1962 Nobel Prize for Physiology or Medicine for this achievement.

After Watson completed his research fellowship he spent the summer of 1953 at Cold Spring Harbor, New York, and became a research fellow in biology at the California Institute of Technology in Pasadena. In 1955 he joined the biology department at

James Watson, a geneticist, one of the discoverers of the structure of DNA, was director of the Cold Spring Harbor Laboratory and instrumental in the formation of the Human Genome Project. (From: *Indiana Alumni Magazine* [1962]. Image courtesy of Indiana University Archives, Bloomington, Indiana.)

Harvard University. In 1968 he was appointed director of Cold Spring Harbor Laboratory but did not give up his Harvard post until 1976. Watson built the laboratory into a world-renowned center for molecular biology and remained its full-time director until 1988. In 1994 he was named president and in 2004 chancellor of the laboratory.

Watson's research and his leadership laid the foundation of the new eugenics. He received many honors and authored numerous publications. His *The Molecular Biology of the Gene* (1965) became a classic text in molecular biology. Watson's autobiographical *The Double Helix: A Personal Account of the Discovery of the Structure of DNA* (1968) was popular, but also controversial due to his treatment of Franklin and disagreement with fellow researchers concerning the process of the DNA discovery. From 1988 to 1992 he was the first director of the National Institute of Health (NIH) component of the Human Genome Project. He resigned due to potential conflicts of interests and differing opinions concerning policies and management of the project. In his semiretirement he remained active on the lecture circuit, but his pronouncements were often controversial. Watson married Elizabeth Lewis (1968) and the couple had two children.

FURTHER READING: Watson, James D., *DNA* (2003); Watson, *The Double Helix* (1980).

WHITNEY, LEON FRADLEY (March 29, 1894–April 11, 1973)

Executive secretary of the **American Eugenics Society** during its most active years (1924–1934), Whitney launched many projects for the society. Born in Brooklyn Heights, New York, the son of an electrical contractor, he received a B.S. (1916) in agriculture at what is now the University of Massachusetts, Amherst. Whitney became a wealthy farmer, animal breeder, and businessman (1916–1924). After being removed as executive secretary of the American Eugenics Society he attended Yale University, then graduated with a D.V.M (1940) from Auburn University, Auburn, Alabama. He practiced as a veterinarian and worked as a clinical instructor in pathology (1946–1963) at the School of Medicine, Yale University. He was also director of the Whitney Laboratory, Orange, Connecticut. Whitney was a prolific writer of educational animal books, mostly during the 1950s, including the popular *How to Breed Dogs*. He founded the Whitney Collection of Dogs at the Peabody Museum of Natural History, Yale University.

Hired in 1924 as executive secretary of the newly formed American Eugenics Society to direct its fund-raising and educational programs, Whitney was overly enthusiastic about eugenics but lacked scientific training in genetics. He launched the society's journal *Eugenics*, promoted and helped to establish **Fitter Family** and eugenics sermon contests, and was actively involved in the **Third International Congress of Eugenics** (1932). He was a member of several organizations including the **Eugenics Research Association**. Whitney was a prolific writer of articles and several eugenics books. For the society he prepared a pamphlet, *A Eugenics Catechism* (1926), and coauthored, with **Ellsworth Huntington**, *The Builders of America* (1927), a **positive eugenics** tract to encourage old stock **Anglo Americans** and educated women to have more children. His most noted work was *The Case for Sterilization* (1934), in which he lauded **Nazi Germany**'s eugenics sterilization program. His over-enthusiasm, fiscal mismanagement, and pro-Nazi publications embarrassed some of the scientific members of the society and resulted in Whitney's being removed from his position in 1934. He married Katharine Carroll Sackett (1916) and they had two children. He died in Orange, Connecticut.

FURTHER READING: Haller, Mark H., *Eugenics* (1984); Reilly, Philip R., *The Surgical Solution* (1991).

WIGGAM, ALBERT EDWARD (October 8, 1871–April 26, 1957)

A noted lecturer and science writer, Wiggam helped to popularize eugenics among the middle class through a variety of publications. He promoted both **positive** and **negative eugenics** and addressed concerns of **race suicide** and **racial degeneration**. Born on a farm in Austin, Indiana, he was the son of a lay preacher and farmer. He attended local schools but was often sick with **tuberculosis**, which delayed his schooling. He received a B.S. (1893) and an M.S. (1903) from Hanover College, near Madison, Indiana. From 1893 to 1899 he held various jobs, including working at a Louisiana sugar refinery and serving as a mining supervisor in Colorado, along with being a bookseller, printer, and florist. He studied philosophy at Colorado College in Denver (1894) and was an editorial writer for the Minneapolis *Journal* (1899–1900). From 1901 to 1919 he was a popular lecturer and was among the first to lecture on heredity and eugenics. Wiggam had the ability to interpret scientific problems and topics for the general public and was a member of many scientific and writing societies. Over his career he wrote for and was on the editorial staff of many magazines. From 1925 until his death he authored a nationally syndicated column, "Let's Explore Your Mind."

Wiggam took an active role in the eugenics movement. He was a member of the **American Breeders Association** and, in its later years, the **Galton Society**, and served on the advisory committee of the **American Eugenics Society** for much of his life. After the creation of the **Eugenics Record Office** in 1910 he helped build its pedigree file by distributing family-record forms to lecture attendees. He was on the editorial board of *Eugenics* through its short life and was a member of the eugenics section of the central committee of the **Third National Conference on Race Betterment** (1928). He also served as vice president of the **Eugenics Research Association** (1934) and was a member of its advisory council (1928–1935) and board of directors (1935–1940). He wrote numerous articles and books with eugenics themes. His most popular works included *The New Decalogue of Science* (1922), **The Fruit of the Family Tree** (1924), and *The Next Age of Man* (1927). Like other eugenicists, he believed that the function of eugenics was "to prevent the race from slipping backward biologically" by supporting **eugenic segregation** and **sterilization** of the mentally and emotionally **unfit**. Wiggam decried **public health** and medical advances for saving the life of "defectives" who "in the good old days of natural selection would have died young." He suggested that education, healthy environments, and moral teaching were critical to raise fit children. Unlike many eugenicists he supported making **birth control** available for everyone. Wiggam married Elisabeth M. Jayne (1902), who died in 1943; he married Helen Scott Holcombe the following year. Although he advocated that the "fit" reproduce, Wiggam fathered no children. He died at his Santa Monica, California, home.

FURTHER READING: Haller, Mark H., *Eugenics* (1984); Kevles, Daniel J., *In the Name of Eugenics* (1985).

WORLD WAR I (1914–1918)

The first major war of the twentieth century, World War I was also called the *Great War*, the *First World War*, and the *War to End All Wars*. While the war was being fought the eugenics movement came to a standstill on both sides of the Atlantic. World War I embroiled most western and eastern European nations, the United States, the Middle East, and other regions. Two opposing factions had emerged by 1910: the Central Powers, including Germany, Austria-Hungary, and Turkey, in opposition to the Allies, including France, Great Britain, Russia, Italy, Japan, and, in 1917, the United States. The assassination of Archduke Francis Ferdinand of Austria at Sarajevo on June 28, 1914, led to a general war between these two alliances. The war ended with the defeat of the Central Powers on November 11, 1918. In the prewar years, **war** increasingly began to be viewed as a **racial poison**. In the United States educator **David Starr Jordan** wrote *Blood of the Nation* (1911), a eugenics and antiwar tract. During the conflict numerous eugenic supporters including **Irving Fisher**, **Paul Popenoe**, **Roswell H. Johnson**, and **Charles Eliot** in the United States wrote about the **dysgenic** effects of war. They feared that the "best of the stock was being killed," while the **unfit** and the weak who did not qualify for military duty stayed home to reproduce, thus leading to **racial degeneracy**.

During the war years, due to racial degeneracy concerns, **public health** and eugenics supporters in the United States campaigned for anti**prostitution** regulations

Eugenicists considered World War I a cause of "race suicide" because the brightest and fittest, such as these university students, were often killed in battle while people mentally and physically unfit for military duty remained at home to reproduce. (Courtesy Indiana University Archives, Bloomington, Indiana.)

around military bases to prevent military personnel from being infected with **venereal diseases**. **Intelligence tests** were developed to screen for the best and brightest recruits. In the aftermath of the war, realization that the fittest young men had been killed led to a resurgence of the eugenics movement in **Britain, Germany**, the **United States**, and other countries. The conflict was viewed as depleting the **Nordic** and **Anglo-Saxon races**, leading to **race suicide**. Many eugenicists agreed that another war should be prevented. To replace the great loss of life and increase overall health and vitality, particularly in Germany, **positive eugenic** measures to encouraging the fit to reproduce were recommended. Nativist eugenicists in the United States, like **Lothrop Stoddard**, and German race theorist **Hans F. K. Günther** concluded in their writings that World War I was eugenically disastrous to the Western world. Stoddard called it a "white civil war." Eugenicists on both sides of the Atlantic in the interwar period (1918–1939), including **Alfred Ploetz**, a pioneer of the German eugenics movement, unsuccessfully campaigned against another war.

FURTHER READING: Fromkin, David, *Europe's Last Summer* (2004).

WORLD WAR II (1939–1945)

The second major war of the twentieth century, World War II engaged the Allies, including Britain, Canada, Australia, France, New Zealand, and the United States, against the Axis, including Germany, Italy, and Japan. Unlike during **World War I**, few eugenic concerns were raised that the conflict was leading to the **race suicide** of northern Europeans. World War II had its origins in the defeat of the German Empire by the British, French, and American forces in World War I. The resulting social and economic discord in postwar Germany fostered an environment that led to the rise to power of Adolf Hitler (1889–1945) and his **Nazi German** regime, who invaded Poland on September 1, 1939. England and France declared war on Germany two days later, but the United States remained neutral until the Japanese bombing of Pearl Harbor, Hawaii, on December 7, 1941. The following day the United States and Britain declared war on Japan, and Hitler declared war on the United States three days after that. The war against Germany ended May 8, 1945. The war against Japan ended September 2, 1945, after the United States dropped atomic bombs on the cities of Hiroshima and Nagasaki.

During the war—and even prior to it—only a few scientists with roots in the earlier eugenics movement spoke out against war as being dysgenic. The major influence of World War II on the waning eugenics movement in the **United States**, in particular, was that all eugenics measures, including **positive eugenics**, were discredited inasmuch as the term became associated with the genocide of **Jews**, Gypsies, and other groups that later became known as the Holocaust. The term *eugenics*, and its ideology, were largely unknown by the postwar baby boom and later generations. Positive eugenics evolved into marriage counseling with emphasis on keeping marriages together for the development of healthy children. To this end, eugenics supporter **Paul Popenoe** wrote a feature, "Can This Marriage Be Saved?" in the *Ladies Home Journal*, a popular magazine, for a number of years. The study of **races** and racial differences evolved into population studies, and concerns by the 1960s had shifted to the population explosion and the earth's limited resources, with many social welfare and health professionals advocating restraint in childbearing.

FURTHER READING: Boyce, Robert W. D., and Joseph A. Maiolo (eds.), *The Origins of World War Two* (2003).

Selected Chronology

1853   Arthur de Gobineau publishes *Essai sur l'inégalité des races humaines* (1853–1855).

1859   Charles Darwin publishes *On the Origin of Species by Means of Natural Selection; or, The Preservation of Favoured Races in the Struggle for Life.*

1865   Austrian monk Gregor Mendel presents his paper "Versuche über Pflanzen-Hybriden" at the February 8 and March 8 meetings of the Naturforschenden Vereins in Brünn (Brno); in 1866 he publishes "Versuche über Pflanzen-Hybriden" in the *Verhandlungen des naturforschenden Vereines in Brünn.*

1869   Francis Galton publishes *Hereditary Genius* with the theme that great abilities are inherited.

1871   American plant breeder Luther Burbank develops the Russet Burbank potato.

       Charles Darwin publishes *The Descent of Man and Selection in Relation to Sex.*

1872   British novelist, essayist, and critic Samuel Butler publishes his satire on social Darwinism and eugenic utopias, *Erewhon; or, Over the Range.*

1875   Francis Galton publishes "The History of Twins, as a Criterion of the Relative Powers of Nature and Nurture" in the November issue of *Fraser's Magazine.*

1876   Charles Darwin publishes *The Effects of Cross and Self Fertilisation in the Vegetable Kingdom.*

1877   Richard Dugdale publishes *The Jukes: A Study in Crime, Pauperism, Disease, and Heredity.*

1879   William James Beal, professor of botany in the State Agricultural College of Michigan, experiments with crossbreeding inbred lines and achieves a significant increase in corn yields.

1882 The U.S. immigration law of August 3, 1882, bars the immigration of "undesirables": idiots, lunatics, convicts, and persons likely to become public charges; it also levies a fifty-cent tax on immigrants landing at U.S. ports.

Congress passes the Chinese Exclusion Act over President Chester Arthur's veto; the law prohibits Chinese laborers from entering the United States for ten years.

1883 Francis Galton introduces the term *eugenics* in *Inquiries into the Human Faculty and Its Development*.

1884 In a letter entitled "Artificial Impregnation" to the journal *Medical World* in 1909, Addison Davis Hard, a Minnesota physician, reports donor insemination of a woman carried out in 1884 by Dr. William Pancoast at the Jefferson College of Medicine in Philadelphia; the sperm was donated by the "best looking" student in the medical class Pancoast taught.

1885 German biologist August Weisman publishes *Die Continuität des Keimplasmas als Grundlage einer Theorie der Vererbung*.

1887 In Clinton, Iowa, Henry Bowers founds the American Protective Association, a secret society concerned about the growth in population and political power of cities with large numbers of immigrants.

1888 Frederick Wines publishes *Report on The Defective, Dependent and Delinquent Classes of the Population of the United States, as Returned at the Tenth Census (July 1, 1880)*.

Victoria Woodhull, an eccentric American reformer given to many causes, publishes *Stirpiculture; or, The Scientific Propagation of the Human Race*.

1889 Francis Galton publishes *Natural Inheritance*.

1891 An act amending various immigration and alien labor acts excludes from admission to the United States: idiots, insane persons, paupers or persons liable to become a public charge, persons suffering from a loathsome or a dangerous contagious disease, polygamists, and felons.

Wilhelm Schallmayer publishes *Über die drohende körperliche Entartung der Culturmenschheit und die Verstaatlichung des ärztlichen Standes*.

1892 In his presidential address to the Association of Medical Officers of American Institutions for Idiotic and Feeble-Minded Persons Dr. Isaac Kerlin speaks about "surgery for the relief of idiotic conditions" and asks, "Whose state shall be the first to legalize oophorectomy and orchitomia for the relief and cure of radical depravity?"

August Weisman, an opponent of the theory of inheritance of acquired traits and known for his germ-plasm concept, a forerunner of DNA theory, publishes *Das Keimplasma; eine Theorie der Vererbung* (translated in 1893 as *The Germ-Plasm: A Theory of Heredity*).

1893   F. Hoyt Pilcher is appointed superintendent of the Kansas State Asylum for Idiotic and Imbecile Children; he begins a program of castration to control, among other things, masturbation; in 1895, after his actions become known the new governor removes him from his position, but he returns in 1897; Kansas legalizes sterilization in 1913, and in 1928 the state Supreme Court declares the law unconstitutional.

1894   A group of Boston professionals including Prescott Hall, Charles Warren, and Robert DeCourcey Ward found the Immigration Restriction League in response to concern about increasing immigration from southern and eastern Europe; the league urges literacy testing on the theory that it would keep out many of the new immigrants.

The *Gobineau-Vereinigung* (Gobineau Association) is founded at Freiburg im Breisgau; Professor Ludwig Schemann is its first president.

1895   Alfred Ploetz publishes *Die Tüchtigkeit unsrer Rasse und der Schutz der Schwachen: Ein Versuch über Rassenhygiene und ihr Verhältniss zu den humanen Idealen, besonders zum Socialismus.*

1896   Connecticut adopts a marriage law that states "no man or woman either of whom is epileptic, imbecile or feeble-minded" may "inter-marry, or live together as man and wife, when the woman is under forty-five years of age"; other states, particularly in the North and the West, soon enact similar eugenic marriage laws.

1897   A bill for eugenic sterilization is introduced in the Michigan legislature; it fails, but the state enacts a sterilization law in 1913; after it is found unconstitutional in 1918 the state passes a revised law in 1923.

1899   The first meeting of the International Conference on Hybridisation (the Cross-Breeding of Species) and on the Cross-Breeding of Varieties, later the International Congress of Genetics, is held in London.

A. J. Ochsner publishes "Surgical Treatment of Habitual Criminals" in the *Journal of the American Medical Association.*

1900   Friedrich Krupp sponsors an essay contest (*Preisausschreiben*) to answer the question "What can we learn from the theory of evolution about internal political development and state legislation?" Wilhelm Schallmayer wins the competition; his work will be published in 1903.

1901   Pennsylvania passes legislation permitting eugenic sterilization; the governor vetoes the bill; Martin W. Barr, chief physician of the Pennsylvania Training School for Feeble-Minded Children at Elwyn, however, continues the practice; the governor vetoes a similar bill in 1905.

Karl Pearson, W.F.R Weldon, and Francis Galton found *Biometrika*, a "Journal for the Statistical Study of Biological Problems."

Gregor Mendel's 1866 study on inheritance in garden peas ("Versuche über Pflanzen-Hybriden") is translated and published as "Experiments in Plant Hybridization," in the *Journal of the Royal Horticultural Society*.

1902   Harry Sharp reports his use of the vasectomy with inmates of the Indiana State Reformatory, Jeffersonville, in "The Severing of the Vasa Deferentia and Its Relation to the Neuropsychopathic Constitution," published in the *New York Medical Journal*.

David Starr Jordan publishes *The Blood of the Nation: A Study of the Decay of Races through Survival of the Unfit*.

1903   A committee of the American Association of Agricultural Colleges meets in St. Louis and founds the American Breeders Association, later the American Genetic Association.

Wilhelm Schallmayer publishes *Vererbung und Auslese im Lebenslauf der Völker: Eine staatswissenschaftliche Studie auf Grund der neueren Biologie* (Heredity and Selection in the Life Process of Nations).

1904   With funding from the Carnegie Institution of Washington Charles Davenport establishes the Station for Experimental Evolution at Cold Spring Harbor, Long Island, New York.

Francis Galton endows a fellowship for the promotion and study of "national eugenics" at University College London. He also establishes the Eugenics Records Office, later called the Galton Laboratory.

Alfred Ploetz and others found the journal *Archiv für Rassen- und Gesellschaftsbiologie*; the journal ceases publication in 1944.

1905   The British parliament passes the Aliens Act (5 Edw. VII. c.130); undesirable aliens include paupers, lunatics, vagrants, prostitutes, the criminal and the diseased, who could be refused entry into Britain.

Alfred Ploetz and Ernst Rudin found the Gesellschaft für Rassenhygiene, in Berlin; the name later becomes Gesellschaft für Rassenhygiene (Eugenik).

1906   The American Breeders Association establishes a committee on eugenics; its duties are "investigation, education, legislation."

John Harvey Kellogg creates the Race Betterment Foundation in Battle Creek, Michigan.

William Bateson introduces the term *genetics* in the June 14 issue of *Nature*.

Robert Reid Rentoul publishes *Race Culture or Race Suicide? A Plea for the Unborn*.

1907   The Eugenics Education Society is founded in Britain; it becomes the Eugenics Society in 1926 and the Galton Institute in 1989; its aim is the popular promotion of eugenics and awareness of heredity in social responsibility.

Indiana passes the first U.S. eugenic sterilization law; it permits sterilization for "all confirmed criminals, idiots, rapists and imbeciles" confined in state institutions if their condition is pronounced incurable by three physicians; other states will enact similar legislation.

The Immigration Act of 1907 authorizes the president to deny entry into the United States of citizens of any county if they are coming "to the detriment of labor conditions therein"; the provision clearly applies especially to Japanese laborers; the act also bars from entry idiots, imbeciles, feebleminded persons, epileptics, insane persons, paupers, professional beggars, polygamists, anarchists, and prostitutes.

In Britain the Galton Laboratory for National Eugenics is established by the merger of the biometric laboratory and the Eugenics Records Office.

1908   The first Better Babies Contest is held at the Louisiana State Fair.

Henry Goddard introduces the Binet-Simon intelligence tests to the United States.

The British Royal Commission on the Care and Control of the Feeble-Minded issues its 8-volume *Minutes of Evidence, Appendices, and Reports* (Cd. 4215–21; 4202).

1909   California's sterilization act of April 26, 1909, provides for the sterilization of the mentally retarded in state hospitals and institutions; it also authorizes sterilization of certain state prison inmates who show evidence of moral or sexual perversion; additional sterilization laws are enacted in 1913 and 1917.

The British Royal Commission on the Poor Laws and Relief of Distress, chaired by Lord George Hamilton, issues its *Report* (Cd. 4499); among its observations: "Cases are not wanting to show that pauperism is hereditary—two generations being quite common, and third generations generally occur."

C. S. Saleeby delivers a lecture to the Sociological Society, "The Obstacles to Eugenics"; in it, he states the *Times* of London reports that negative eugenics proposed to kill nobody. It "distinguishes between the right to live and the right to become a parent."

Francis Galton publishes *Essays in Eugenics*.

Dr. Harry Sharp publishes "Vasectomy as a Means of Preventing Procreation in Defectives" in the *Journal of the American Medical Association*.

J. M. Murdoch, superintendent of the State Institution for Feeble-Minded of Western Pennsylvania, publishes "Quarantine Mental Defectives," a committee report, in the *Proceedings of the National Conference of Charities and Correction*.

Irving Fisher publishes *A Report on National Vitality, Its Wastes and Conservation*.

C. W. Saleeby publishes *Parenthood and Race Culture: An Outline of Eugenics*.

1910   With financial help from Mary Harriman, Charles Davenport establishes the Eugenics Record Office, an expansion of the Station for Experimental Evolution, in Cold Spring Harbor, New York, with himself as director and Harry H. Laughlin as superintendent.

American Breeders Magazine begins publication; in 1914 it becomes *Journal of Heredity*, "the official journal of the American Genetic Association."

The Dillingham Commission (the Immigration Commission, formed in 1907 and chaired by Senator William P. Dillingham of Vermont) ends its investigation; it issues between 1909 and 1911 its 42-volume report containing masses of statistical and sociological data on immigrants; in the final report the commission worries that the "new immigration," comprised of people who cannot be assimilated, is diluting the Anglo-Saxon stock and lowering the standard of living; among other things, it recommends limiting immigration from southern and eastern Europe and excluding Asians.

Karl Pearson publishes *Nature and Nurture: The Problem of the Future*, the text of the presidential address at the annual meeting of the Social and Political Education League.

1911   The New Jersey law (1911 Acts, Chapter 190) of April 21, 1911, authorizes the sterilization of insane, epileptic, and retarded persons as well as certain criminals; in 1913 the New Jersey Supreme Court rules the legislation unconstitutional.

Francis Galton dies; his will provides for the endowment of the Galton Professorship of Eugenics.

Karl Pearson is appointed the first Galton Professor of Eugenics at University College, London.

David Starr Jordan publishes *The Heredity of Richard Roe: A Discussion of the Principles of Eugenics.*

Charles Davenport publishes *Heredity in Relation to Eugenics.*

1912   The First International Eugenics Congress is held at the University of London.

In New York the Public Health Law (Ch. 445 of the Laws of 1912) establishes a Board of Examiners to examine the condition of criminals and others deemed "defective," such as inmates in state institutions, and to determine if they should be sterilized; in 1918, in the case *In re Thomson*, the law is found unconstitutional.

Harvey Ernest Jordan publishes *Eugenics: The Rearing of the Human Thoroughbred.*

Florence H. Danielson and Charles Davenport publish *The Hill Folk: Report on a Rural Community of Hereditary Defectives.*

Arthur H. Estabrook and Charles Davenport publish *The Nam Family; A Study in Cacogenics.*

Henry H. Goddard publishes *The Kallikak Family: A Study in the Heredity of Feeble-Mindedness.*

New York physician Prince Morrow publishes *Eugenics and Racial Poisons.*

1913   The British parliament passes the Mental Deficiency Act, 1913; its provisions are mild versions of recommendations contained in the 1908 report of the Royal Commission on the Care and Control of the Feeble-Minded.

The Ärztliche Gesellschaft für Sexualwissenschaft und Eugenik (Medical Society for Sexology and Eugenics) is founded in Berlin.

Governor Francis E. McGovern of Wisconsin signs a sterilization bill into law; the law, providing for the sterilization of criminals and the insane, is seen as a progressive measure, modern and scientific.

Prominent American attorney Charles A. Boston publishes "A Protest against Laws Authorizing the Sterilization of Criminals and Imbeciles," in issue 3 of the *Journal of the American Institute of Criminal Law and Criminology.*

Edith Spaulding and William Healy publish "Inheritance as a Factor in Criminality" in the *Journal of the American Institute of Criminal Law and Criminology;* their study finds no "evidence of direct inheritance of criminalistic traits."

1914   The first National Conference on Race Betterment is held, in Battle Creek, Michigan.

The Ärztliche Gesellschaft für Sexualwissenschaft und Eugenik begins publication of its journal *Zeitschrift für Sexualwissenschaft*.

Archiv für Frauenkunde und Eugenetik begins publication.

Harry H. Laughlin publishes *Report of the Committee to Study and to Report on the Best Practical Means of Cutting off the Defective Germ-Plasm in the American Population*, under the auspices of the Eugenics Record Office.

Henry Goddard publishes *Feeble-Mindedness: Its Causes and Consequences*.

F. Scott Fitzgerald writes the song "Love or Eugenics" for the Princeton University Triangle Club musical *Fie! Fie! Fi-Fi!*

D. W. Griffith's film *The Escape*, based on a 1913 play, is released; it explores the effects of unwise selection of human mates.

Harry H. Laughlin publishes *Legal, Legislative, and Administrative Aspects of Sterilization*.

1915   The Eugenics Registry is formed under the auspices of the Race Betterment Foundation in partnership with the Eugenics Record Office; its mission is to collect data on natural inheritance and to "combat race decay."

The Second National Conference on Race Betterment is held in San Francisco.

Concern about the falling birthrate in Germany prompts the founding of the Deutsche Gesellschaft für Bevolkerungspolitik (German Society for Population Policy).

Chicago surgeon Harry Haiselden provokes controversy when he publicizes his practice of allowing "defective" newborns to die by intentionally withholding treatment; his actions in the case of "Baby Bollinger" are not unprecedented in medicine, but Haiselden makes his view public in articles and lectures, emphasizing the dangers to society of "lives of no value"; a fictionalized account of the incident is filmed in 1916 as *The Black Stork*.

1916   *Eugenical News*, the organ of the Eugenics Record Office and, later, of the Eugenics Research Association and the American Eugenics Society, begins publication.

American anthropologist Alfred Kroeber criticizes eugenics in his article "Inheritance by Magic," published in *American Anthropologist*.

Lewis M. Terman publishes *The Measurement of Intelligence: An Explanation of and a Complete Guide for the Use of the Stanford Revision and Extension of the Binet-Simon Intelligence Scale*; scoring on the test determines "intelligence quotient," or "IQ."

American geneticist William E. Castle publishes *Genetics and Eugenics: A Textbook for Students of Biology and a Reference Book for Animal and Plant Breeders*, the most widely used college text of its time.

Madison Grant publishes *The Passing of a Great Race*.

Arthur H. Estabrook publishes *The Jukes in 1915*.

1917　The Immigration Act, requiring among other things a literacy test for immigrants, becomes law over President Woodrow Wilson's veto.

To aid in placement of soldiers the U.S. Army begins to administer alpha and beta intelligence tests to recruits under a program headed by Robert Yerkes.

Birth Control Review begins publication.

William Le Baron's play *The Very Idea*, a eugenics farce, opens in New York; it opens in London in 1919; the play becomes a movie in 1920 and again in 1929.

The Black Stork (alternate title *Are You Fit to Marry?*), a film based on the 1915 "Baby Bollinger" case in Chicago, is released; two other films with strong eugenic themes, *The Garden of Knowledge* and *Married in Name Only*, appear in theaters.

1918　Charles Davenport, Madison Grant, Henry Fairfield Osborn, and others organize the Galton Society in New York City.

Paul Popenoe and Roswell H. Johnson publish *Applied Eugenics*.

1919　Raymond Pearl publishes "Sterilization of Degenerates and Criminals Considered from the Standpoint of Genetics" in the April issue of *Eugenics Review*.

Arthur C. Rogers and Maud A. Merrill publish *Dwellers in the Vale of Siddem: A True Story of the Social Aspect of Feeble-Mindedness*.

Thomas Hunt Morgan publishes *The Physical Basis of Heredity*.

1920　Fitter Families contests start at the Kansas Free Fair in Topeka, Kansas.

Karl Binding, a German legal scholar specializing in criminal law, and Alfred Hoche, a German psychiatrist, publish *Die Freigabe der Vernichtung lebensun-*

werten Lebens. Ihr Mass und ihre Form (variously translated as Permission to Destroy Life Devoid of Value or Permitting the Destruction of Life Unworthy of Life).

Lothrop Stoddard publishes *Rising Tide of Color against White World-Supremacy*.

1921   The first American Birth Control Conference is held, in New York City.

The Second International Congress of Eugenics is held, at the American Museum of Natural History in New York City.

The U.S. House Committee on Immigration and Naturalization publishes *Biological Aspects of Immigration*; based upon the statements of Harry H. Laughlin.

British mystery novelist R. Austin Freeman, who trained as a physician, publishes *Social Decay and Regeneration*; he worries about the "sub-man," and how "for many years there has been flowing into this country a steady stream of men and women of the lowest type—the very dregs of inferior populations."

Erwin Baur, Fritz Lenz, and Eugen Fischer publish the 2-volume *Grundriss der menschlichen Erblichkeitslehre und Rassenhygiene*. Bd. 1, *Menschliche Erblichkeitslehre*; Bd. 2, *Menschliche Auslese und Rassenhygiene*.

1922   Prominent eugenicists found the Eugenics Committee of the United States of America, with Irving Fisher as its first president; in 1925 it becomes the American Eugenics Society.

Harry H. Laughlin publishes *Eugenical Sterilization in the United States*; the volume contains a "model sterilization law."

Hans F. K. Günther publishes *Rassenkunde des deutschen Volkes*; by 1930 fourteen editions of the work have been published.

Lothrop Stoddard publishes *The Revolt against Civilization: The Menace of the Under Man*.

1923   Munich University establishes a chair for racial hygiene; the first holder is Fritz Lenz.

Earnest Sevier Cox publishes *White America*.

The U.S. House Committee on Immigration and Naturalization publishes its hearings, *Analysis of America's Modern Melting Pot*, a review of the physical and mental health of immigrants; the document is the statement of the sole witness, Harry H. Laughlin.

The Second International Congress of Eugenics publishes its scientific papers in two volumes: *Eugenics, Genetics and the Family* and *Eugenics in Race and State.*

Carl Brigham, a professor of psychology at Princeton University, publishes *A Study of American Intelligence.*

1924   The Johnson-Reed Immigration Restriction Act of 1924 (also called the Johnson Act or the Johnson-Reed Act) sets the annual immigration quota of any nationality at 2 percent of the foreign-born of that nationality as recorded in the 1890 census; it limits immigration, starting with the fiscal year beginning July 1, 1927, to 150,000 at a ratio based on the number of people of that nationality as recorded in the census of 1920; it also prohibits from immigration "aliens ineligible for citizenship," which meant the Chinese and the Japanese, under existing law.

An act of Virginia, approved March 20, 1924, provides for the sterilization of the feebleminded.

Albert Edward Wiggam publishes *The Fruit of the Family Tree.*

1925   The Deutscher Bund fur Volksaufartung und Erbkunde (German Federation for Population Betterment and Heredity) is founded; in 1931 it will merge with the German Society for Race Hygiene (Eugenics).

The Galton Laboratory for National Eugenics and the Eugenics Society (London, England) begin publication of *Annals of Eugenics*; in 1954 the title changes to *Annals of Human Genetics.*

Lewis Terman publishes the first volume of the 5-volume set *Genetic Studies of Genius*; the final volume appears in 1959.

Adolf Hitler publishes the first volume of *Mein Kampf*; the second volume appears in 1927.

1926   E. S. Gosney establishes the Human Betterment Foundation (incorporated in 1928) "to foster and aid constructive and educational forces for the protection and betterment of the human family in body, mind, character and citizenship;" he also states that "its first major problem is to take over the investigation of the possibilities of race betterment by eugenic sterilization"; Gosney is president of the organization until his death in 1942.

Carl Brigham, a psychologist at Princeton University, begins administering the Scholastic Aptitude Test (SAT) to test groups as a predictor for academic success; the test is based on work Brigham had done with Robert Yerkes in testing soldiers for the U.S. Army in World War I; the SAT eventually becomes a standard for college admissions.

The Eugenics Education Society is renamed the Eugenics Society.

Henry Pratt Fairchild publishes *The Melting-Pot Mistake*.

American attorney Clarence Darrow publishes an antieugenics article, "The Eugenics Cult," in the June issue of *American Mercury*; he concludes, "Among the schemes for remolding society this is the most senseless and impudent that has ever been put forward by irresponsible fanatics to plague a long-suffering race."

Leonard Darwin publishes *The Need for Eugenic Reform*.

Arthur H. Estabrook and Ivan E. McDougle publish *Mongrel Virginians: The Win Tribe*.

1927   In *Buck v. Bell* the U.S. Supreme Court upholds a Virginia law that permits the sterilization of the feebleminded; writing for the Court, Oliver Wendell Holmes says, "It is better for all the world, if instead of waiting to execute degenerate offspring for crime or to let them starve for their imbecility, society can prevent those who are manifestly unfit from continuing their kind;" he adds, commenting on the family history of Carrie Buck, "Three generations of imbeciles is enough"; later scholarship will reveal that neither Buck nor her daughter were feebleminded and that the out-of-wedlock pregnancy was the result of rape, not promiscuity or degeneracy.

The Kaiser Wilhelm Institut für Anthropologie, Menschliche Erblehre und Eugenik is established in Berlin; Eugen Fischer is its first director.

Raymond Pearl publishes "The Biology of Superiority" in the November issue of *American Mercury*; he laments certain trends in eugenics, especially the outdated ideas of "orthodox eugenists" and concludes that "it would be high time that eugenics cleaned house, and threw away the old-fashioned rubbish which has accumulated in the attic."

British writer Charles Wicksteed Armstrong publishes *The Survival of the Unfittest*.

1928   The third National Conference on Race Betterment is held, in Battle Creek, Michigan.

Wycliffe Draper offers $5,000 to the Eugenics Research Association to establish prizes for a study of the relative fecundity of Nordic and other races and, if Nordics are inferior in fecundity, additional prizes for essays on the causes and prevention of the lower birthrate.

Eugenics: A Journal of Race Betterment, the official journal of the American Eugenics Society, begins publication.

The U.S. House Committee on Immigration and Naturalization publishes its hearings *Eugenical Aspects of Deportation*; the hearing is primarily the statement of Harry H. Laughlin.

1929 In Britain a joint committee of the Board of Education and Board of Control appointed in 1924 and chaired by Arthur Henry Wood issues its report, *Report of the Mental Deficiency Committee.*

E. S. Gosney and Paul Popenoe publish *Sterilization for Human Betterment: A Summary of Results of 6,000 Operations in California, 1909–1929.*

1930 Pope Pius XI's encyclical *Casti Connubi* condemns eugenics as well as divorce, birth control, and companionate marriage.

Madison Grant and Charles Stewart Davison edit and publish *The Alien in our Midst; or, "Selling our Birthright for a Mess of Pottage."*

The journal *Eugenik: Erblehre Erbpflege* begins publication.

R. A. Fisher publishes *The Genetical Theory of Natural Selection.*

1931 Major A. G. Church, a Labour member of Parliament and member of the Eugenics Society's Committee for Legalizing Eugenic Sterilization, asks leave to introduce a private member's bill to legalize eugenic sterilization; the request is voted down.

1932 The Third International Congress of Eugenics is held, at the American Museum of Natural history in New York; Henry Fairfield Osborn gives the principal address, "Birth Selection versus Birth Control."

Aldous Huxley publishes the novel *Brave New World.*

1933 The new National Socialist government in Germany enacts the Law for the Prevention of Hereditary Disease in Posterity, or Sterilization Law; it takes effect in January 1934; among those considered for sterilization for heritable diseases are people with schizophrenia, epilepsy, alcoholism, manic depression, hereditary deafness or blindness, severe hereditary physical deformity, Huntington's chorea, and congenital feeblemindedness.

The German government takes control of the Gesellschaft für Rassenhygiene (Eugenik); it restores the organization's former name and moves its headquarters from Berlin to Munich and makes the society responsible to the Ministry of the Interior.

Madison Grant publishes *The Conquest of a Continent; or, The Expansion of Races in America.*

Ernest Lidbetter publishes *Heredity and the Social Problem Group.*

1934   The German Ministry of the Interior announces measures for the suppression of hereditary disease by sterilization with the establishment of special courts, consisting of a judge and two physicians; the hearings would be private, fair and impartial, according to the government.

In Britain the Board of Control's Committee on Sterilization, chaired by Laurence Brock, issues *Report of the Departmental Committee on Sterilisation* (Cmd. 4485); the Brock Committee favors voluntary sterilization, but its work is largely forgotten after news of Germany's sterilization programs reaches Britons.

Leon Whitney, executive secretary of the American Eugenics Society, publishes *The Case for Sterilization*.

The film *Tomorrow's Children* is released; in the melodrama the adoptive daughter of a family of alcoholic degenerates faces forced sterilization.

1935   In Germany the Marital Health Law of October 18, 1935, bans marriage between the hereditarily healthy and hereditarily ill and requires a medical certificate from the local health office attesting to the fitness of the partners.

In Germany the National Socialist government enacts the Law for the Protection of German Blood and Honor (*Schutz des Blutes und der Ehre*); the law forbids marriage and sexual intercourse between Jews and gentiles.

Heinrich Himmler establishes the *Lebensborn* (Well of Life) program; it aims to assist SS families with a large number of children and to provide maternity facilities to expectant mothers, married or not, of racially desirable children; the first home opens late in 1936; the initiative, the subject of lurid speculation, fails in its objective to increase birthrates substantially.

Hermann Muller publishes *Out of the Night: A Biologist's View of the Future*.

Columbia University psychologist Otto Klineberg publishes *Negro Intelligence and Selective Migration*.

French-American Nobel Prize winner Dr. Alexis Carrel publishes *Man the Unknown*.

1936   The Human Betterment Foundation publishes a twelve-page pamphlet, *Human Sterilization*; it opens, "Strong, intelligent, useful families are becoming smaller and smaller. Irresponsible, diseased, defective parents, on the other hand, do not limit their families correspondingly. There can be but one result. The result is race degeneration."

The American Neurological Association's Committee for the Investigation of Eugenical Sterilization, chaired by Abraham Myerson, publishes its report

Eugenical Sterilization: A Reorientation of the Problem; the report states, among other things, that "so far as mental disease is concerned, the race is not rapidly going to the dogs, as has been the favorite assertion for some time."

1937   The Pioneer Fund Inc. is founded with backing from Wickliffe Draper; its stated purpose is to "conduct or aid in conducting study and research into the problems of heredity and eugenics in the human race"; it also may provide aid for education to children "descended from white persons who settled in the original thirteen colonies prior to the adoption of the Constitution"; founders include Harry H. Laughlin, Frederick Osborn, Wickliffe Draper, Malcolm Donald, and Vincent R. Smalley.

Adolf Hitler orders the sterilization of the Rheinlandbastarde, children of German mothers and African fathers who were part of the French army occupying the Rhineland after World War I.

1939   On August 18 the German Ministry of the Interior issues a decree requiring physicians, nurses, and midwives to register children under age three who are retarded or deformed.

In an October order, backdated to September 1, Adolf Hitler authorizes a program of euthanasia of the mentally and physically handicapped and patients deemed incurable, primarily in institutions, state, private, or church-run; the program becomes known as T-4 after the address of its headquarters, Tiergartenstrasse 4, Berlin.

The Carnegie Institution of Washington ceases funding the Eugenics Record Office.

Hermann Muller, J.B.S. Haldane, and twenty-one other prominent scientists sign the "Geneticists' Manifesto"; the document, a reply to the question posed by the Science Service of Washington, D.C., "How could the world's population be improved most effectively genetically?" appears as "Social Biology and Population Improvement" in the September 16 issue of *Nature*.

Harvard anthropologist Earnest A. Hooton publishes *Crime and the Man*; he concludes, "So it behooves us to learn our human parasitology and human entomology, to practice an artificial and scientific selection with intelligence, if we wish to save our skins."

1940   British geneticist J. A. Fraser Roberts publishes *An Introduction to Medical Genetics*.

Frederick Osborn publishes *Preface to Eugenics*.

1941   The Film *Ich klage an* (I accuse) premiers in Berlin in August 1941, in which a physician whose wife suffers multiple sclerosis kills her with her consent; the Nazis use this film to justify killing the mentally ill and retarded.

1942 In *Skinner v. Oklahoma* (316 U.S. 535) the Supreme Court strikes down Oklahoma's Habitual Criminal Sterilization Act because of "its failure to meet the requirements of the equal protection clause of the Fourteenth Amendment."

U.S. psychiatrist Foster Kennedy publishes "The Problem of Social Control of the Congenital Defective," in the July issue of the *American Journal of Psychiatry*; he observes "Good breeding begets good brains; with no good brains there can be no good mind"; he favors euthanasia for "those hopeless ones who never should have been born—Nature's mistakes."

1944 With the publication of their "Studies on the Chemical Nature of the Substance Inducing Transformation of Pneumococcal Types" in the January issue of the *Journal of Experimental Medicine*, Oswald Avery, Colin MacLeod, and Maclyn McCarty announce their discovery that DNA, rather than protein, as some believed, is the hereditary material in living organisms.

1945 F. O. Butler, Superintendent of the Sonoma State Home, publishes "A Quarter of a Century of Experience in Sterilization of Mental Defectives in California" in the April issue of *American Journal of Mental Deficiency*; his experience confirms his "faith in the universal need of a law permitting sterilization."

1947 On trial before the Nuremberg Tribunal for his role in Nazi euthanasia and sterilization programs, Karl Brandt introduces into evidence as a portion of his defense the writings of Madison Grant, particularly *The Passing of the Great Race*; Alexis Carrel's *Man the Unknown*; and Indiana's sterilization law of 1907.

1948 The American Society of Human Genetics is founded; in 1949 Hermann Muller becomes its first annual president.

1949 The American Society of Human Genetics begins publication of the *American Journal of Human Genetics*.

Lionel Penrose publishes *The Biology of Mental Defect*.

1952 British physician Douglas Bevis describes how amniocentesis can be used to test for Rh-factor incompatibility in "The Antenatal Prediction of Haemolytic Disease of the Newborn," published in *The Lancet*; the prenatal test has implications for later use in screening for genetic disorders.

1953 Biochemist James Watson and British biophysicist Francis Crick announce their discovery of the structure of DNA in "A Structure for Deoxyribose Nucleic Acid," published in the April 25 issue of *Nature*.

1954 *Annals of Eugenics* changes its name to *Annals of Human Genetics*.

1956 The first International Congress of Human Genetics is held, in Copenhagen, Denmark.

1960   *Mankind Quarterly* begins publication.

1962   Francis Crick, James Watson, and Maurice Wilkins receive the Nobel Prize for Physiology or Medicine for determining the molecular structure of DNA; Rosalind Franklin, who might have shared the prize, died in 1959.

1963   The Galton Professorship of Eugenics becomes the Galton Professorship of Human Genetics and the Francis Galton Laboratory of National Eugenics becomes the Galton Laboratory of the Department of Human Genetics and Biometry.

1967   C. B. Jacobson and R. H. Barter publish "Intrauterine Diagnosis and Management of Genetic Defects" in the November 15 issue of the *American Journal of Obstetrics and Gynecology*.

1969   The British journal *Eugenics Review* changes title to the *Journal of Biosocial Science*.

The U.S. journal *Eugenics Quarterly* changes title to *Social Biology*.

Albert Rosenfeld, science writer and science editor of *Life* magazine publishes *The Second Genesis: The Coming Control of Life*.

Arthur R. Jensen publishes "How Much Can We Boost IQ and Scholastic Achievement?" in the *Harvard Educational Review*.

1970   The National Academy of Sciences rejects (again) William Shockley's proposal that the academy encourage research in "dysgenics"; Shockley fears the "retrograde evolution" of the U.S. population through the increasing number of African Americans of low IQ; Shockley will suggest a program of voluntary sterilization in which the government would pay people who consent to being sterilized up to $1,000 for each point they test below 100 on the IQ scale.

Paul Ramsey, professor of Christian Ethics at Princeton University, publishes *Fabricated Man: The Ethics of Genetic Control*.

1971   Harvard professor Richard Herrnstein publishes "I.Q." in the September issue of *Atlantic Monthly*.

1972   Arthur R. Jensen publishes *Genetics and Education*.

1973   The American Eugenics Society is renamed the Society for the Study of Social Biology.

1975   Concerned about the safety of recombinant DNA techniques, a group of scientists, including Nobelists James Watson and Paul Berg, organizes a meeting to discuss the issues troubling them; the International Congress on Recombinant DNA Molecules, better known as Asilomar, gathers at the Asilomar Conference Center in Pacific Grove, California.

1976   Venture capitalist Robert Swanson and molecular biologist Herbert Boyer found Genentech, the first genetic engineering company, in South San Francisco.

1978   In Britain Louise Brown, the first "test tube baby," is born July 25.

In addition to other reports and studies, the U.S. National Commission for the Protection of Human Subjects of Biomedical and Behavioral Research publishes *The Belmont Report: Ethical Principles and Guidelines for the Protection of Human Subjects of Research.*

1979   Robert K. Graham establishes the Repository for Germinal Choice in Escondido, California; Graham hoped to begin by offering sperm of Nobel Prize winners but expanded the donor pool to include other healthy, highly intelligent, and talented men; Robert Shockley acknowledged being a donor; the repository closes in 1999.

1980   In California the Court of Appeals rules in the case of *Curlender v. Bio-Science Laboratories* (165 Cal. Rptr. 477), a case of "wrongful life" involving a child born with Tay-Sachs disease; the court finds that "plaintiff has adequately pleaded a cause of action for punitive damages. We see no reason in public policy or legal analysis for exempting from liability for punitive damages a defendant who is sued for committing a 'wrongful-life' tort."

The Virginia chapter of the American Civil Liberties Union files a class action suit against the Commonwealth in the Federal District Court for the Western district of Virginia on behalf of "Judith Doe" and others who were involuntarily sterilized in state institutions.

1981   Stephen Jay Gould publishes *The Mismeasure of Man*; a revised and expanded edition appears in 1996, after the publication of *The Bell Curve* in 1994.

1982   The first child is born to a mother inseminated with sperm from the Repository for Germinal Choice.

The President's Commission for the Study of Ethical Problems in Medicine and Biomedical and Behavioral Research publishes its report *Splicing Life: A Report on the Social and Ethical Issues of Genetic Engineering with Human Beings.*

1984   William Shockley's libel suit against the *Atlantic Constitution* ends with a judgment in his favor—a $1 award for actual damages but no award for punitive damages; he had sued over a 1980 column that compared his proposals for voluntary sterilization to Nazi experiments and had, he alleged, labeled him as a racist.

1988   James Watson becomes director of the Office of Human Genome Research in the National Institutes of Health; in 1990 the Human Genome Project, a consortium of institutions, officially begins the attempt to map and sequence the entire human genome.

1989 In Britain the Eugenics Society changes its name to the Galton Institute.

J. Philippe Rushton, Professor of Psychology at the University of Western Ontario, presents his paper "Evolutionary Biology and Heritable Traits (With Reference to Oriental-White-Black Difference)" at the Symposium on Evolutionary Theory, Economics and Political Science, AAAS Annual Meeting in San Francisco; the paper is published in the December 1992 issue of *Psychological Reports*.

Richard Herrnstein publishes "IQ and Falling Birthrates," in the May issue of *The Atlantic*.

1990 American geneticist W. French Anderson leads a team at the National Institutes of Health that performs the first approved gene therapy on four-year-old Ashanti DeSilva by infusing her with her own white blood cells treated to contain the ADA gene she is missing, the source of her immune deficiency condition.

Human Genome Project begins to sequence the human genome.

1992 Faced with complaints of racism, the National Institutes of Health withdraws funding for a conference on genetics and crime, "Genetic Factors in Crime: Findings, Uses and Implications," which was to be held at the University of Maryland.

1993 Dean H. Hamer, chief of the Section on Gene Structure and Regulation in the Laboratory of Biochemistry of the National Cancer Institute, and colleagues publish "A Linkage between DNA Markers on the X-chromosome and Male Sexual Orientation" in the July issue of *Science*; the research shows that many gay men share a common marker in the X chromosome, suggesting the existence of a "gay gene."

The play *The Twilight of the Golds* by Jonathan Tolins (made into a film in 1997) opens in New York City; it concerns a couple who, after genetic testing reveals their unborn son may be gay, debate ending the pregnancy with the wife's family, which includes her gay brother.

1994 The made-for-television movie *Against Her Will: The Carrie Buck Story* airs.

Richard J. Herrnstein and Charles A. Murray publish *The Bell Curve: Intelligence and Class Structure in American Life*.

Dean H. Hamer and Peter Copeland publish *The Science of Desire: The Search for the Gay Gene and the Biology of Behavior*.

1995   President Bill Clinton creates the National Bioethics Advisory Commission; the American Life League responds by forming the American Bioethics Advisory Commission; in 1997 the two bodies publish their separate reports.

The University of Maryland conference on links between genes and criminal or violent behavior, postponed from 1992, takes place at the Aspen Institute in rural Maryland.

J. Philippe Rushton publishes *Race, Evolution, and Behavior: A Life History Perspective*.

1996   Led by embryologist Ian Wilmut, researchers at the Roslin Institute in Scotland clone a sheep they name Dolly; they report her birth in the February 27, 1997, issue of *Nature*.

University of Edinburgh psychologist Chris Brand publishes *The g Factor; General Intelligence and Its Implications*; he is suspended from teaching and his publisher withdraws the book, which discusses race and inherited intelligence.

1997   Dr. Raymond B. Cattell is selected to receive the American Psychological Foundation Gold Medal Award for Lifetime Achievement in Psychological Science at the American Psychological Association Convention; critics accuse him of racism and of being aligned with "fascist and eugenics causes" in some of his writings; the award is postponed and Cattell declines it, removing his name from consideration for the distinction.

The film *Gattaca* is released; it explores a future where genetic engineering determines one's makeup and place in society before birth.

1998   Swedish director Peter Cohen releases the film *Homo Sapiens 1900*, a documentary history of eugenics; it opens in New York on March 3, 2000.

1999   Ron Harris, a fashion photographer, offers models as egg donors to the highest bidders, auctioning their ova on the Internet at www.ronsangels.com.

Advertising appears in campus newspapers at Princeton, Harvard, Yale and other Ivy League schools offering to pay up to $50,000 for eggs of a woman who is intelligent, athletic, and at least 5 feet 10 inches tall, with an SAT score of 1400 or more.

2000   President Bill Clinton announces, in the presence of Craig Venter of Celera, Francis Collins of the Human Genome Project, and James Watson, the completion of a draft sequence of the human genome.

The show "Perfecting Mankind: Eugenics and Photography" runs at the International Center of Photography in New York from January 11 through March 18.

Greg Cox publishes *The Rise and Fall of Khan Noonien Singh*, Vol. 1 of *The Eugenics Wars*, based on characters and premises created by Gene Roddenberry for the television show and movies *Star Trek*.

2002 Oregon governor John Kitzhaber apologizes to Oregonians who were sterilized by castration, vasectomy, tubal ligation, or hysterectomy under provisions of the state's 1913 eugenics legislation.

The Virginia General Assembly passes a resolution honoring the memory of Carrie Buck; a historical marker is erected in Charlottesville, where she was born; Governor Mark R. Warner offers the "Commonwealth's sincere apology for Virginia's participation in eugenics."

Gregory Stock publishes *Redesigning Humans: Our Inevitable Genetic Future.*

2003 California governor Gray Davis issues an apology for sterilizations conducted under the state's 1909 (and later) eugenics legislation.

Governor Mike Easley of North Carolina signs legislation removing involuntary sterilization from state law; he says, "To the victims and families of this regrettable episode in North Carolina's past, I extend my sincere apologies."

2004 In Britain the Human Fertilization and Embryology Authority grants a license to a team at the University of Newcastle for human therapeutic cloning to produce stem cells for research; Ian Wilmut, who helped create "Dolly" the sheep, applies for a cloning license to study motor neuron disease.

Selected Bibliography

Books

Alcott, William A., *The Physiology of Marriage*. New York: Arno Press and New York Times, [1866], 1972.

American Academy of Political and Social Science, *Race Improvement in the United States*. Philadelphia: 1909.

American MENSA Limited, *A History of Mensa: Commemorating American Mensa's 30th Anniversary, 1960–1990*. Brooklyn, NY: American Mensa, Ltd., 1990.

Bachrach, Susan D. (project director), *Deadly Medicine: Creating the Master Race*. Washington: United States Holocaust Memorial Museum, 2004.

Bannister, Robert C., *Social Darwinism: Science and Myth in Anglo-American Social Thought*. Philadelphia: Temple University Press, [1979], 1988.

Banton, Michael, "Galton's Conception of Race in Historical Perspective," in Milo Keynes, (ed.), *Sir Francis Galton, FRS*, 1993.

Bates, Barbara, *Bargaining for Life: A Social History of Tuberculosis, 1876–1938*. Philadelphia: University of Pennsylvania Press, 1992.

Baur, Erwin, Eugene Fischer, and Fritz Lenz, *Human Heredity*. Trans. Eden and Cedar Paul. New York: Macmillan Company, 1931.

Begleiter, Henri, and Benjamin Kissin, *The Genetics of Alcoholism*. New York: Oxford University Press, 1995.

Bennett, J. H., *Natural Selection, Heredity, and Eugenics*. New York: Clarendon Press; Oxford University Press, 1983.

Biddiss, Michael Denis, *Father of Racist Ideology: The Social and Political Thought of Count Gobineau*. New York: Weybright and Talley, 1970.

Billington, Ray Allen, *The Origins of Nativism in the United States, 1800–1844*. New York: Arno Press, [1933], 1974.

Black, Edwin, *War Against The Weak: Eugenics and America's Campaign to Create a Master Race*. New York: Four Walls Eight Windows, 2003.

Blackwell, Elizabeth, *Pioneer Work in Opening the Medical Profession to Women: Autobiographical Sketches*. New York: Source Book Press, [1895], 1970.

Blocker, Jack S., Jr., *American Temperance Movements: Cycles of Reform*. Boston: Twayne Publishers, 1989.

Box, Joan Fisher, *R. A. Fisher: The Life of a Scientist*. New York: Wiley, 1978.

Boyce, Robert W. D., and Joseph A. Maiolo, (eds.), *The Origins of World War Two: The Debate Continues*. New York: Palgrave Macmillan, 2003.

Boyle, T. Coraghessan, *The Road to Wellville*. New York: Viking, 1993.

Brandt, Allan M., *No Magic Bullet: A Social History of Venereal Disease in the United States since 1880*. New York: Oxford University Press, 1985.

Broder, Sherri, *Tramps, Unfit Mothers, and Neglected Children: Negotiating the Family in Nineteenth-Century Philadelphia*. Philadelphia: University of Pennsylvania Press, 2002.

Brookes, Martin, *Extreme Measures: The Dark Visions and Bright Ideas of Francis Galton*. New York: Bloomsbury, 2004.

Bruce, Robert V., *Bell: Alexander Graham Bell and the Conquest of Solitude*. Boston: Little, Brown, 1973.

Buchanan, Allen, Dan W. Brock, Norman Daniels, and Daniel Wikler, *From Chance to Choice: Genetics and Justice*. New York: Cambridge University Press, 2000.

Burt, Elizabeth V., *The Progressive Era: Primary Documents on Events from 1890 to 1914*. Westport, CT: Greenwood Press, 2004.

Bush, Lester E., Jr., *Health and Medicine among the Latter-day Saints: Science, Sense, and Scripture*. New York: Crossroad, 1993.

Campbell, Persia, *Mary Williamson Harriman*. New York: Columbia University Press, 1960.

Carey, Gregory, *Human Genetics for the Social Sciences*. Thousand Oaks, CA: Sage Publications, 2003.

Carlson, Elof Axel, *Genes, Radiation, and Society: The Life and Work of H. J. Muller*. Ithaca, NY: Cornell University Press, 1981.

———. *The Unfit: A History of a Bad Idea*. Cold Spring Harbor, NY: Cold Spring Harbor Laboratory Press, 2001.

Carson, Gerald, *Cornflake Crusade*. New York: Rinehart, 1957.

Cashman, Greg, *What Causes War? An Introduction to Theories of International Conflict*. New York: Lexington Books, 2000.

Castle, William E., *Genetics and Eugenics: A Text-book for Students of Biology and a Reference Book for Animal and Plant Breeders*. Cambridge, MA: Harvard University Press, 1932.

Cate, Curtis, *Friedrich Nietzsche*. London: Hutchinson, 2002.

Cavalli-Sforza, L. L., *Genes, Peoples, and Languages: A Picture of Recent Human Evolution*. Trans. Mark Seilstad. New York: North Point Press, 2000.

Cavalli-Sforza, L. L., Paolo Menozzi, and Alberto Piazza, *The History and Geography of Human Genes*. Princeton, NJ: Princeton University Press, [1994], 1996.

Cecil, Robert, *The Myth of the Master Race: Alfred Rosenberg and Nazi Ideology*. New York: Dodd, Mead, 1972.

Chesler, Ellen, *Woman of Valor: Margaret Sanger and the Birth Control Movement in America*. New York: Simon & Schuster, 1992.

CIBA Foundation Symposium 194, *Genetics of Criminal and Antisocial Behaviour*. New York: Wiley, 1996.

Clark, William R., and Michael Grunstein, *Are We Hardwired? The Role of Genes in Human Behavior*. New York: Oxford University Press, 2000.

Clarke, Charles Walter, *Taboo: The Story of the Pioneers of Social Hygiene*. Washington, DC: Public Affairs Press, 1961.

Clay, Catrine, and Michael Leapman, *Master Race: The Lebensborn Experiment in Nazi Germany*. London: Hodder & Stoughton, 1995.

Clayton, Julie, and Carina Dennis, (eds.), *50 Years of DNA*. New York: Palgrave Macmillan, 2003.

Connelly, Mark Thomas, *The Response to Prostitution in the Progressive Era*. Chapel Hill: University of North Carolina Press, 1980.

Coon, Carleton S., Stanley M. Garn, and Joseph B. Birdsell, *Races: A Study of the Problems of Race Formation in Man*. Westport, CT: Greenwood Press, [1950], 1981.

Coon, Carleton Stevens, *The Origin of Races*. New York: Knopf, [1962], 1971.

Copeland, Dale C., *The Origins of Major War*. Ithaca, NY: Cornell University Press, 2000.

Cordasco, Francesco, *The White Slave Trade and the Immigrants: A Chapter in American Social History*. Detroit, MI: Blaine Ethridge Books, 1981.

Corsi, Pietro, *The Age of Lamarck: Evolutionary Theories in France, 1790–1830*. Trans. Jonathon Mandelbaum. Berekley: University of California Press, 1988.

Cotton, Edward H., *The Life of Charles W. Eliot*. Boston: Small, Maynard & Company, 1926.

Cutler, Robert W. P., *The Mysterious Death of Jane Stanford*. Stanford, CA: Stanford General Books, 2003.

Danielson, Florence H., and Charles B. Davenport, *The Hill Folk: Report on a Rural Community of Hereditary Defects*. Cold Spring Harbor, NY: Press of the New Era Print Co., 1912.

Darwin, Charles, *The Descent of Man and Selection in Relation to Sex*. New York: A. L. Burt, [1871], 1990.

———. *On the Origin of Species by Means of Natural Selection; or, The Preservation of Favoured Races in the Struggle for Life*. New York: Penguin, [1859], 1985.

Darwin, Leonard, *The Need for Eugenic Reform*. New York: [Appleton, 1926], Garland Publishing, 1984.

———. "Preface," *Problems in Eugenics*. Vol. 2, *Report of Proceedings of the First International Eugenics Congress, University of London, July 24th to 30th 1912*. London: Eugenics Education Society, 1913.

Davenport, Charles, *Heredity in Relation to Eugenics*. New York: H. Holt and Company, 1911.

Dawkins, Richard, *The Extended Phenotype: The Long Reach of the Gene*. New York: Oxford University Press, [1982], 1999.

DiBerardino, Marie A., *Genomic Potential of Differentiated Cells*. New York: Columbia University Press, 1997.

Dorey, Annette K. Vance, *Better Baby Contests: The Scientific Quest for Perfect Childhood Health in the Early Twentieth Century*. Jefferson, NC: McFarland & Co., 1999.

Dover, Gabriel A., *Dear Mr. Darwin: Letters on the Evolution of Life and Human Nature*. Berkeley: University of California Press, 2000.

Dreyer, Peter, *A Gardner Touched with Genius: The Life of Luther Burbank*. New York: Coward, McCann & Geoghegan, 1975.

Drlica, Karl, *Understanding DNA and Gene Cloning: A Guide for the Curious*. Hoboken, NJ: Wiley, 2004.

Dubos, René, and Jean Dubos, *The White Plague: Tuberculosis, Man and Society*. New Brunswick, NJ: Rutgers University Press, [1952], 1987.

Duffy, John, *The Sanitarians: A History of American Public Health*. Urbana, IL: University of Illinois Press, [1990], 1992.

Dugdale, Richard Louis, *"The Jukes": A Study in Crime, Pauperism, Disease, and Heredity*. New York: Arno Press, [1877], 1970.

Duster, Troy, *Backdoor to Eugenics*. New York: Routledge, 2003.

Edwards, Robert G., and Patrick C. Steptoe, *A Matter of Life: The Story of a Medical Breakthrough*. New York: Morrow, 1980.

Endler, John A., *Natural Selection in the Wild*. Princeton, NJ: Princeton University Press, 1986.

Engs, Ruth C., *Alcohol and Other Drugs: Self-Responsibility*. Bloomington, IN: Tichenor Publishers, 1987.

———— (ed.), *Controversies in the Addiction Field*. Kendal-Hunt: Dubuque, IA, 1990.

Engs, Ruth Clifford, *Clean Living Movements: American Cycles of Health Reform*. Westport, CT: Praeger, 2001.

————. *The Progressive Era's Health Reform Movement: A Historical Dictionary*. Westport, CT: Praeger, 2003.

Epstein, Barbara, *The Politics of Domesticity: Women, Evangelism, and Temperance in Nineteenth-Century America*. Middletown, CT: Wesleyan University Press, [1981], 1986.

Estabrook, Arthur H., *The Jukes in 1915*. Washington, DC: Carnegie Institution of Washington [1916], 1977.

Estabrook, Arthur H., and Charles B. Davenport, *The Nam Family: A Study in Cacogenics*. Lancaster, PA: New Era Printing Company, [1912], 1968.

Estabrook, Arthur H., and Ivan E. McDougle, *Mongrel Virginians: The Win Tribe*. Baltimore, MD: The Williams & Wilkins Company, 1926.

Farrall, Lyndsay Andrew, *The Origins and Growth of the English Eugenics Movement, 1865–1925*. New York: Garland, 1985.

Ferrence, Roberta G., *Deadly Fashion: The Rise and Fall of Cigarette Smoking in North America*. New York: Garland, 1989.

Fienberg, Stephen E., and D. V. Hinkley, (eds.), *R. A. Fisher: An Appreciation*. New York: Springer-Verlag, 1980.

Finlayson, Anna Wendt, *The Dack Family: A Study in Hereditary Lack of Emotional Control*. Cold Springs Harbor, NY: Eugenics Record Office, 1916.

Fisher, Irving Norton, *My Father, Irving Fisher*. New York: Comet Press Books, 1956.

Fisher, Ronald Aylmer, Sir, *Natural Selection, Heredity, and Eugenics: Including Selected Correspondence of R. A. Fisher with Leonard Darwin and Others*. Ed. J. H. Bennett. New York: Oxford University Press, 1983.

Foote, Robert H., *Artificial Insemination to Cloning: Tracing 50 Years of Research*. Ithaca, NY: Cornell University, 1998.

Fowler, Orson, *Love and Parentage: Including Important Directions and Suggestions to Lovers, Etc.* New York: Fowler and Wells, 1847.

Fromkin, David, *Europe's Last Summer: Who Started the Great War in 1914?* New York: Knopf, 2004.

Gallagher, Nancy L., *Breeding Better Vermonters: The Eugenics Project in the Green Mountain State*. Hanover, NH: University Press of New England, 1999.

Galton, Francis, Sir, "Herbert Spencer Lecture Delivered before the University at Oxford, June 5, 1907,"*Essays in Eugenics*. London: Eugenics Education Society, 1909.

———. *Hereditary Genius: An Inquiry into its Laws and Consequences*. London: Macmillan, 1869.

———. *Inquiries into Human Faculty and Its Development*. London: Macmillan & Co., 1883.

———. *Memories of My Life*. New York: AMS Press, [1908], 1974.

Gessler, Bernhard, *Eugen Fischer (1874–1967): Leben und Werk des Freiburger Anatomen, Anthropologen und Rassenhygienikers bis 1927*. Frankfurt am Main: Peter Lang, 2000.

Gillham, Nicholas W., *A Life of Sir Francis Galton: From African Exploration to the Birth of Eugenics*. New York: Oxford University Press, 2001.

Gobineau, Arthur, Comte de, *The Moral and Intellectual Diversity of Races*. New York: Garland, [1856], 1984.

Goddard, Henry H., *The Kallikak Family: A Study in the Heredity of Feeble-mindedness*. New York: Arno Press, [1912], 1972.

Gordon, Linda, *Woman's Body, Woman's Right: Birth Control in America*. New York: Penguin Books, 1990.

Gosney, Ezra S., and Paul Popenoe, *Sterilization for Human Betterment: A Summary of Results of 6,000 Operations in California, 1909–1929*. New York: Macmillan, 1929.

Gould, Stephen Jay, *The Structure of Evolutionary Theory*. Cambridge, MA: Belknap Press of Harvard University Press, 2002.

Graham, Sylvester, *A Lecture to Young Men*. New York: Arno Press, [1833], 1974.

Grant, Madison, *The Passing of the Great Race; or, The Racial Basis of European History*. New York: Charles Scribner's Sons, [1916], 1970.

Gray, Madeline, *Margaret Sanger: A Biography of the Champion of Birth Control*. New York: R. Marek, 1978.

Greeley, Andrew M., *The American Catholic: A Social Portrait*. New York: Basic Books, 1977.

Green, Harvey, *Fit for America: Health, Fitness, Sport, and American Society*. New York: Pantheon Books, 1986.

Grittner, Frederick K., *White Slavery: Myth, Ideology, and American Law*. New York: Garland Publishers, 1990.

Grosvenor, Edwin S. and Morgan Wesson, *Alexander Graham Bell: The Life and Times of the Man Who Invented the Telephone*. New York: Harry Abrams, 1997.

Günther, Hans F. K., *The Racial Elements of European History*. Trans. G. C. Wheeler. Wayne, PA: Landpost Press, [1927], 1992.

Guyer, Michael F., *Being Well-Born: An Introduction to Eugenics*. Indianapolis, IN: Bobbs-Merrill Company, 1916.

Hale, Christopher, *Himmler's Crusade: The Nazi Expedition to Find the Origins of the Aryan Race*. Hoboken, NJ: Wiley, 2003.

Hall, Ruth E., *Passionate Crusader: The Life of Marie Stopes*. New York: Harcourt Brace Jovanovich, 1977.

Haller, Mark H., *Eugenics: Hereditarian Attitudes in American Thought*. New Brunswick, NJ: Rutgers University Press, [1963], 1984.

Harris, John, *Clones, Genes, and Immortality: Ethics and the Genetic Revolution*. New York: Oxford University Press, 1998.

Hawkins, Mike, *Social Darwinism in European and American Thought, 1860–1945: Nature as Model and Nature as Threat*. New York: Cambridge University Press, 1997.

Hawley, R. Scott, and Catherine A. Mori, *The Human Genome: A User's Guide*. San Diego, CA: Academic Press, 1999.

Henig, Robin Marantz, *The Monk in the Garden: The Lost and Found Genius of Gregor Mendel, the Father of Genetics*. Boston, MA: Houghton Mifflin, 2000.

———. *Pandora's Baby: How the First Test Tube Babies Sparked the Reproductive Revolution*. Boston, MA: Houghton Mifflin, 2004.

Henry, Clarissa, and Marc Hillel, *Children of the SS*. London: Hutchinson, 1976.

Herrnstein, Richard J., and Charles Murray, *The Bell Curve: Intelligence and Class Structure in American Life*. New York: Free Press, 1994.

Higham, John, *Strangers in the Land: Patterns of American Nativism, 1860–1925*. New Brunswick, NJ: Rutgers University Press, 1955.

Hofstadter, Richard, (ed.), *The Progressive Movement, 1900–1915*. New York: Simon & Schuster, [1963], 1986.

———. *Social Darwinism in American Thought*. New York: George Braziller, [1955], 1959.

Holmes, Samuel J., *Trend of the Race: A Study of Present Tendencies in the Biological Development of Mankind*. New York: Harcourt, Brace and Company, 1921.

Holt, Marilyn Irvin, *Linoleum, Better Babies, and the Modern Farm Woman, 1890–1930*. Albuquerque: University of New Mexico Press, 1995.

Hooper, Judith, *Of Moths and Men: An Evolutionary Tale; The Untold Story of Science and the Peppered Moth*. New York: Norton, 2002.

Howe, Louise Kapp, *The White Majority: Between Poverty and Affluence*. New York: Random House, 1971.

Hubbard, Ruth, and Elijah Wald, *Exploding the Gene Myth: How Genetic Information Is Produced and Manipulated by Scientists, Physicians, Employers, Insurance Companies, Educators, and Law Enforcers*. Boston, MA: Beacon Press, 1993.

Huxley, Aldous, *Brave New World*. London: The Folio Society, [1932], 1971.

Ingalls, Robert P., *Mental Retardation: The Changing Outlook*. New York: Collier Macmillan, 1986.

Jacobson, Matthew Frye, *Special Sorrows: The Diasporic Imagination of Irish, Polish, and Jewish Immigrants in the United States*. Cambridge, MA: Harvard University Press, 1995.

Jacoby, Kerry N., *Souls, Bodies, Spirits: The Drive to Abolish Abortion since 1973*. Westport, CT: Praeger, 1998.

Jacoby, Russell, and Naomi Glauberman, *The Bell Curve Debate: History, Documents, Opinions*. New York: Times Books, 1995.

James, Henry, *Charles W. Eliot, President of Harvard University, 1869–1909*. 2 vols. Boston: Houghton Mifflin Company, 1930.

Jensen, Arthur R., *The G Factor: The Science of Mental Ability*. Westport, CT: Praeger, 1998.

Jones, Helen, *Health and Society in Twentieth-Century Britain*. New York: Longman, 1994.

Jones, Steve, *Darwin's Ghost: the Origin of Species Updated*. New York: Random House, 2000.

Jordan, David Starr, *The Blood of the Nation: A Study of the Decay of Races Through the Survival of the Unfit*. Boston, MA: American Unitarian Association, 1902.

———. *The Days of a Man, Being Memories of a Naturalist, Teacher, and Minor Prophet of Democracy*. 2 vols. Yonkers-on-Hudson, NY: World Book Company, 1922.

———. *The Heredity of Richard Roe: A Discussion of the Principles of Eugenics*. Boston: American Unitarian Association, 1911.

Jordan, Harvey Ernest, "Eugenics: Its Data, Scope and Promise, As Seen by the Anatomist." In *Eugenics: Twelve University Lectures*. 107–138. Ed. Lucy James Wilson. New York: Dodd, Mead and Company, 1914.

Jordanova, L. J., *Lamarck*. New York: Oxford University Press, 1984.

———. "Needed—A New Human race." In *Proceedings of the First National Conference on Race Betterment, January 8, 9, 10, 11, 12, 1914*. Ed. Emily F. Robins. Battle Creek, MI: Race Betterment Foundation, 1914.

Kellogg, John H., "The Eugenics Registry," In *Official Proceedings of the Second National Conference on Race Betterment, August 4, 5, 6, 7 and 8, 1915; Held in San Francisco, California*. Battle Creek, MI: Race Betterment Foundation, (1915), 76–87.

———. *Plain Facts for Old and Young*. Buffalo, NY: Heritage Press, [1879], 1974.

Kennedy, David M., *Birth Control in America: The Career of Margaret Sanger*. New Haven, CT: Yale University Press, 1970.

Kerr, Anne, and Tom Shakespeare, *Genetic Politics: From Eugenics to Genome*. Cheltenham, England: New Clarion Press, 2002.

Kühl, Stefan, *The Nazi Connection: Eugenics, American Racism, and German National Socialism*. New York: Oxford University Press, 1994.

Kevles, Daniel J., *The Code of Codes: Scientific and Social Issues in the Human Genome Project*. Cambridge, MA: Harvard University Press, 1992.

———. *In the Name of Eugenics: Genetics and the Uses of Human Heredity*. New York: Alfred A. Knopf, 1985.

Keynes, Milo, (ed.), *Sir Francis Galton, FRS: The Legacy of His Ideas/Proceedings of the Twenty-Eighth Annual Symposium of the Galton Institute, London, 1991*. Houndmills, Basingstoke, Hampshire: Macmillan Press, in association with the Galton Institute, 1993.

Klaw, Spencer, *Without Sin: The Life and Death of the Oneida Community*. New York: Allen Lane, 1993.

Kline, Wendy, *Building a Better Race: Gender, Sexuality, and Eugenics from the Turn of the Century to the Baby Boom*. Berkeley: University of California Press, 2001.

Klotzko, Arlene Judith, (ed.), *The Cloning Sourcebook*. New York: Oxford University Press, 2001.

Kluger, Richard, *Ashes to Ashes: America's Hundred-Year Cigarette War, the Public Health, and the Unabashed Triumph of Philip Morris*. New York: Alfred A. Knopf, 1996.

Kolata, Gina Bari, *Clone: The Road to Dolly, and the Path Ahead*. New York: W. Morrow and Co., 1998.

Kraut, Alan M., *Silent Travelers: Germs, Genes, and the "Immigrant Menace."* New York: Basic Books, 1994.

Laughlin, Harry Hamilton, *Eugenical Sterilization in the United States.* Chicago: F. Klein Company, 1922.

Levin, Jerome D., *Alcoholism: A Bio-psycho-social Approach.* New York: Hemisphere, 1990.

Littlewood, Roland, *Pathologies of the West: An Anthropology of Mental Illness in Europe and America.* Ithaca, NY: Cornell University Press, 2002.

Lukas, Richard C., *The Forgotten Holocaust: The Poles under German Occupation, 1939–1944.* Lexington: University Press of Kentucky, 1986.

Lynn, Richard, *Eugenics: A Reassessment.* Westport, CT: Praeger, 2001.

Magnello, Eileen, *The Road to Medical Statistics.* New York: Rodopi, 2002.

Marchione, Margherita, *Americans of Italian Heritage.* Lanham, MD: University Press of America, 1995.

Markel, Howard, *Quarantine! East European Jewish Immigrants and the New York City Epidemics of 1892.* Baltimore, MD: Johns Hopkins University Press, 1997.

Marks, Jonathan, *Human Biodiversity: Genes, Race, and History.* New York: Aldine De Gruyter, 1995.

Martin, Geoffrey J., *Ellsworth Huntington: His Life and Thought.* Hamden, CT: Archon Books, 1973.

Mauss, Armand L., *All Abraham's Children: Changing Mormon Conceptions of Race and Lineage.* Urbana: University of Illinois Press, 2003.

Mazumdar, Pauline M. H., *Eugenics, Human Genetics, and Human Failings: The Eugenics Society, Its Source, and Its Critics in Britain.* New York: Routledge, 1992.

McClain, Charles J., *In Search of Equality: The Chinese Struggle against Discrimination in Nineteenth-Century America.* Berkeley: University of California Press, 1994.

McGee, Glenn, *The Perfect Baby: A Pragmatic Approach to Genetics.* Lanham, MD: Rowman & Littlefield Publishers, 1997.

McLoughlin, William G., *Revivals, Awakenings, and Reform: An Essay on Religion and Social Change in America, 1607–1977.* Chicago: University of Chicago Press, 1978.

Mehler, Barry, "The History of the American Eugenics Society, 1921–1940" (Ph.D. Dissertation, University of Illinois, Urbana-Champaign, 1988).

Menand, Louis, *The Metaphysical Club.* New York: Farrar, Straus and Giroux, 2001.

Money, John, *The Destroying Angel: Sex, Fitness & Food in the Legacy of Degeneracy Theory, Graham Crackers, Kellogg's Corn Flakes & American Health History.* Buffalo, NY: Prometheus Books, 1985.

Moorehead, Alan, *Darwin and the Beagle.* Harmondsworth, England: Penguin, 1971.

Morris, Charles, *The Aryan Race: Its Origins and its Achievements.* Chicago: S. C. Griggs and Company, [1888], 1988.

Muller, H. J., *Out of the Night: A Biologist's View of the Future.* London: V. Gollancz, [1936], 1984.

———. *Studies in Genetics: The Selected Papers of H. J. Muller.* Bloomington: Indiana University Press, 1962.

Nietzsche, Friedrich Wilhelm, *Thus Spoke Zarathustra: A Book for All and None.* Trans. Walter Arnold Kaufmann. New York: Modern Library, 1995.

Nissenbaum, Stephen, *Sex, Diet, and Debility in Jacksonian America: Sylvester Graham and Health Reform*. Chicago: Dorsey Press, [1980], 1988.

Noyes, John Humphrey, *Male Continence or Self Control in Sexual Intercourse*. Oneida, NY: Oneida Community Mansion House, [1872], 1992.

Ordover, Nancy, *American Eugenics: Race, Queer Anatomy, and the Science of Nationalism*. Minneapolis: University of Minnesota Press, 2003.

Osborn, Frederick, *Preface to Eugenics*. New York: Harper & Brothers, 1940.

Ostling, Richard N., and Joan K. Ostling, *Mormon America: The Power and the Promise*. San Francisco, CA: Harper SanFrancisco, 1999.

Paul, Diane B., *Controlling Human Heredity: 1865 to the Present*. Atlantic Highlands, NJ: Humanities Press, 1995.

Pearson, Karl, *The Life, Letters and Labours of Francis Galton*. 3 vols. Cambridge: Cambridge University Press, 1914–1930.

Perkins, Harry F., (ed.), *A Decade of Progress in Eugenics: Scientific Papers of the Third International Congress of Eugenics*. Baltimore, MD: The Williams & Wilkins Company, 1934.

Pernick, Martin S., *The Black Stork: Eugenics and the Death of "Defective" Babies in American Medicine and Motion Pictures since 1915*. New York: Oxford University Press, 1996.

Pickens, Donald K., *Eugenics and the Progressives*. Nashville, TN: Vanderbilt University Press, 1968.

Pinker, Steven, *The Blank Slate: The Modern Denial of Human Nature*. New York: Viking, 2002.

Pivar, David J., *Purity Crusade: Sexual Morality and Social Control, 1868–1900*. Westport, CT: Greenwood Press, 1973.

———. *Purity and Hygiene: Women, Prostitution, and the "American Plan," 1900–1930*. Westport, CT: Greenwood Press, 2001.

Popenoe, Paul, and Roswell H. Johnson, *Applied Eugenics*. New York: Macmillan Company, 1918.

Porter, Theodore M., *Karl Pearson: The Scientific Life in a Statistical Age*. Princeton, NJ: Princeton University Press, 2004.

Rafter, Nicole Hahn, *Creating Born Criminals*. Urbana: University of Illinois Press, 1997.

———. (ed.), *White Trash: The Eugenic Family Studies, 1877–1919*. Boston: Northeastern University Press, 1988.

Reed, James, *From Private Vice to Public Virtue: The Birth Control Movement and American Society since 1830*. New York: Basic Books, 1978.

Regal, Brian, *Henry Fairfield Osborn: Race, and the Search for the Origins of Man*. Burlington, VT: Ashgate, 2002.

Reilly, Philip R., *Abraham Lincoln's DNA and Other Adventures in Genetics*. Cold Spring Harbor, NY: Cold Spring Harbor Laboratory Press, 2000.

———. *The Surgical Solution: A History of Involuntary Sterilization in the United States*. Baltimore, MD: The Johns Hopkins University Press, 1991.

Renfrew, Colin, *Archaeology and Language: The Puzzle of Indo-European Origins*. New York: Cambridge University Press, 1988.

Richardson, Ken, *The Making of Intelligence*. New York: Columbia University Press, 2000.

Ridley, Matt, *Nature via Nurture: Genes, Experience, and What Makes Us Human*. New York: HarperCollins, 2003.

Rifkin, Jeremy, *The Biotech Century: Harnessing the Gene and Remaking the World*. New York: Jeremy P. Tarcher/Putnam, 1998.

Robbins, Emily F., (ed.), *Proceedings of the First National Conference on Race Betterment, January 8, 9, 10, 11, 12, 1914*. Battle Creek, MI: Gage Printing Company, 1914.

Robertson, Constance N., *The Oneida Community: An Autobiography, 1851–1876*. Syracuse, NY: Syracuse University Press, 1970.

Rose, June, *Marie Stopes and the Sexual Revolution*. Boston: Faber and Faber, 1992.

Rosen, Christine, *Preaching Eugenics: Religious Leaders and the American Eugenics Movement*. New York: Oxford University Press, 2004.

Rosen, Ruth, *The Lost Sisterhood: Prostitution in America, 1900–1918*. Baltimore, MD: Johns Hopkins University Press, 1982.

Rosenberg, Charles E., *No Other Gods: On Science and American Social Thought*. Baltimore, MD: The Johns Hopkins University Press, 1976.

Ruse, Michael, *The Darwinian Revolution: Science Red in Tooth and Claw*. Chicago: University of Chicago Press, 1999.

Russell, Cheryl, *Racial and Ethnic Diversity: Asians, Blacks, Hispanics, Native Americans, and Whites*. Ithaca, NY: New Strategist Publications, 2000.

Rydell, Robert W., *World of Fairs: The Century-of-Progress Expositions*. Chicago: University of Chicago Press, 1993.

Saleeby, C. W., *Parenthood and Race Culture: An Outline of Eugenics*. New York: Moffat, Yard & Company, 1909.

Saleeby, Caleb Williams, *The Progress of Eugenics*. New York: Funk & Wagnalls Company, 1914.

Sanger, Margaret, *Margaret Sanger: An Autobiography*. New York: Dover Publications, [1938], 1971.

———. *My Fight for Birth Control*. New York: Maxwell Reprint Co., [1931], 1969.

———. *The Selected Papers of Margaret Sanger*. Ed. Esther Katz. Urbana: University of Illinois Press, 2003.

Saunderson, Henry Hallam, *Charles W. Eliot; Puritan Liberal*. New York: Harper & Brothers, 1928.

Scher, Steven J., and Frederick Rauscher, *Evolutionary Psychology: Alternative Approaches*. Boston: Kluwer Academic Publishers, 2002.

Schwarz, Richard William, *John Harvey Kellogg, M.D.* Berrien Springs, MI: Andrews University Press, 1981.

Searle, G. R., *Eugenics and Politics in Britain, 1900–1914*. Leyden, Netherlands: Noordhoff International Publishing, 1976.

Segal, Nancy L., *Entwined Lives: Twins and What They Tell Us about Human Behavior*. New York: Dutton, 1999.

Segers, Mary C., and Timothy A. Byrnes, (eds.), *Abortion Politics in American States*. Armonk, NY: M.E. Sharpe, 1995.

Selden, Steven, *Inheriting Shame: The Story of Eugenics and Racism in America*. New York: Teachers College Press, 1999.

Singer, Peter, and Deane Wells, *Making Babies: The New Science and Ethics of Conception*. New York: Charles Scribner's Sons, 1985.

Sloan, Don M., with Paula Hartz, *Abortion: A Doctor's Perspective/A Woman's Dilemma*. New York: D. I. Fine, 1992.

Smith, J. David, *The Sterilization of Carrie Buck*. Far Hills, NJ: New Horizon Press, 1989.

Sokolow, Jayme A., *Eros and Modernization: Sylvester Graham, Health Reform, and the Origins of Victorian Sexuality in America*. Rutherford, NJ: Fairleigh-Dickinson University Press, 1983.

Soloway, Richard A., *Demography and Degeneration: Eugenics and the Declining Birthrate in Twentieth-Century Britain*. Chapel Hill: University of North Carolina Press, 1990.

Stacey, Meg, (ed.), *Changing Human Reproduction: Social Science Perspectives*. Newbury Park, CA: Sage Publications, 1992.

Steen, Grant R., *DNA and Destiny: Nature and Nurture in Human Behavior*. New York: Plenum Press, 1996.

Stern, Madeleine Betting, *Heads & Headlines: The Phrenological Fowlers*. Norman: University of Oklahoma Press, 1971.

Stewart, Elizabeth A., *Exploring Twins: Towards a Social Analysis of Twinship*. New York: St. Martin's Press, 2000.

Stock, Gregory, *Redesigning Humans: Our Inevitable Genetic Future*. Boston: Houghton Mifflin, 2002.

Stoddard, Lothrop, *The Rising Tide of Color against White World-Supremacy*. Westport, CT: Negro Universities Press, [1920], 1971.

Stone, Dan, *Breeding Superman: Nietzsche, Race and Eugenics in Edwardian and Interwar Britain*. Liverpool, England: Liverpool University Press, 2002.

Strauss, William, and Neil Howe, *The Fourth Turning: An American Prophecy*. New York: Broadway Books, 1997.

Sturtevant, A. H., and E. B. Lewis, *A History of Genetics*. Cold Spring Harbor, NY: Cold Spring Harbor Press, 2001.

Swain, Carol M., *The New White Nationalism in America: Its Challenge to Integration*. New York: Cambridge University Press, 2002.

Teller, Michael E., *The Tuberculosis Movement: A Public Health Campaign in the Progressive Era*. Westport, CT: Greenwood Press, 1988.

Trefil, James, and Margaret Hindle Hazen, *Good Seeing: A Century of Science at the Carnegie Institution of Washington*. Washington, DC: Joseph Henry Press, 2002.

Troyer, Ronald J., and Gerald E. Markle, (eds.), *Cigarettes: The Battle over Smoking*. New Brunswick, NJ: Rutgers University Press, 1983.

Vercollone, Carol, Heidi Moss, and Robert Moss, *Helping the Stork: The Choices and Challenges of Donor Insemination*. New York: Macmillan, 1997.

Walters, Ronald C., *American Reformers 1815–1860*. New York: Hill and Wang, 1978.

Watson, James D., *DNA: The Secret of Life*. New York: Alfred A. Knopf, 2003.

———. *The Double Helix: A Personal Account of the Discovery of the Structure of DNA*. Ed. Gunther S. Stent; critical edition. New York: Norton, 1980.

Weindling, Paul, *Health, Race, and German Politics between National Unification and Nazism, 1870–1945*. New York: Cambridge University Press, 1989.

Weismann, August, *The Germ-Plasm: A Theory of Heredity*. Bristol, England: Thoemmes Press, [1893], 2003.

Weiss, Sheila Faith, *Race Hygiene and National Efficiency: The Eugenics of Wilhelm Schallmayer*. Berkeley: University of California Press, 1987.

Whitney, Leon, *The Case for Sterilization*. New York: Frederick A. Stokes Company, 1934.

Whorton, James C., *Crusaders for Fitness: The History of American Health Reformers*. Princeton, NJ: Princeton University Press, 1982.

Wiggam, Albert E., *The Fruit of the Family Tree*. Indianapolis, IN: Bobbs-Merrill, 1924.

Wilkins, Maurice, *The Third Man of the Double Helix: Memoirs of a Life in Science*. Oxford: Oxford University Press, 2003.

Wilson, Edward O., *On Human Nature*. New York: Bantam Books, 1978.

Winters, Paul A., (ed.), *Cloning*. San Diego, CA: Greenhaven Press, 1998.

Witkowski, Jan, *Illuminating Life: Selected Papers from Cold Spring Harbor (1903–1969)*. Cold Spring Harbor, NY: Cold Spring Harbor Laboratory Press, 2000. http://www.eugenicsarchive.org/eugenics

Yount, Lisa, (ed.), *Cloning*. San Diego, CA: Greenhaven Press, 2000.

Yudell, Michael, and Robert DeSalle, (eds.), *The Genomic Revolution: Unveiling the Unity of Life*. Washington, DC: Joseph Henry Press, with the American Museum of Natural History, 2002.

Zenderland, Leila, *Measuring Minds: Henry Herbert Goddard and the Origins of American Intelligence Testing*. Cambridge: Cambridge University Press, 1998.

Articles

Allen, Garland E., "Science Misapplied: The Eugenics Age Revisited," *Technology Review* 99 (August–September 1996), 22–32.

———. "The Eugenics Record Office at Cold Spring Harbor, 1910–1940," *Osiris*, 2nd Series 2 (1986), 225–264.

Banton, Michael, "Galton's Conception of Race in Historical Perspective," in Milo Keynes, (ed.), *Sir Francis Galton, FRS*, 1993.

Bigelow, Maurice A., "Brief History of the American Eugenics Society," *Eugenical News* 31 (December 1946), 49–51.

Bix, Amy Sue, "Experiences and Voices of Eugenics Field-workers: 'Women's Work' in Biology," *Social Studies of Science* 27 (August 1997), 625–668.

Dann, Kevin, "From Degeneration to Regeneration: The Eugenics Survey of Vermont, 1925–1936," *Vermont History* 59 (Winter 1991), 5–29.

Devlin, Dennis S., and Colleen L. Wickey, " 'Better Living Through Heredity,' Michael F. Guyer and the American Eugenics Movement," *Michigan Academician* 16 (Winter 1984), 199–208.

Engs, Ruth Clifford, "Do Traditional Western European Drinking Practices Have Origins in Antiquity?" *Addiction Research* 2 (1995), 227–239.

Estabrook, A. H., "The Indiana Survey," *Journal of Heredity* 8 (April 1917), 156–159.

Field, Geoffrey G., "Nordic Racism," *Journal of the History of Ideas* 38 (July–September 1977), 523–540.

Fogel, Robert W., "The Fourth Great Awakening and the Political Realignment of the 1990s," *Brigham Young University Studies* 35(3) (1995), 31–43.

Galton, Francis, "Eugenics: Its Definition, Scope, and Aims," *American Journal of Sociology* 10 (July 1904), 1–25.

Getz, Lynne M., "Biological Determinism in the Making of Immigration Policy in the 1920s," *International Science Review* 70 (October 1995), 26–33.

Gould, Stephen Jay, "Does the Stoneless Plum Instruct the Thinking Reed?" *Natural History* 101 (April 1992), 16–25.

Gugliotta, Angela, "'Dr. Sharp with His Little Knife': Therapeutic and Punitive Origins of Eugenic Vasectomy: Indiana, 1892–1921," *Journal of the History of Medicine and Allied Science* 53 (October 1998), 371–406.

Hirshbein, Laura Davidow, "Masculinity, Work and the Fountain of Youth: Irving Fisher and the Life Extension Institute, 1914–31," *Canadian Bulletin of Medical History* 16 (January 1999), 89–124.

Johnson, Hildegarde Walls, "Fitter Families for Future Firesides," *Journal of Heredity* 16 (December 1925) 457–460.

Jordan, Harvey Ernest, "Eugenics: Its Data, Scope and Promise, As Seen by the Anatomist." In *Eugenics: Twelve University Lectures*, Ed. Lucy James Wilson, 107–138. New York: Dodd, Mead and Company, 1914.

Kellogg, John H., "The Eugenics Registry." In *Official Proceedings of the Second National Conference on Race Betterment, August 4, 5, 6, 7 and 8, 1915; Held in San Francisco, California*. Battle Creek, MI: Race Betterment Foundation, (1915), 76–87.

———. "Needed—A New Human Race." In *Proceedings of the First National Conference on Race Betterment, January 8, 9, 10, 11, 12, 1914*. Ed. Emily F. Robins. Battle Creek, MI: Race Betterment Foundation, 1914.

Kenny, Michael G., "Toward a Racial Abyss: Eugenics, Wickliffe Draper, and the Origins of the Pioneer Fund," *Journal of History of the Behavioral Sciences* 38 (Summer 2002), 259–283.

Lenz, Fritz, "Eugenics in Germany," *The Journal of Heredity* 15 (May 1924), 223–231.

Leon, Sharon M, "'Hopelessly Entangled in Nordic Pre-suppositions': Catholic Participation in the American Eugenic Society in the 1920s," *Journal of the History of Medicine and Allied Sciences* 59 (2004), 3–49.

Love, Rosaleen, "'Alice in Eugenics-land': Feminism and Eugenics in the Scientific Careers of Alice Lee and Ethel Elderton." *Annals of Science* 36 (1979), 45–158.

Osborn, Frederick, "History of the American Eugenics Society," *Social Biology* 21 (Summer 1974), 115–126.

Penrose, L. S., "The Galton Laboratory: Its Work and Aims," *The Eugenics Review* 41 (1949–1950), 17–27.

Rodwell, Grant, "Dr. Caleb Williams: The Complete Eugenicist," *History of Education* 26 (1997), 23–40.

Ryan, Patrick J., "Unnatural Selection: Intelligence Testing, Eugenics, and American Political Cultures," *Journal of Social History* 30 (1997), 669–685.

Weingart, Peter, "German Eugenics Between Science and Politics." *Osiris* 5 (1989), 260–282.

Weiss, Sheila Faith, "Race and Class in Fritz Lenz's Eugenics," *Medizinhistorisches Journal* 27 (1992), 5–25.

———. "The Race Hygiene Movement in Germany," *Osiris* 3 (1987), 193–236.

———. "Wilhelm Schallmayer and the Logic of German Eugenics," *Isis* 77 (1986), 33–46.

Wiseman, James, "In Vino Veritas," *Archaeology* 89 (September/October 1997), 12–17.

Selected Encyclopedias and Biographies

Buenker, John D., and Edward R. Kantowicz, *Historical Dictionary of the Progressive Era, 1890–1920.* Westport, CT: Greenwood Press, 1988.

Friedman, Howard, *Encyclopedia of Mental Health.* New York: Academic Press, 1998.

Garraty, John A., and Mark C. Carnes, (eds.), *American National Biography.* New York: Oxford University Press, 1999.

Knight, Jeffrey A., (ed.), *Encyclopedia of Genetics, Vols. 1 and 2.* Pasadena, CA: Salem Press, 1999.

Ng, Franklin, *The Asian American Encyclopedia.* New York: Marshall Cavendish, 1995.

Salzman, Jack, David Lionel Smith, and Cornel West, (eds.), *Encyclopedia of African-American Culture and History.* New York: Macmillan Library Reference, 1996.

Sternberg, Robert J., *Encyclopedia of Human Intelligence.* New York: Simon & Schuster, 1995.

United States, Bureau of the Census, *The Statistical History of the United States from Colonial Times to the Present.* New York: Basic Books, 1976.

Who Was Who among North American Authors, 1921–1939. Detroit, MI: Gale Research Company, 1976.

Who Was Who in America, 15 vols. to date. Chicago, IL: A. N. Marquis Company.

Zentner, Christian, and Friedemann Bedüürftig, (eds.), *The Encyclopedia of the Third Reich.* New York: Da Capo Press, 1997.

Index

Index pages for main entries are provided in bold type.

About the Author

RUTH CLIFFORD ENGS is Professor of Applied Health Science at Indiana University, Bloomington. She has published numerous articles and books, including *Clean Living Movements: American Cycles of Health Reform* (Praeger, 2000) and *The Progressive Era's Health Reform Movement: A Historical Dictionary* (Praeger, 2003).